Phase Transformation of Kaolinite Clay

Akshoy Kumar Chakraborty

Phase Transformation of Kaolinite Clay

Springer

Akshoy Kumar Chakraborty
Retired Scientist, Refractory Division
Central Glass and Ceramic Research Institute
Jadavpur, West Bengal
India

ISBN 978-81-322-1153-2 ISBN 978-81-322-1154-9 (eBook)
DOI 10.1007/978-81-322-1154-9
Springer New Delhi Heidelberg New York Dordrecht London

Library of Congress Control Number: 2013937192

Printed on acid-free paper

Springer is part of Springer Science+Business Media (www.springer.com)

Dedicated to my Revered
Sadhan-Siddha Sadguru Swami Nigamananda
A Renowned Spiritual Personality
A Devotee of Spiritual Seeker

Akshoy Kumar Chakraborty

Preface

The importance of Kaolinite Clay in the development of modern ceramic science can best be appreciated by considering its widespread influence on ceramic, material science, and mineralogy. The vision for this book is to upgrade the knowledge of the phase transformation process in ceramic science.

The application of basic theory on the mechanism of decomposition of kaolinite during heating leads to ample opportunities and challenges in the development of clay-based technologies. For example, refractory-brick making industries, production of synthetic mullite industries, white ware industries, production of earthenware, tile, insulators, sanitary wares, cement-mortar production industries, in production of filter and pollutant adsorbing materials, etc.

Excepting some books on clay mineralogy and ceramics which depicts a brief summaries of the results as a chapter covering compilation of studies on thermal effects of clay by various researches for about a century.

This situation and circumstance incited the author to present comprehensive studies on thermal decomposition of kaolinite by different physicochemical methods carried out by various authors with special reference to characterization of controversial spinel phase, mullitization path, and full explanation of thermal events of kaolinite.

The present volume consists of two parts. Part I includes a review of the previous research papers relevant to kaolinite to mullite reaction series. Part II presents the critical analysis of the published thoughts and findings with the recent experimental observations of the present authors with a view to put forward a new mechanism of K–M Reaction Series which explains the various earlier controversies existing in thermal decomposition processes of kaolinite. It is necessary to keep the chapters as concise as possible to accommodate a large voluminous publication of about 100 years. Readers may consult a large content of additional references that have been presented in the book as and when necessary for any particular subject in the clay-mullite application field.

It is hoped that this publication will be received by a large section of clay researchers, clay mineralogists and teaching professors, ceramic technocrats, and various end users of clay products throughout the world with keen interest. This book is planned to serve as a text for both undergraduate and graduate students. Researchers of clay mineralogist and students of material science at university are the primary audience for this book. Both researchers and students can realize the new mechanism of transformation of kaolinite to mullite. It will be a source of inspiration and encouragement for further study and research for the teaching faculty, with renewed applications in various clay-mullite fields by production managements.

First of all, I must separately thank the publishers of various journals for the figures and tables cited from their scientific journals in writing this book.

I have been fortunate enough to be a part of the clay-mullite research field in the Central Glass and Ceramic Research Institute, India and have had a number of scientific, library, and administrative opportunities for a span of 30 years in the XRD Section.

I take this opportunity to thank my revered late parents, my beloved wife MunMun, and brothers, late Rajendranath and Sanatkumar, for their constant encouragements to study the disputed clay-mullite research subject at a stretch and bring it to a concrete conclusion by forgoing a few industry jobs offering much higher perks.

My special thanks go to Mr. Dilip Kumar Ghosh, the then Head of the XRD Section as the coauthor of a large number of publications in SCI journals. My sincere thanks to my colleagues, Mr. Nirmal Kr. Ghosh, Mr. Schidananda Sengupta, and others for generous assistances given by them.

Finally, I would like to thank Basu, a sole work provider of Guru for his help in preparing the manuscript.

The author is especially indebted to SPRINGER for this opportunity to give valuable suggestions for improvement as a whole and to publish this book.

Kolkata Akshoy Kumar Chakraborty

Contents

Part I

Part I

Chapter 1
Introduction

1.1 Theoretical and Industrial Importance of the Study of Thermal Decomposition of Kaolinite

A pure kaolinite crystal has the chemical formulae $3Al_2O_32SiO_22H_2O$ or $Al_2Si_2O_5(OH)_4$. This would give a theoretical composition Al_2O_3–39.8 %, SiO_2–46.38 %, and H_2O–13.9 %. This clay mineral is of considerable interest to the production of various types of ceramic materials for having certain valuable properties.

(i) Plasticity and binding power useful for fabrication of green ceramic products.
(ii) White color after burning.
(iii) Fine texture imparts to finished articles.
(iv) Most important one is the formation of considerable amount of mullite crystals during sintering ceramic bodies which necessarily makes the body hard of good mechanical strength.

Mullite ($3Al_2O_32SiO_2$) is one of the most stable ingredients developed in refractory and pottery articles. The important properties contributed by mullite as follows.

(i) High refractoriness.
(ii) High thermal shock resistance.
(iii) High resistance to spalling.
(iv) High resistance to corrosive agents like glass, slags, etc.
(v) Elongated needle shaped crystals intergrowth in the matrix and thus improve the rigidity of the fired ceramic.

The qualities mentioned above are mostly dependent upon the size and quantity of mullite developed in the matrix. Therefore, the technology is growing immensely to produce it from various aluminosilicate raw materials. One of the most important raw materials is the clay mineral containing kaolinite. During gradual heat treatment to higher temperature it forms mullite via two intermediate phases namely metakaolinite and spinel. The compositions of these phases have

A. K. Chakraborty, *Phase Transformation of Kaolinite Clay*,
DOI: 10.1007/978-81-322-1154-9_1, © Springer India 2014

been the subject of extensive investigations due to its theoretical importance and industrial value as well. In this book, major emphasis is given on the evolution of structural changes of kaolinite to mullite via two intermediate phases.

1.2 A Short Structural View of Kaolinite

A preliminary structure of kaolinite has been presented in this chapter so as to understand the reader the basic course of its thermal decomposition. An ideal structural scheme for kaolinite has been suggested first by Pauling (1930). It is a layer structure consisting of two sheets namely one tetrahedral silica sheet and the other octahedral gibbsite layer.

1.2.1 Tetrahedral Si–O Network

Three oxygens of each tetrahedron SiO_4 are connected at three corners with three similar adjacent units and form a hexagonal ring. The resultant sheet extends along a and b crystallographic axes. The bases of all tetrahedrons with their electrically satisfied oxygens lie on the same plane i.e., on the plane of the paper and tips of them point in the same direction. One hexagonal ring has six silicon atoms and each silicon is connected to three rings. Six oxygen atoms of each ring are shared between two rings. Another six oxygen atoms which are above each silicon atom, still require one valence share, are shared between three rings. The composition of the Si–O layer is : $6Si^{+4}/3 + 6O^{-2}/2 + 6O^{-2}/3 = (Si_2O_5)^{-2}$. Ideally, Si–O hexagonal network is referred to as a rectangular cell with two sides of lengths a and b with $b/a = \sqrt{3}$. The value of Si–O distance is 0.160 nm. The calculated value for b is 0.905 nm, which is close to observed value about 8.95 Å for kaolinite. The hexagonal network is considered to be composed of three strings of oxygen atoms intersecting at angles of 120°. The O–O distance in silica tetrahedral sheet is 0.255 nm.

1.2.2 Octahedral Network

It constitutes gibbsite which has a layer structure of formulae $Al_2(OH)_6$ unit. Aluminum atoms in octahedral co-ordination are embedded in between two close-packed layers of hydroxyls. The O–O distance for a regular octahedron is 0.267 nm. The expected ideal value for b parameter 0.864 nm which is far short of observed b parameter of kaolinite, 0.895 nm. Idealized projections of those are shown by Brindley (1961).

1.2.3 Kaolin Layer

The tetrahedral and octahedral sheets fit together to form a composite tetrahedral octahedral basic unit layer which is called a "kaolin layer". Therefore, to fit two layers together and to form composite kaolin layer, decrease in b value Si–O network is necessary. This will occur by deformations of the above network. The decrease in b will be obtained from the deformation of the hexagonal Si–O network, alternate left and right handed rotations of Si–O tetrahedral and shortening of the shared edges of Al–O(OH) octahedral.

The distorted and deviated two sheets now fit together to form a composite kaolin layer structure with following changes.

Si–O distance increases from 0.160 to 0.163 nm. Al–O distance (0.132 + 0.57) = 0.189 nm in octahedral layer and Al–O(OH) comes to 0.193 nm.

$$(Si_2O_5)^{-2} + [Al(OH)_4]^{+2} ------\rightarrow [Al_2(Si_2O_5)(OH)_4]$$

During joining $4OH^-$ groups per unit cell of octahedral layer have been replaced by four apex of the tetrahedral layer. Thus a common layer constituting of oxygen and hydroxyl ions has formed. The charge distribution of the composite layer is given below.

$6O^{-2}$	−12
$4Si^{+4}$	+16
$4O^{-2} + 2OH^-$	−10 (Layer common to tetrahedral and octahedral)
$4Al^{+3}$	+12
$6OH^-$	−6

The charges within the structural unit are balanced and the structural formulae is $(OH)_8\ Si_4\ Al_4O_{10}$. The kaolinite layer is continuous in a- and b-directions and is stacked one above the other in c-direction and in this way variations between kaolin group minerals occur. The following minerals are recognized as belonging to the kaolin group : Nacrite, Dickite, Kaolinite, Livesite, and Halloysite.

1.2.4 Relationship Between Adjacent Layers

The (001) of the Si–0 network of one layer is projected on the hydroxyl sheet of the next adjacent Al–O(OH) layer. Oxygen and hydroxyl atoms are grouped in pairs and forming bonds. In ideal layer structure, the O–OH distance is equal. In triclinic kaolinite structure $\alpha = 91.6°$, $\beta = 104.8°$, and $\gamma = 89.9°$ with distortion of layer structure results to unequal O–OH distances. Average value of it is = 0. 289 nm with a, b, and c, as per Brindley and Robinson (1946), are 0. 0.515, 0.895, and 0,739 nm, respectively. Details of structural modifications of kaolin minerals are not the subject matter of this book and these are reviewed earlier by Grim (1968), and Grimshaw (1971).

Naturally occurring kaolin clays during X-ray diffraction study show considerable variations in reflection intensities. This arises the occurrence of different order of kaolinites.

(1) *Highly crystalline/high degree of structural regularity*
Brindley and Robinson (1946) *have* shown the d spacings versus intensities of well- and poorly crystallized kaolinites. Well-crystallized kaolinite shows resolution of doublet at 0.418 and 0.413 nm spacing's. Moreover, it shows four clear reflections between 002 (d = 0.357) and $20\bar{1}$ (d = 0.256 nm).

(2) *Poorly crystallized kaolinite* (*highly "b—axis disordered*")
It shows no reflection in the spacing range 0.35–0.25 nm. A group of reflections produces blurring (see diffractogram of poorly crystallized Georgia kaolinite). Thus, in place of sharp hkl reflections, only one hk band is noted having indices 02, 11 or $1\bar{1}$.

Murray and Lyons (1956) showed the variation of X-ray diagram from well-crystallized kaolinite to b–axis disordered kaolinite (Fig. 1.1). The other definition regarding ordering of kaolinite is called as Hinckley index as proposed by Hinckley (1962, Hinckley 1963). Brindley and Robinson (1947) called this disordered kaolinite as "the fire clay mineral".

Out of the minerals of the kaolin group, particular emphasis has been given to kaolinite mineral. A pure kaolinite crystal has the overall composition $2[Al_2(Si_2O_5)(OH)_4]$ per unit cell which conforms to the oxide formulae Al_2O_3 $2SiO_2$ $2H_2O$.

Naturally occurring clays consist of mineral kaolinite as basic mineral. There are three major variety among the large domain of kaolinite clay. Fire clay is on one hand to china clay on the other extreme.

(1) China Clay is a white burning clay and the backbone of white ware industries.
(2) Fire Clay imparts slight yellowish color after burning and is largely used in refractory industries,
(3) Ball Clay is a sedimentary origin, and of fine grain size distribution and usually contains organic matter. It develops white to cream color during heating. It imparts more workable plasticity and green strength to the fabricated green tri axial ceramic bodies.

1.2.5 The Major Objective of this Book is to Focus on the Followings

(1) What is happening during gradual heating in kaolin layer? There are two possibilities. Kaolin layer may disintegrate into free oxides and then recombine in due course of heating or the Si–O–Al bonds in kaolin layer may persist during successive phase transformation processes up to mullite.

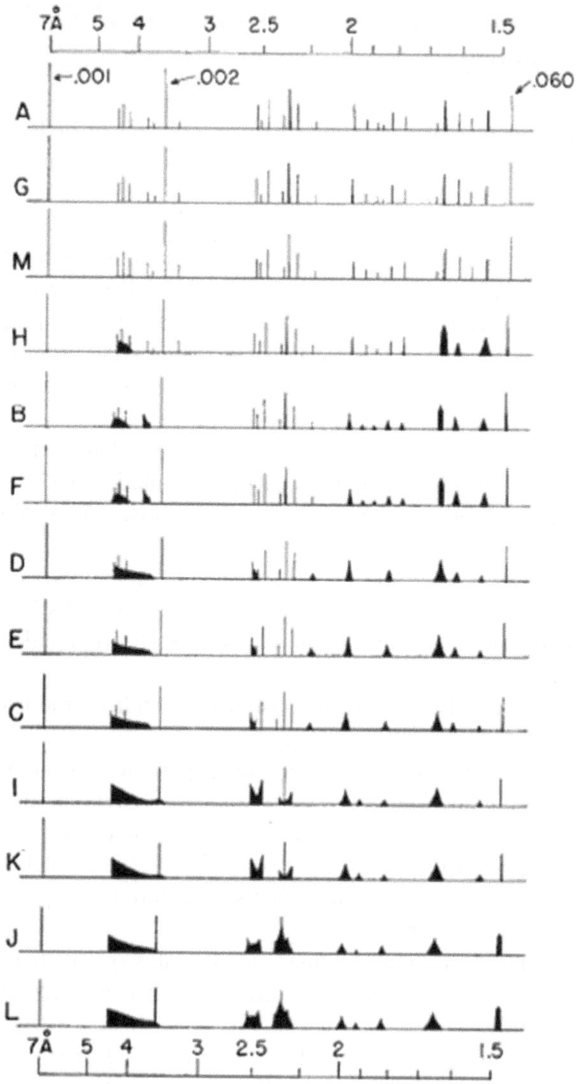

Fig. 1.1 X-ray powder diagram of Kaolinites arranged in order of crystalinity (After Murray and Lyons 1956) Free available

(2) During dehydroxylation process of kaolinite (–O–OH) pairing perhaps lost. What results bring forth upon this dehydroxylation?

The present book summarizes the detail accounts of the above two views, and finally substantiates a new mechanism of phase transformation of kaolinite.

1.3 Chemical and XRD Analysis of Kaolinites of Different Origins

Chemical analyses of different clays are given in Table A Appendix which shows how those analysis change with different origin and source of clay minerals. Quartz is found to be an universal impurity. Besides quartz, other impurities e.g., iron bearing oxides and hydroxides, TiO_2 are also associated with kaolinite. X-ray diffraction patterns of kaoinites of different origins are shown in figures below. Some of the features arise during XRD characterization of clays.

(1) XRD pattern changes with degree of crystallinity of kaolinite. Some of the peaks are missing, some become broad when order of crystallinity changes. e.g., highly crystalline Georgia to medium crystalline Georgia and to poorly crystalline Georgia clays (Fig. 1.2).

(2) Titania bearing phase cannot be identified easily in original clay when present in small quantity. Clay sample marked HGMS _ORG supplied by M/S. 20 Microns shows common peaks of anatase and kaolinite at 0.351 nm (Figs. 1.3, 1.4).

Clay is decomposed on heating at 700° C. Sample mark ORG heat shows XRD peaks of both rutile and anatase and this conforms association of later phase in raw clay. Similar to this technique, anatase becomes identified when MP kaolinite is heated to 600° C.

(3) Likely, identification of quartz and hematite is easier when clay is heated at 600° C by the above procedure.

(4) Orientation effect. Kaolinite particles are flaky or platy in nature. As such preferential orientation occurs and (00l) reflections are enhanced. Accordingly, characteristic XRD pattern of it depends upon shaping of the particles in a powder compact. Chen et al. (2000), Chen and Tuan (2002) have shown usual powder pattern in comparison with their textured patterns obtained with two different processing techniques.

(1) During die-pressing, kaolin plates orient themselves on the plane perpendicular to the die-pressing direction. This compact shows the strongest intensities of {001} planes.

Contrary, planes parallel to die-pressing direction shows even less intensities (Fig. 1.5).

(2) Using tape casting of kaolin slurry, kaolin flakes is aligned parallel and forms a highly textured green tape which eventually demonstrates strong XRD peaks of (001) planes.

(3) Clay slip cast on one surface of plaster of paris mould also shows preferred orientation. Clay plates on the mould surface are aligned parallel to it, at the inside clay plates are randomly oriented.

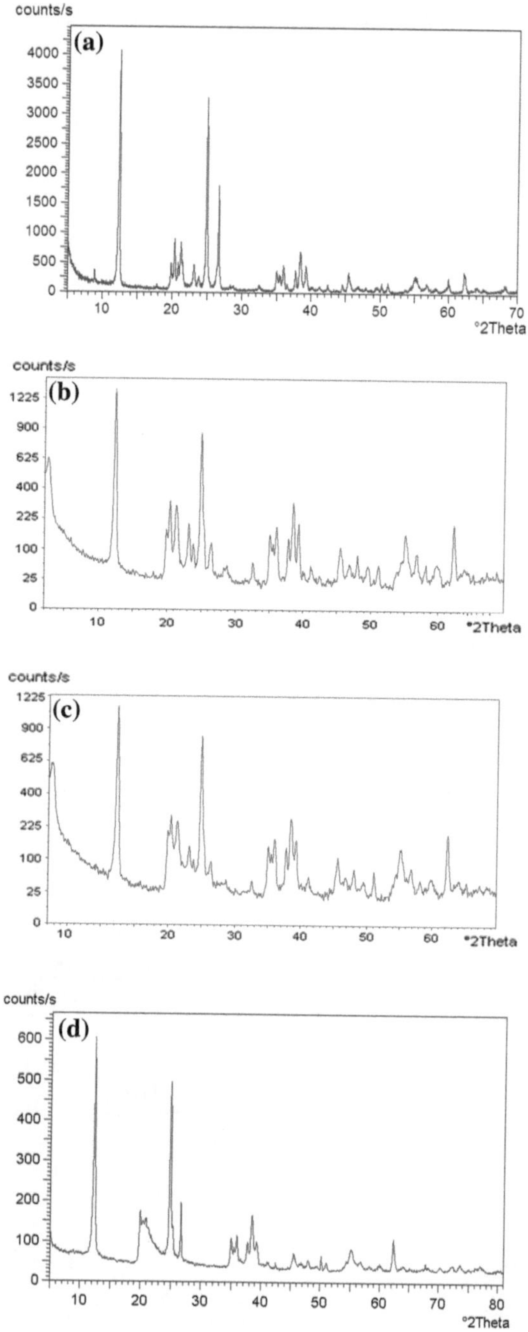

Fig. 1.2 **a** XRD pattern of raw Rajmohol china clay. **b** Georgia highly crystalline. **c** Georgia medium crystalline. **d** Georgia poorly crystalline. **e** English China clay. **f** Karlovy vary clay. **g** New Zealand China clay. **h** New Zealand Halloysite. **i** Fire clay (Makdamnagar)

Fig. 1.2 continued

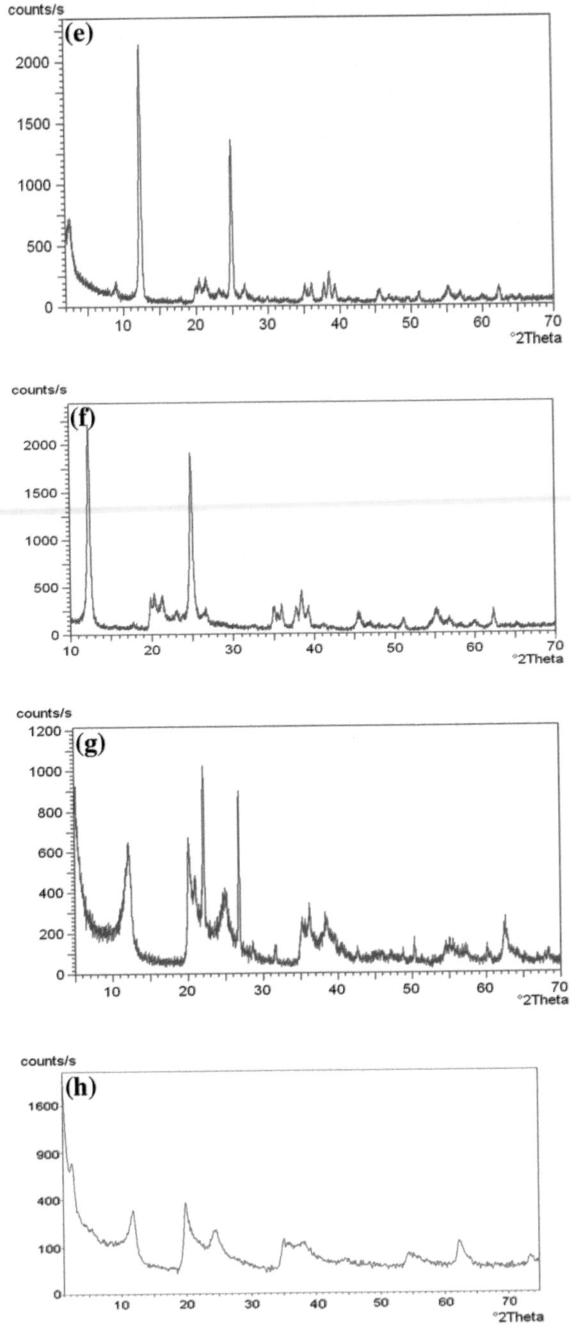

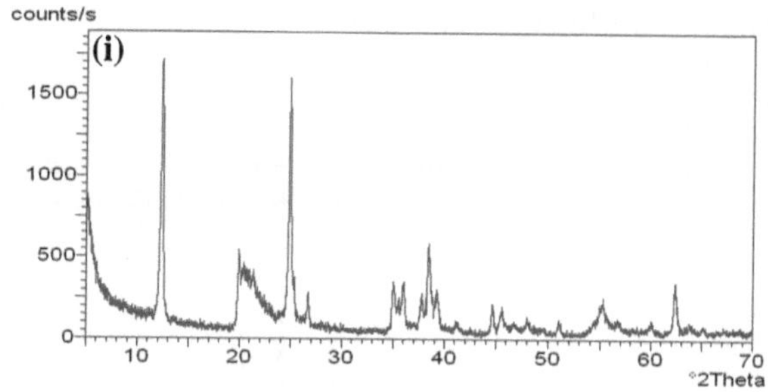

Fig. 1.2 continued

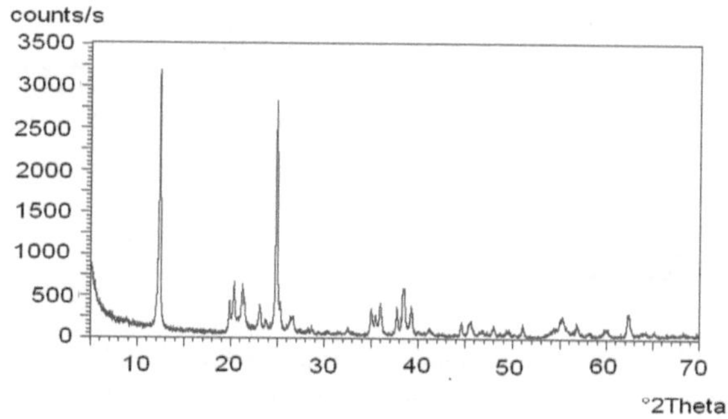

Fig. 1.3 Raw clay sample mark (HGMS-ORG) received from M/s. 20 MICRONS

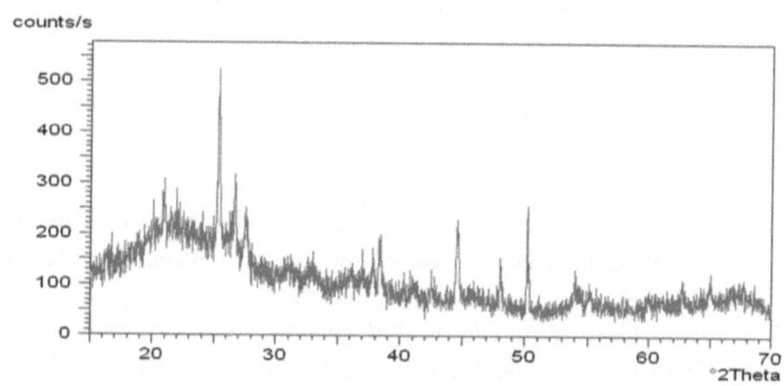

Fig. 1.4 Raw clay sample mark (HGMS-ORG) heat treated to 700 °C

Electron microscopic studies of raw kaolinite shows hexagonal outline by various researchers working with TEM. Eitel et al. (1939) has first shown the morphology of kaolinite as pseudo-hexagonal platelets in the transmission electron microscopic study. I R Pattern of kaolinite is shown in Chap. 7.

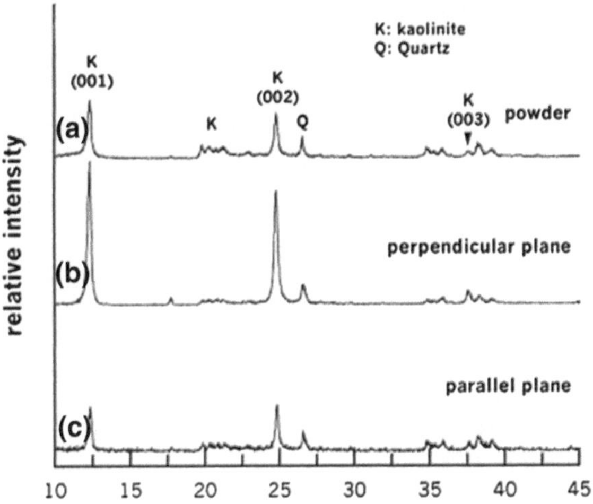

Fig. 1.5 The XRD patterns of (**a**) the kaolin powder and **b**, **c** the powder compact. In (**b**) and (**c**) the plane taken is perpendicular and parallel to the die-pressing direction, respectively (After Chen et al. 2000) Reprinted by permission of Ceram. International

References

G.W. Brindley, K. Robinson, Structure of Kaolinite. Miner. Mag. **27**, 242–253 (1946)

G.W. Brindley and K. Robinson. X-ray studies of some kaolinite fireclays.Trans. Brit. Ceram. Soc. **46**, 49–62 (1947)

G.W. Brindley, Kaolin, serpentine, and kindered minerals Ch. II, in *The X-ray Identification and Crystal Structures of Clay Minerals*, ed. by G. Brown. (The Mineralogical Society, London, 1961), pp. 51–132

C.Y. Chen, G.S. Lan, W.H. Tuan, Micro structural evolution of mullite during the sintering of kaolinite powder compacts. Ceram. Int. **26**(7) 715–20 (2000)

C.Y. Chen, W.H. Tuan, Evolution of mullite texture on firing tape-cast kaolin bodies. J. Am. Ceram. Soc. **85**(5) 1121–1126 (2002)

W. Eitel, H.O. Miiller, O.E. Radczewski, Ubermikroskopische Untersuchungen an Tonminerdlien (Ultramicroscopic examination of clay minerals). Ber. Deut. Keram. Ges. **20**(4) 165–180 (1939)

R.E. Grim, *Clay Mineralogy*, 2nd edition. (McGraw-Hill, New York, 1968)

R.W. Grimshaw, in *The Chemistry and Physics of Clays,* 4th edn. (*Wiley-Interscience, New York, 1971*)

D.N. Hinckley, Variability in 'crystallinity' values among the kaolin deposits of the coastal plain of Georgia and South Carolina, in *Proceedings of the 11th National Conference on Clays and Clay Minerals* (Ot-tawa, 1962), pp. 229–235

D.N. Hinckley, Variability in crystallinity values among the kaolin deposits of the coastal plain of Georgia and South Carolina. Clays Clay Miner. **11**, 229–235 (1963)

H. Murray, S.C. Lyons, Degree of crystal perfection of kaolinite. Natl. Acad. Sci. Publ **456**, 31–40 (1956)

L. Pauling, Structure of the chlorites. Proc. Natl. Acad. Sci. Wash. **16**, 578–582 (1930)

Chapter 2
Thermal Evolution of Kaolinite

2.1 Various Method Used for Thermal Evolution of Kaolinite

Methods used by various investigators for ascertaining thermal decomposition of kaolinite are chronologically submitted in a comprehensive manner and then a summary is given. Following different experimental techniques (Table 2.1) have been chosen by various investigators using kaolinites of different localities varying in (i) degree of crystallinities, (ii) impurity oxides contents, and (iii) particle sizes, etc.

Table 2.1 Different experimental tools used to characterize phase developments in heated kaolinite

Study	Objective
1. Several physical changes—D.T.A., Dilatometry, Density, and Surface area	To ascertain the probable steps of thermal transformation of kaolinite
2. Solubility of SiO_2 and Al_2O_3 components	To reveal the nature and state of intermediate metakaolinite phase
3 Diffraction: Powder and single crystal X-ray diffraction/Electron diffraction	To identify crystalline phases formed on heating. Structural parameters e.g., unit cell dimensions of spinel, mullite, and cristobalite phases. To ascertain the phase relationship among phases
4. Microscopy: mainly electron microscopy (TEM), (HREM), and (FE-TEM)	To exhibit the morphology of parent kaolinite to heated phases and to establish interrelationships between them
5. Spectroscopy: IR, XRF, and MAS NMR	To note the changes in co-ordination number of Al among successive phases during thermal process. Simultaneously to note the changes in the character of the Si environment
6. Radial Electron Function (RED)	To correlate the theoretical RDF curves of possible transformation equations of metakaolinite to experimentally drawn RDF pattern

(continued)

A. K. Chakraborty, *Phase Transformation of Kaolinite Clay*,
DOI: 10.1007/978-81-322-1154-9_2, © Springer India 2014

Table 2.1 (continued)

Study	Objective
7. Thermodynamic approach	To calculate theoretically the changes of free energy, heat of formation, lattice energy of several possible reaction paths of kaolinite decomposition. and finally prediction there of
8. Alkali extraction of siliceous phase (A)	To estimate the amount of silica/or siliceous phase liberated during heating of kaolinite
9. QXRD	To estimate the quantities of spinel, mullite. and cristobalite phase formed during heating kaolinite
10. Alkali leaching followed by XRD, IR, HREM, EDS MAS NMR, EF-TEM, and Studies of leached residues	To evaluate the nature of intermediate spinel phase

Method used by various investigators for proper ascertaining Kaolinite–Mullite (K–M) Reaction Series are chronologically submitted in comprehensive manner in the following chapters.

Chapter 3
Thermal Methods

3.1 Introduction

Heating a substance causes a variety of changes. According to International Confederation on thermal analysis (ICTA), a group of techniques are being used to correlate the temperature variation with some physical properties of the substance. Thermal analysis technique shows some curves which is a measure of property as a function of temperature and that is characteristic of a particular substance.

Kaolinite transforms ultimately to mullite and β-cristobelite during heating via two intermediate steps. Each step of transition exhibits some physical changes which are detected during its thermal analysis test.

A number of thermal methods utilizing various equipments are available for investigation. These are based upon some fundamental changes of properties.

(i) **DTA and DDTA (Differential DTA)** It is based on energy changes. Accordingly, it can detect changes in heat content occurring during decomposition, recrystallization, and polymorphic transformation phenomenon during heating of a kaolinite at a specified rate with respect to a reference substance. As such temperature difference between them occurs.

(ii) **TG and DTG** It is based on weight change. As such it can detect loss of moisture content during decomposition of kaolinite when heated at a specific rate. The weight loss of most kaolinite sample is measured as a function of temperature.

(iii) **TMA and DTMA** It is based on dimensional change. It can detect contraction and expansion behavior during structural transformation of kaolinite when heated at a specified rate. The length of kaolinite specimen is measured as a function of temperature.

A short survey of the earlier studies of DTA, TGA and TMA of kaolinites of different origins and procedures is presented.

A. K. Chakraborty, *Phase Transformation of Kaolinite Clay*,
DOI: 10.1007/978-81-322-1154-9_3, © Springer India 2014

3.2 DTA Technique

Le Chatelier (1887) first investigated the behavior of clays on heating. He noted directly the rate of change of temperature (dts/dt) of clay samples in terms of temperatures (Ts). The temperatures were measured by means of a thermocouple consisting of pure Pt and 10 % Rh in Pt in connection with galvanometer. Sample was heated at a uniform rate of heating about 2 °C/min. A photographic apparatus was employed for recording of heating curve data. A brilliant induction spark was made to pass at an interval of 2 s before silt and gave on the plate after reflection from the galvanometer. The spacing of the image of the silt was a measure of rate of heating. The calibration of the thermocouple was made by taking as fixed points of the boiling or fusing of following substances e.g., Water at 100 °C; S at 448 °C and Se at 1,045 °C, respectively. At the moment of fusion or vaporization, the temperature remained stationary. This led to the superposition of several consecutive images of the spark. Results show that crystalline kaolin observes a vary well-marked deceleration which ended at 700 °C and a slower acceleration around 1000 °C. Halloysite shows first deceleration between 150 and 200 °C, a second very important deceleration ending at 700 °C corresponding to dehydration and finely a sharp acceleration indicating a crystallization phenomenon accompanied by a releasing of heat at 1,000 °C. Allophane shows a distinct deceleration between 150 and 220 °C and then precedes a sharp acceleration at 1,000 °C.

He concluded that free SiO_2 cannot be present in pure clays as hydrated SiO_2 showed an endotherm between 100 and 200 °C. The evaluation of heat is due to molecular changes of the alumina to the insoluble form. Free alumina does not exist in clays, but is liberated by their decomposition and dehydration reactions. He came to this conclusion on the basis that alumina prepared by calcinations of aluminum nitrate showed the same endothermic reaction followed by a sudden acceleration in the rise of temperature at 850 °C.

Wohlin (1913) after Le Chatelier improved the thermal analysis technique by inserting a second thermocouple in the furnace. He plotted the temperatures of sample (Ts) and that of the furnace or neutral body (Tr) at definite intervals of time and then compared. Plot shows a straight line when there is no thermal effect occurred during heating of the sample. When a reaction took place, energy would either be absorbed or evolved, as such the plot will change its slope at some temperature. If tangent of the slope decreased in comparison to original, the reaction is called as endothermic and inverse is termed an exothermic process.

Thereafter, the temperature of the test material relative to that of an adjacent inert material was measured, instead of to that of furnace. To do that a thermocouple embedded in the test sample and another in the inert material were connected in opposition. During the run the temperature of the inert material increases regularly as the temperature of furnace increases. When a thermal reaction takes place in a sample, unbalance of temperature between the inert and reference occurs. The temperature of the sample may be greater or lesser than that of inert material. It depends on whether the reaction is exothermic or endothermic.

Consequently, an e.m.f. is generated at the junction of the differential thermocouple circuit. This e.m.f. value/difference of temperature (ΔT) is plotted against temperature of inert material. When no thermal reaction is taking place in the specimen, the temperatures at both the junctions of the difference couple are the same and no e.m.f. is setup. This technique is the basis for modern method of thermal analysis. For endothermic reaction, the DTA trace moves generally downwards and it goes upwards for exothermic reactions due to change in the direction of e.m.f. Otherwise, the trace moves more or less in a straight line and is termed as base line. Thus in DTA technique, one can detect and measure the temperature difference ΔT (°C) or unbalance of temperature between the reference substance and sample which absorbs or generates enthalpy during heating them at a constant rate inside a furnace.

Insley and Ewell (1935) used a platinum wire resistance furnace, a divided Pt cylinder, a Pt–Rh to Pt to Pt–Rh differential thermocouple, and a type K potentiometer as constituent of DTA apparatus. The normal heating rate of 6 °C/min was selected. Readings of the temperature of the sample and that of the galvanometer deflection due to the temperature difference between the two sides of cylinder were made at intervals of 1 min or less and plotted. They noted three heat effects which are in order of the occurrence of increasing temperature difference.

(i) A large endothermic effect which occurs over a broad temperature range.
(ii) A large exothermic effect which is very sharp and intense and occurs in a narrow temperature range.
(iii) A very small exothermic effect which occurs over a broad temperature range and which is frequently difficult to detect. They also noted that all of these reactions causing these heat effects are irreversible.

Contrary to previous work, Norton (1939) designed a DTA apparatus where he took kanthal resistance furnace and measured the temperature of the specimen by a separate thermocouple imbedded in the center of the nickel block (with a view to neutralize the thermal gradients) which is connected directly to the potentiometer with the cold junction in ice. Thermal curves for some clay minerals showed that all these minerals are characterized by a strong exothermic peak in the region of 980 °C. The beginning of the endothermic effect is found in all of them at ~475 °C. Nacrite shows a comparatively long peak running from 610 to 670 °C. Dickite shows a rather long range of slight heat absorption and then a rather sudden peak at 700 °C. Kaolinite has a definite peak at 610 °C. Anauxite gives a curve substantially same as kaolinite. Halloysite shows two low-temperature peaks which are sharp and distinct, first at 150 °C and the smaller at 325 °C. The main absorption peak at 580 °C in halloysite comes at a lower temperature than for kaolinite. Allophane shows a large endothermic effect at low temperature at 180 °C and a small exotherm at 600 °C.

Grim and Rowland (1944) investigated several clay minerals and noted that hydrated halloysite shows similar thermal reactions as kaolinites with an additional sharp endothermic reaction at 100–150 °C accompanying the loss of 2 molecules of water and with transition to dehydrated halloysite as given below.

$$Al_2O_32SiO_24H_2O \longrightarrow Al_2O_32SiO_22H_2O + 2H_2O \tag{3.1}$$

Thus, they pointed out that DTA results are very much useful in the identification of clay minerals. In addition to characteristic endothermic and exothermic peaks of kaolinite between 450 and 600 °C and between 900 and 1,000 °C, Yamauchi and Kato (1943) also noted one endotherm at ~100 °C and another exothermic peak at 150–300 °C. These two peaks were explained due to the burning out of organic substance associated in the original clay.

Grimshaw et al. (1945) studied a wide range of British clays by standardization of DTA equipment and with greater refinements in procedures. In the apparatus developed by them, a fixed weight of clay sample (0.8 g) is packed in one of the two cubical compartment of refractory specimen holder made of sintered alumina. The other compartment is being filled with inert material usually calcined Al_2O_3. A chromel-to-alumel thermocouple junction is located at the center of both clay and alumina. This thermocouple shows greater sensitivity due to showing higher e.m.f. than Pt/Pt–Rh thermocouple. It is however satisfactory up to 1,000 °C instead of 1,400 °C, where Pt/Pt–Rh thermocouple shows greater sensitivity. The specimen holder is placed inside the cavity of refractory block made of alumina or mullite. The furnace consists of refractory tube made of alumina or mullite and is wound with nichrome wire. The standard thermal analysis block fits into the tube of the furnace. It is heated up by using a variable transformer. During operation, when a difference in temperature occurs between sample and inert due to thermal transformation of the sample, the e.m.f. in the two thermocouples will differ proportionately and a net potential difference across the leads will be revealed from the galvanometer. The temperature during the run will be noted by a mill voltmeter through a change over switch. The deflection of the galvanometer is recorded either by a continuous pen system or by a chopper bar mechanism or alternatively by a mirror galvanometer cum photographic system. They showed that the characteristic peaks of kaolin minerals are slightly different in nature may be very large-medium-large in their intensity. Peak positions identified by them using DTA show differences.

Bradley and Grim (1951) studied a series of clay minerals with a view toward establishing the significance of various thermal effects by different physical means. They noted the usual large endothermic and exothermic effects for Georgia clay as noted by Insley and Ewell for Zettlitz kaolinite. Moreover, they observed a broad and pronounced second exothermic peak in the temperature range 1,207–1,300 °C.

De keyser (1963a) studied the high-temperature DTA of six kaolinites and showed two exothermic peaks for mullitization around 1,100–1,300 °C with high intensities for Zettlitz kaolinite.

Besides the occurrences of prominent endothermic at 500–600 °C range, a sharp exotherm at 980 °C and a broad exothermic around 1,250 °C, some authors e.g., Glass (1954). Johnson et al. (1982) noted a very small exothermic at 1,300–1,400 °C and designated it for the crystallization of amorphous SiO_2 to β-cristobalite. The details are given in next chapter.

A comprehensive review regarding DTA studies of clay minerals have been done by Smothers et al. (1951), Mackenzie (1957) and West (1970).

Several designs of DTA equipment are coming up nowadays. Old mill volt meter and galvanometer are replaced by modern temperature measuring devices and X–Y recorder, etc. A small temperature difference (ΔT) or mill volt is amplified to achieve more sensitive ΔT measurement. Recent equipments are even fully computerized.

3.3 Dynamic Difference Calorimeter

Usually in DTA methods, thermocouple junctions are embedded in powder samples. During its run, kaolinite powder gradually sinters and shrinks. As a result thermal conductivity between thermocouple and sample powder changes which influence the thermal effects. In DDK methods, kaolinite sample and reference material are placed in two platinum crucibles which rest upon two platinum dishes. Thermocouple junctions are welded at the base of the disks. Sensitivity of Dynamic Difference Calorimeter (DDK) may be less than that of DTA but it gives reproducible results. Formation of any bead with fused mass and the thermocouple junction is avoided. Platinum cups could be washed from fused clay very easily. DDK trace of Georgia kaolinite when loosely packed in the Pt–cup exhibits two exotherms at ~130, ~591 °C, and a sharp exotherm at ~987 °C. At the high-temperature side, it exhibits two peaks of small magnitudes at ~1,250 °C, and lastly at ~1,420 °C, respectively.

Chakraborty (1992) made DTA analysis of two Indian kaolinites using Shamadzu Thermomechanical Analysis System with varying packing density and rate of heating. Figure 3.1 shows DDK curve (a) Rajmahal kaolinite and (b) Bhandak kaolinite. With higher sensitivity run, DTK curves of both kaolinites in the vicinity of 980 °C exhibit two exotherms. The first one preceded by an endotherm is very small and insignificant but the other one is quite sharp and very well-defined. In the mullite formation region, a big and broad exotherm in the temperature range of 1,150–1,400 °C is observed. Within this peak, two exothermic peaks: maximum (T_m) one at 1,250 °C and other at 1,330 °C are noted which tallies with the observations made earlier by De keyser (1963a). Lastly, one more exotherm at about 1,420 °C is well-marked which correspond to previous findings of Glass (1954) and later issue of Johnson et al. (1982).

3.4 D.DTA

The first derivative of a differential thermal curve often provides very useful information (Fig. 3.2). Slight changes in the slope of a gradually drifting curve or reversals in slope, are represented by sharp change in the direction of a curve representing the first derivative of the differential thermal curve.

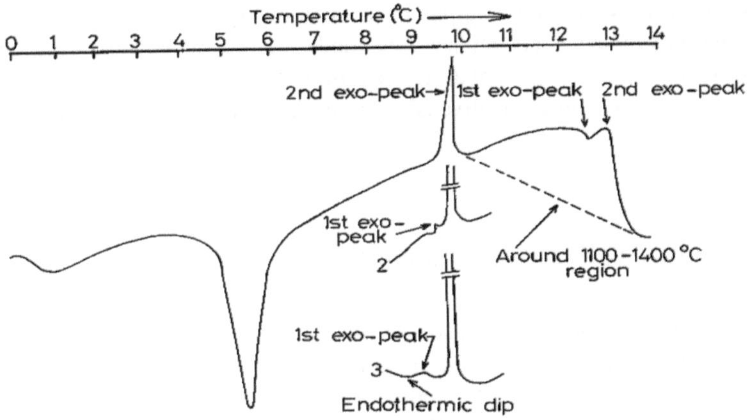

Fig. 3.1 DTA trace of Rajmohol Kaolinite. Sample wt, 40 mg; rate of heating, 10 °C/min; sensitivity, ± 100 μv. *2* Portion of DTA trace of Rajmohol kaolinite with sensitivity, ±10 μv showing one additional exotherm. *3* Portion of DTA trace of Bhandak kaolinite with sensitivity, ±10 μv showing endothermic dip in addition to two exotherms (after Chakraborty 1992). Reprinted by permission of the American Ceramic Society

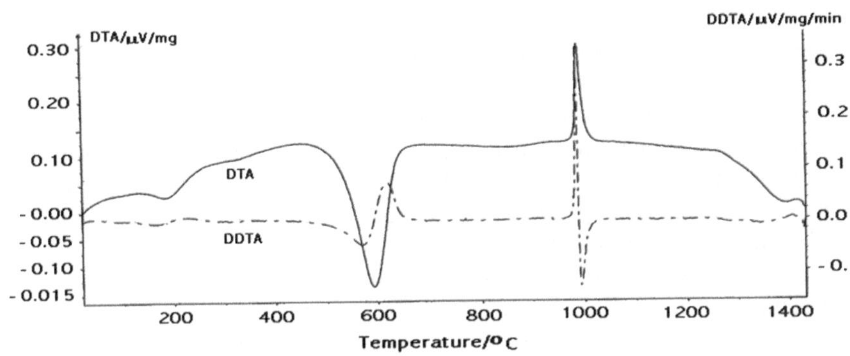

Fig. 3.2 DTA and DDTA curves of Chitarpur Kaolinite

3.5 TGA

Generally, thermogravimetric analysis shows a weight loss due to dehydration of physically or chemically bound water similar to an endothermic phenomenon noticed in DTA study. There are two techniques of measuring this weight change.

(i) Determination of weight loss by equilibrium heating schedule

Sokoloff (1912), Tammann and Papa (1923), Rhode (1927), Van Nieuwenburg and Pieters (1929), Nutting (1933), Ross and Kerr (1930), Kelly and Jenny (1936) determined weight losses of clays by heating until such time so as to attain constancy in weight at the desired temperature. Loss in weight versus temperature plot data of different clays show that dehydration takes place over a considerable temperature range and there is single temperature at which all the combined moisture is expelled.

(ii) Determination of weight loss by nonequilibrium heating schedule

Instead of static heating at the temperature until constancy in weight attains, some authors heated clay continuously at a constant rate e.g., 5 °C/h to 10 °C/min, i.e., in nonequilibrium condition. Loss in weight is recorded continuously up to higher temperature.

De keyser (1953), Erdey et al. (1954) designed a thermogravimetry apparatus. Sample in a cup is suspended from a balance by a wire which is attached with a permanent magnet moving inside a solenoid in an electric furnace. The sample is heated continuously at a uniform rate. As soon as dehydration occurs, change in weight takes place and the beam of the balance connecting the magnet moves as a result induced e.m.f. is generated in the solenoid and this is recorded by a recorder. The starting point of dehydration, the point of maximum degree of dehydration, and lastly the end point of it are recorded. The technique of weight loss versus temperature under uniform rate of heating was developed by De Keyser (1959c) and others colleagues is called as differential thermogravimetry (DTG). TG and DTA curves of Chitarpur Kaolinite is shown in Fig. 3.3.

Kulbicki and Grim (1959) and later De keyser et al. (1963b) performed thermogravimetric differential analysis and confirmed the existence of OH groups in metakaolinite. The former authors designed a sensitive equipment which is able to detect the possible release of small amount of water from clay minerals at higher temperature. The water vapor evolved from the kaolinite specimen during heating is carried by a flow of N_2 previously dried through a water absorber; the absorption of water causes a slight increase in the temperature of the carrying gas which is detected by means of a differential thermocouple and recorded as a function of the temperature of the specimen. The vapor curves are close to that of DTA curves. Each of the curves shows a faint depression of about 950–1,000 °C and this is analogous to DTA curve of each of these minerals which sometimes show a slight

Fig. 3.3 TG and DTA curves of Chirtarpur Kaolinite

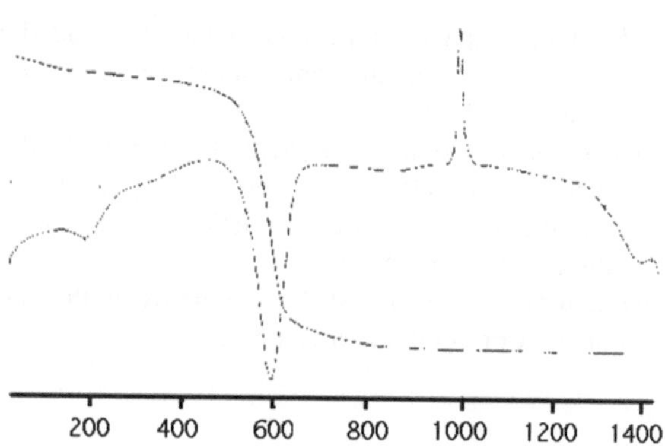

endothermic reaction at this temperature interval (Fig. 3.1, plot 3). They suggested that during very rapid heating, the small amount of water present in the so-called metakaolinite structure is expelled when the structure is destroyed.

Like derivative DTA, differential thermo gravimetry has also come up in detecting inflection point of small reactions and the precise temperature at which the reaction starts. De keyser (1963c) noted a small endothermic peak just before 980 °C exothermic peak of kaolinite by DTG technique. This peak is analogous to that of DTA and the amount of loss in weight of water was measured by De keyser to be 0.4 wt%.

3.6 TMA

Kaolinite minerals show dimensional changes during heating similar to changes in enthalpy and weight occurring in DTA and TGA analysis. The changes are mostly contractions in different transformation stages excepting expansion in one step only. The expansion/contraction behavior of different kaolinites on heating have been studied extensively by various authors.

Hyslop and McMurdo (1938) carried out thermal expansion study using differential thermal expansion apparatus, fabricated earlier by his colleagues, by heating clay samples in the form of cylindrical at 10 °C/min rate of rise temperature in an oxidizing atmosphere. They noted that both kaolinite and halloysite show an initial slight expansion up to 500 °C, thereafter a sudden contraction occurs. Curves then flatten up to 900 °C. Finally a rapid contraction takes place in both cases. Thereafter, the profile in case of kaolinite continues showing contraction but the same for halloysite does not show any further contraction.

Heindl and Meng (1939) studied the effects of heat on changes in length from room temperature to 1,000 °C, and compared them with endothermic and exothermic effects of Flint clays of different origins e.g., Missiouri, Kentuchy, Ohio, Pennsylvania, Washington, and Plastic clays of Tyrene, Georgia kaolin, Zettlitz kaolin, and aluminous clays. Usual interferometer was used for making measurement in changes of lengths. They noted the following events.

(i) Kaolinite expands uniformly at the start and then rather slowly from room temperature up to the temperature range of 470–550 °C and the expansion amounts to 0.3 %.
(ii) Contraction commences thereafter and proceeds relatively slow for approximately up to 600 °C and then very rapidly for relatively short range of temperature of 50 °C. The contraction for this interval of approximately 100 °C amounts to 1.0–1.8 %.
(iii) Contraction occurs slowly in between the temperature 650–920 °C. The amount of contraction is 1.5–2.3 %.

(iv) Contraction proceeds at a very rapid rate in the temperature range of (930–980 °C). The amount of contraction in this temperature interval is 1.0–1.4 %.
(v) Contraction again continues at a slower rate beyond 1,000 °C.

Tsuzuki and Nagasawa (1969) carried out combined dilatometric and DTA analysis of kaolin minerals. They showed that after an intense contraction caused by dehydroxylation between 550 and 650 °C, a moderate contraction occurs up to 850 °C. Thereafter, the rate of contraction suddenly rises at 980 °C exotherm i.e., there are two stages of transformations up to 1,000 °C. First is endothermic and contractile, while the second is exothermic and contractile also.

Structural transformations of kaolinite which are generally accompanied by volume changes/essentially shrinkages were also registered by Schuller and Kromer (1975) by their differential dilatometric studies. In addition to two shrinkage steps noted by several authors one at 550 °C due to escape of OH groups as water/steam by an endotherm and at the other 960 °C during crystallization by an exotherm, they observed sintering of heated kaolinite by two more steps, characterized by a high shrinking rate. Normally, third step with a high shrinking rate was followed by a forth step with a lesser one. In some clay, shrinkage during the fourth step resembles that of the third step and in some clay it even exceeded it. They tried to explain the third step of shrinkage attributable to the formation of mullite. The fourth step was ascribed to the formation of a molten phase influenced by the quantity of flux content of kaolin.

Flank (1979) studied thermomechanical analysis (TMA) in a Dupont Model 943 using 1/16” die auger extruded kaolinite plate. As in previous studies, he noted that TMA curve consists of two fairly sharp shrinkage “step” of roughly comparable magnitude, one at about 450–550 °C and the second at 900–980 °C. He explained the first step to be related to dehydroxylation and second step to be due to mullitization in contrast to the explanation put forward by Schuller and Kromer (1975). The temperature of occurrences is somewhat below than those commonly observed transitions in DTA.

Chakraborty (1992, 2003) studied DTMA of English china clay in a Shimadzu Thermal Analysis System. Dilatometric curve (1) of Fig. 3.4 consists of the following four major contraction steps and these occur at the following temperatures.

(i) First one is at about 500 °C,
(ii) The second is at about 980 °C,
(iii) The third large contraction is in between 1,100 and 1,400 °C and lastly,
(iv) The fourth is at 1,400–1,450 °C respectively.

The corresponding derivative TMA curve (2) shows differential contraction peak at about 980 °C preceded by an asymmetry. DTMA curve (3) in a sensitive run shows two contraction peaks in the asymmetry region on TMA.

Chen et al. (2000) showed that kaolin powder shrinks in anisotropic manner during sintering process (Fig. 3.5). Shrinkage in the direction of thickness is larger than diameter shrinkage.

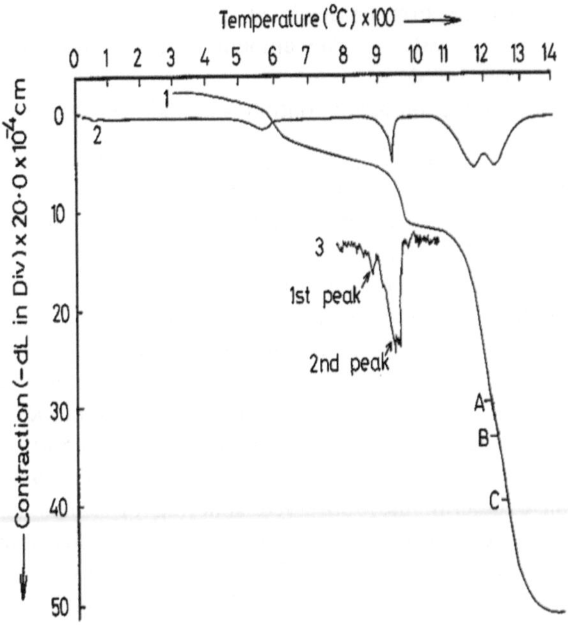

Fig. 3.4 Dilatometric and *2* DTMA curves of English kaolinite. Rate of heating, 10 °C/min; sample size, 5.2 mm in length and 4 mm in dia; sensitivity ± 1000 μm; chart speed, 1.25 mm/min. *3* Portion of DTMA curve of English kaolinite. Sensitivity, ±50 μm (after Chakraborty 1992). Reprinted by permission of the American Ceramic Society

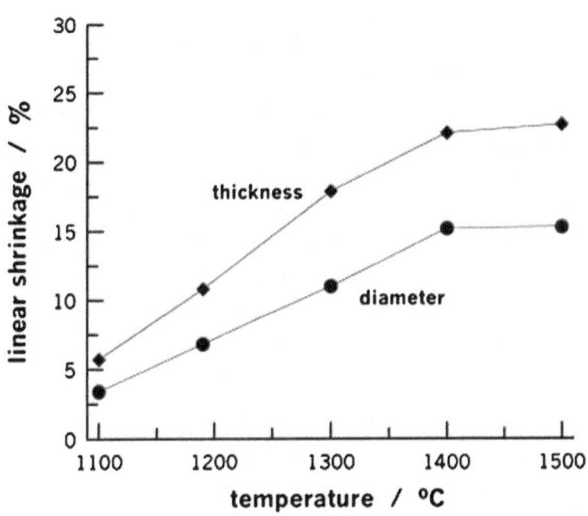

Fig. 3.5 The linear shrinkage of kaolinite powder compact (after Chen et al. 2000). Reprinted by permission of Ceram. International

3.7 Correlation Between DTA, TGA, and TMA Results

As shown above the thermal changes of kaolinite comprises of three major steps, namely dehydroxylation of kaolinite to metakaolinite at 500–600 °C endotherm; sudden crystallization at about 980 °C exotherm; and further recrystallization to mullite at about 1,250 °C exotherm. First transformation is associated with loss in weight. Above all, these three transformations are accompanied by considerable volume changes and also enthalpy changes. It is inferred from the above studies that an interrelationship may exist in the results among the three methods of thermal investigations. Following authors attempted to correlate different thermal studies.

3.7.1 Between Dilatometry and DTA

Heindl and Meng (1939) first compared thermal changes of Georgia kaolinite with DTA. According to them the two very-rapid contractions took place during heating in the temperature range 500–600 °C and 925–980 °C concurred with similar events happened during endothermic and exothermic reactions of kaolinite.

3.7.2 Between DTA, Dilatometer, and TGA

Freund (1960a, b) tried to compare the DTA data of Vaughn (1958, 1955) and Glass (1954) with shrinkage curve of his own to that of weight loss curves given by Mackenzie (1957). It was shown that striking thermal effects in DTA, well correspond with noted volume changes in shrinkage curves.

Udagawa et al. (1969) carried out thermogravimetric analyses and DTA analysis of allophane and kaolinite at the heating rates of 5 °C/min and 10 °C/min, respectively, and then compared.

Tsuzuki and Nagasawa (1969) compared the differential contraction curve with simultaneously recorded DTA curve. Peak temperatures in two cases are almost identical, but the peak on differential dilatometric curve starts at a far lower temperature than that in the DTA curve i.e., contraction begins before crystallization reaction at 980 °C exotherm. According to them, the contraction in this period may be due to the small endothermic peak occurring immediately before the 980 °C exotherm.

3.7.3 Between TGA and DTA

Watanabe (1987) compared the TGA and DTA traces of Georgia kaolinite by performing the two analyses at the same heating rate. Charkraborty (1993) showed that thermal events occurring in the DTK of Rajmohal kaolinite are in good agreement with the effects recorded in DTMA (Fig. 3.6). For example:

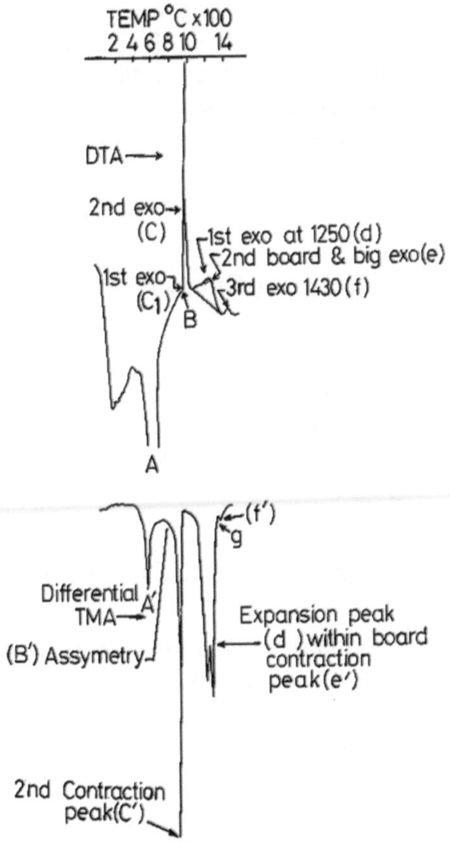

Fig. 3.6 Differential TMA curve of Rajmohol kaolinite compared with its DTA curve. Experimental conditions: In DTA: Rate of heating, 20°/min; sample weight, 40 mg; Sensitivity, 50 μv and chart drive, 1.25 mm/min. In DTMA: Rate of heating, 20°/min; sample size, 2 mm height and 4 mm dia; sensitivity, ±50 μm and chart drive, 1.25 mm/min (after Chakraborty 1993)

(i) During metakaolinite formation, DTK shows an endothermic peak (A) and DTMA records a differential contraction peak (A′)
(ii) During the onset of decomposition of metakaolinite, DTK shows one endothermic dip (B). While, the corresponding DTMA trace exhibits a contraction peak (B′) in the asymmetric portion.
(iii) During crystallization of decomposed metakaolinite, DTK shows a sharp 980 °C exotherm (C) and the DTMA trace exhibits a very sharp differential contraction peak (C′).
(iv) During recrystallization, DTK shows one exotherm at about 1250 °C(D) and DTMA shows a differential expansion peak (D′).
(v) Before and after recrystallization phenomenon, DTK exhibits a broad exotherm in the range of 1,100–1,400 °C with its maximum at 1,330 °C (E). While DTMA shows a large differential contraction peak (E′).

(vi) At the end of crystallization processes, DTK exhibits an exothermic peak (F) in the temperature range 1,410–1,440 °C. This peak coincides with differential contraction peak (F′) in the same temperature range in DTMA record. Therefore, it is inferred that the thermal effects observed during DTK run corroborate well with the dimensional changes noted in DTMA of kaolinite. Hence, the same phenomenon may be responsible for exhibiting two different physical changes in the two methods of analysis.

3.8 Factors Affecting Thermal Events in DTA, TGA and TMA

Naturally occurring kaolinites generally vary to a great extent in their crystalline order of perfection and particle size. Different types of impurities, e.g., quartz, mica, illite, feldspar, TiO_2, and Fe_2O_3 are often associated with clays of different localities. Obviously, all these factors will affect the DTA and TMA characteristics of kaolinites. In this chapter, it has been shown how crystallinity, crystal size, and impurities influence the DTA traces and shrinkage behaviors in TMA, respectively.

3.8.1 Crystallinity and Size of Kaolinite

Grim (1947) studied different kaolinites for DTA and showed that variations in the size and perfection of crystallinity of particles of kaolinite reflected in the variations of intensities of the thermal reaction characteristics of the mineral. The intensity of the large endothermic reaction at ~600 °C is slightly greater for well-crystallized kaolinite than for poorly crystallized variety. Further, endotherm for a well-crystallized kaolinite occurs over a greater temperature range. The intensity of 980 °C exothermic peak is considerably greater for a well-crystallized than for a poorly crystallized kaolinite. The exothermic slope of the curve between 600 and 900 °C is greater for well-crystallized kaolinite.

Well-crystallized kaolinite shows a slight endothermic reaction (called as endothermic dip) immediately before 980 °C exotherm, whereas a poorly crystallized kaolinite does not. He explained that loss of (OH) from the kaolinite lattice at ~600 °C in case of a well-crystallized kaolinite is not accompanied by complete destruction of lattice. It may however occur at the small endothermic dip at ~925–950 °C.

According to Arens (1951) poorly crystallized type of kaolins showed a shifting of the endothermic reaction to lower temperature range with reduction in intensity of the peak. By TEM and DTA, Sudo et al. (1952) showed both higher endothermic and exothermic peaks for well-crystallized kaolinite. The 980 °C exothermic peak was sharp in nature.

Wahl and Grim (1964) carried out DTA study of a well-crystallized Georgia kaolinite, a poorly crystallized kaolinite e.g., Illionis kaolinite and Eureka halloysite (Fig. 3.7). Well-crystallized kaolinite shows complete dehydroxylation at ~600 °C and follows the 1st sharp exothermic reaction at 980 °C. The 2nd exothermic reaction begins at approximate 1,200 °C and shows maximum with its T_m at 1,280 °C. Finally, it exhibits a 3rd exothermic reaction 1,350 °C. In comparison, DTA curve of a poorly crystallized variety of kaolinite shows an endothermic reaction of reduced intensity at 550 °C, the 1st exotherm occurs at a lower temperature of ~950 °C and is of lesser in magnitude as indicated by a subdued peak. Similarly, the T_m of 2nd exotherm also occurs at a low temperature at ~1,200 °C in comparison to 1,280 °C in of a well-crystallized variety. As expected, the 3rd exotherm is exhibited at ~1,250 °C which is lower than for a well-crystallized variety. DTA curve for halloysite is similar to that of poorly crystallized kaolinite up to 950 °C. But the 2nd exotherm occurs at a higher temperature at 1,250 °C than in poorly crystallized kaolinite. The 3rd exotherm is not noticed in case of halloysite. By examining the phases developed during heating, Wahl and Grim (1964) demonstrated that structural differences in the original clay mineral may be responsible for the formation of different phases. Graf et al. (1962) also showed that structural inheritance is often a major factor in controlling new mineral crystallization.

Bellotto (1994) took two well-crystallized kaolinite namely KAOSAR from Italy; KGa-1 from Georgia and a poorly crystallized kaolinite named as KGa-2 from other locality of Georgia for TG–DTA study. Regarding the occurrence of dehydroxylation peak, he noted that the maximum of the DTA peak shifted toward higher temperature from 511 °C for KGa-2 to 572 °C for KAOSAR and thus concluded that dehydroxylation of kaolinite is slowed down as the order of kaolinite increases. Regarding the occurrence of 1st exothermic peak, he also observed a similar shifting toward higher temperature from 986 °C for KGa-2 to 993 °C for KAOSAR.

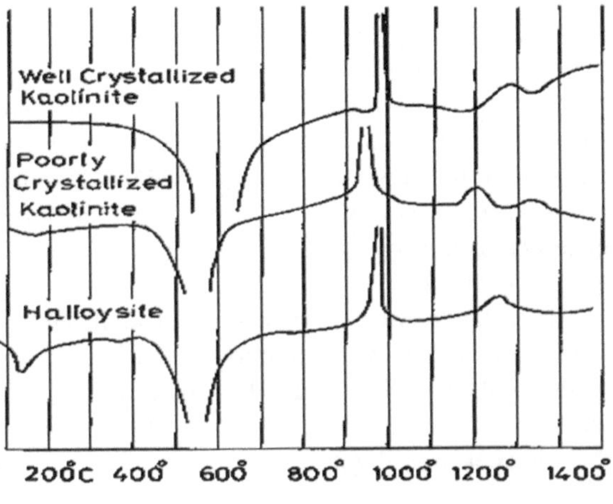

Fig. 3.7 Differential thermal curves of well-crystallized kaolinite, poorly crystallized kaolinite, and hydrated halloysite (after Wahl and Grim 1964) free available

3.8.2 *Impurities, Particle Size, and Crystalline Order of Kaolinite*

Glass (1954) investigated a selected group of kaolinite and halloysite clays of varying particle sizes, structural variations, and varying impurity content, e.g., mica by DTA. Characteristics of the halloysite thermal curve in comparison to kaolinite are the followings.

(i) There is a large initial endothermic peak at ~100 °C.
(ii) The large endothermic peak near 600 °C occurs at approximately the same temperature as in case of a poorly crystallized kaolinite but it is of reduced intensity.
(iii) The peak for halloysite returns to base line is faster than it departs. Where as a well-crystallized kaolinites show about equal rates of speed.
(iv) The rise in base line from 600 °C onwards (exothermic slope ~600 °C) is similar to the shift of base line (slope) for mica bearing poorly crystallized kaolinite.
(v) The small endothermic reaction at about 950 °C is less pronounced than it is for well-crystallized kaolinite.
(vi) The exothermic reaction below 1,000 °C is similar to that of well-crystallized kaolinite.
(vii) The curve above 1,000 °C is similar to that of a well-crystallized kaolinite. The peak at 1,250 °C is greater in intensity. Well-crystallized kaolinite shows one exotherm at 1,250 °C. The poorly crystallized kaolinite shows two distinct exothermic peaks. Mica bearing variety shows peaks at 1,200 °C and 1,300 °C. On the other hand the mica free clay shows peak temperature at ~1,200 °C and 1,240 °C, respectively.

The thermo grams obtained by DTA of kaolin samples mixed with different amounts of micrometric CaF_2 are shown in Fig. 3.8. Lomeli et al. (2009) suggested that the temperature of the endothermic peaks associated with dehydroxylation reaction (2) were not modified by the mineralizing additive.

The temperature of the exothermic peaks occurring at 989 °C related to the metakaolinite transformation was slightly shifted to a lower value as the amount of micrometric CaF_2 was increased. Even with the addition of conventional CaF_2 displaced the temperature of the reaction by only 20 °C.

3.8.3 *Particle Size of Kaolinite*

Norton (1939) fractioned Edger plastic kaolin in the centrifuge and prepared a series of monodispersed fraction. The heating curve for this set of fractions showed that the initial part of the endothermic curve occurs at the same temperature and the peak comes at about the same temperature in all cases, but the finer the particle

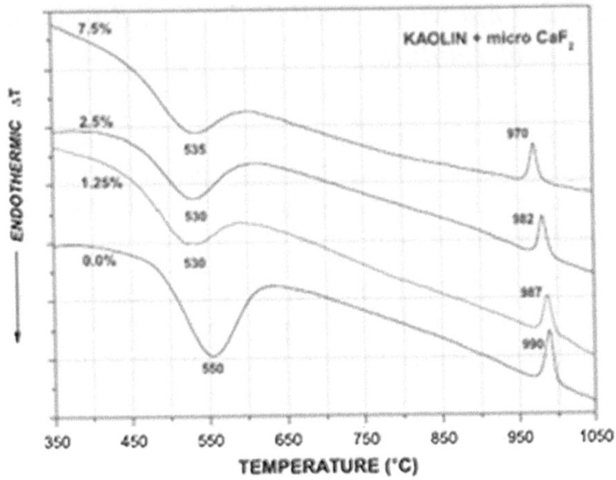

Fig. 3.8 DTA thermo grams of kaolin sample with different amounts of micrometric CaF_2 (after Lomeli 2009). Reprinted by permission of the American Ceramic Society

the more rapidly the curve backs to the zero line. The fine fraction shows a double exothermic peak between 900 °C and 1,000 °C.

Harman and Fraulini (1940) showed that occurrence of 600 °C endothermic peak is a function of particle size and cationic form of kaolinite.

Spiel et al. (1945) showed the variation of DTA curve with particle size. Both the intensity of the thermal reaction and the temperature of occurrence of the peaks decrease as the particle size decreases.

3.8.4 Heating Schedule

3.8.4.1 Dynamic Heating

Samoilov (1915) observed the influence of heating rate on DTA analysis. He observed that a lower heating rate lowers down the temperature of exothermic reaction in kaolinite. Houldsworth and Cobb (1923) showed that greater the rate of heating, more delayed the endothermic reaction and the 980 °C exothermic reactions, respectively. Spiel et al. (1945) stated that the area of a DTA peak should be independent of the rate of rise of temperature. They also showed that slower the heating rate, the broader the peak and lower the temperature of the peak.

Arens (1951) showed how the position of endothermic peak changes with rate of heating. The change in area was much greater than that shown by Spiel et al. (1945). With increase of rate heating from 1.75 to 12 °C per minute, Gerad-Hirne and Lamy (1951) noted the increase in the base-line shift. On heating kaolin at various rates, Spinedi and Franciosi (1952) found that the area under the endothermic peak passed through a maximum.

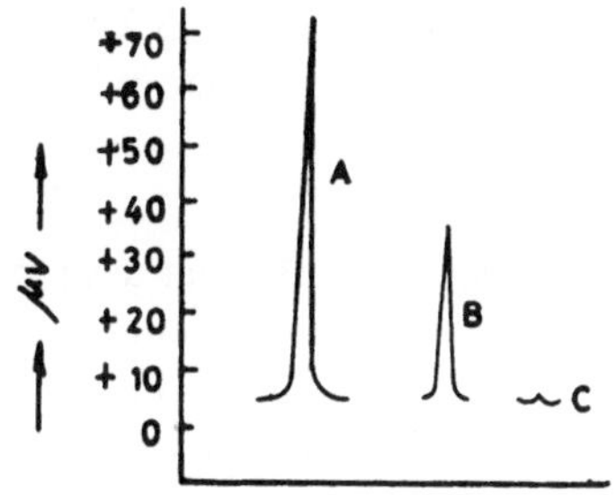

Fig. 3.9 Intensity of 980 °C DTA curves of Bhandak Kaolinite Soaked at 850 °C for Various Duration. $A = 1$ h; $B = 10$ h; $C = 30$ h (after Chakraborty)

3.8.4.2 Static Heating

Long/static heating effects 980 °C exotherm. Insley and Ewell (1935), heat-treated Zettlitz kaolinite for a different duration of time prior to DTA analysis. Apparently, it shows that size of the exothermic effect decreases with increase in temperature of heat treatment for a definite duration. Dekeyser (1963e) showed that on prolonged heating above 800 °C the exothermal peak decreases notably and vanishes after a heating of 24 h at 900 °C.

Chakaborty (1976) showed that at a fixed temperature, the intensity of DTA exotherm decreases with increase of soaking period (Fig. 3.9).

3.8.5 Atmosphere

The effect of water vapor on the decomposition temperature of kaolinite is expressed in a Clausius-Clapeyron type equation derived by combining the Van't Hoff equation with the equilibrium constant.

$$\text{Kaolinite} \underset{K_r}{\overset{K_f}{\longleftrightarrow}} \text{Metakaolinite} + 2H_2O \tag{3.2}$$

$$K' \,(\text{Equilibrium constant}) = \frac{K_f}{K_r} = P^2H_2O$$

$$\text{Or } \ln_{K'} = 2\ln_p.H_2O$$

$$\text{The Vant Hoff equation is } d\ln_{kp} = \frac{H}{RT^2}\,dT$$

$$\text{Or } \ln_{kp} = \frac{-H}{RT} + C$$

$$K' \sim Kp \tag{3.3}$$

By combining Eq. (3.2), results Eq. (3.4) which show the relationship of the pressure of water vapor, p.H_2O with the temperature i.e.,

$$\ln p.H_2O = H/2RT + C' \tag{3.4}$$

It thus indicates that a simple increase of water pressure should result in an increase of the temperature of dehydroxylation. Stone (1952) carried out DTA studies of different kaolinite samples at four values of partial pressure of water vapor. It is noted that increase of the partial pressure of water ranging from 5 to 760 mm, the occurrence of dehydroxylation temperature of kaolin group of minerals increases. The exothermic peak temperature is lowered by increasing the water vapor content (Fig. 3.10).

3.8.6 *Effect of Grinding on DTA Analysis*

The mechano-chemical effects produced by grinding kaolinite have been studied by Laws and Page(1946), Dragsdorf et al. (1951), Haase and Winter (1959), Miller and Oulton (1970), Korneva and Yusupov (1976), Kawai, Yoshida and Hashizume (1990), Iglesium and Anzar (1996), Gonza′lez Garcı′a et al. (1991), in different aspects, e.g., crystal structural break down of kaolinite by XRD, textural, and morphological changes by SEM, sorption behavioral studies (Suraj 1997), decrease in particle size and its distribution vis-a-vis increase in surface area by BET and finally thermomechanical effects by DTA, etc. Aglietti et al. (1986), Kristo'f et al. (1993), and Sa'nchez-Soto et al. (2000) compared the high-temperature transformation behavior of ground kaolinite versus unground variety.

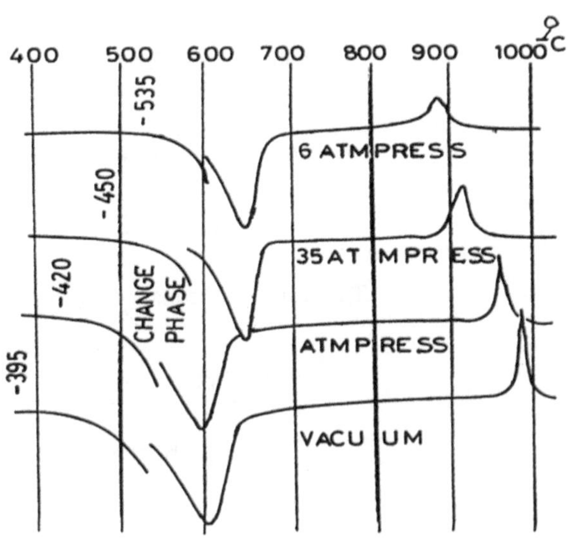

Fig. 3.10 Thermograms of kaolinite under different pressures (after Stone and Rowland 1955) free available

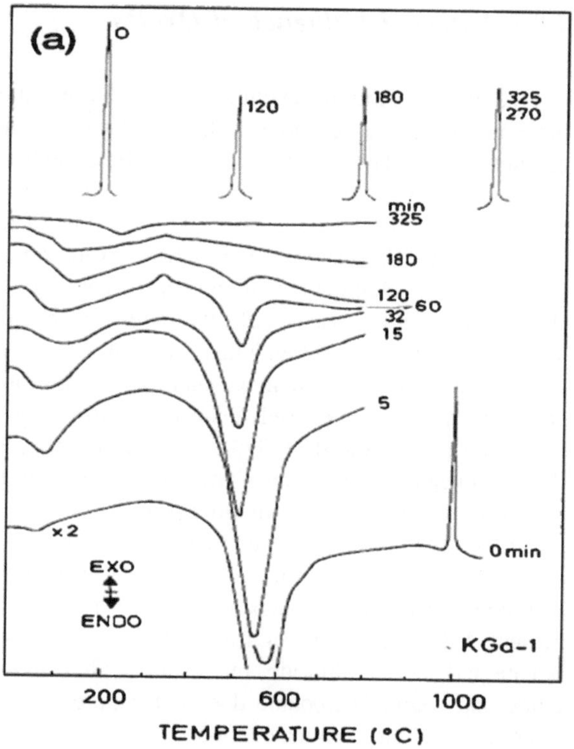

Fig. 3.11 DTA curves of the original kaolinite samples and of samples ground for different times: **a** KGa-1 (heating rate, 10 °C/min (after Sa'nchez-Soto 2000). Reprinted by permission of the American Ceramic Society

The reported results of these authors yield differently. There may be various reasons namely crystallinity, impurity content of original kaolinite and its induced reactivity generated when grinding is produced by impact and friction forces, such as vibratory, oscillating, and planetary mills.

The amorphization of crystalline kaolinite is evident after grinding KGa-1 at 120–180 min. Reflections at 001, 002, 003, and 006 gradually become broad and finally vanishes. Peaks related to b axis become random and structure transforms to disordered state. However, no information is gathered regarding the breakdown of tetrahedral silica sheet, octahedral aluminum hydroxyl sheet, even decomposition of –OH groups, and rupture of SI–O–Al linkages. DTA analysis of these ground clays is shown in (Fig. 3.11). The result shows that when grinding time was increased, the endothermic dehydroxylation peak in both samples shifted slightly to lower temperatures. The temperature of the characteristic first exothermic effect shifted slightly to lower temperatures with grinding, although the DTA effects did not increase with grinding time in either kaolinite sample, at least up to 325 min Fig. 3.12.

3.8.7 External Influence of Oxides

The kaolinite to mullite evolution processes are affected by presence of oxides inherent in natural resources. These induce low melting glassy phase during heating process. The nature of the liquid phase modify the exhibition of exotherms, formation temperatures of spinel and mullite phases and their quantities, and finally the densification process by viscous flow of it.

Belyankin (1932) showed that 980 °C exothermic effect could be screened by the presence of alumina containing substances, e.g., boehmite or hydrargillite. He showed that Chasorvyarsk (USSR) kaolin did not exhibit 980 °C exotherm but showed only the 600 °C endothermic peak. X-ray study of this clay showed to contain kaolinite. The chemical analysis of it showed a less Al_2O_3 content in comparison to kaolinite, Belyankin called this clay mineral as "monothermite". On analysis by Sedletski (1949), it was found that this kaolinite is admixture of illite or mica. The presence of these minerals is probably responsible for nonappearance of the 950 °C exothermic peak.

Gruver et al. (1949) showed that fluxes decrease or eliminate thermal effect of clay, e.g., admixture of iron oxide and soda. It is shown that the suppression of the heat effect of kaolinite is dependent on both the amount and type of contaminant present. Effective suppression is caused by fusible constituent, e.g., sodium carbonate and sodium chloride. Interestingly, they showed that combine use of iron oxide and sodium carbonate does not suppress the formation of γ-Al_2O_3 much more than sodium carbonate alone. They are of the opinion that oxides may react with clay when heated during thermal analysis. On DTA studies on the effect of CaO on kaolinite, it was noted that the characteristic 980 °C exothermic peak shift at a considerably lower temperature of 800 °C. Gruver et al. (1949) observed that certain impurities and active oxides may react with clay during heating and thus effect heat of transition at 600 °C and 980 °C, respectively (Fig. 3.13).

Sudo et al. (1952) was also noted that impurity oxides produced a downward shift of the peak temperature and also decrease the amplitude of the peak when a hydrated halloysite was analyzed by DTA (Fig. 3.13).

Dekeyser (1963a) had shown that 2 weight percent Na_2O addition on Zettlitz kaolinite followed by heating for 24 h at 700 °C, the exothermic peak is notably reduced and the starting temperature of the exothermal reaction is lowered. He also added 2.5 weight percent AlF_3 to Zettlitz kaolinite. The Al_2O_3 liberated by the reaction of AlF_3 with SiO_2 acted in the same way as the Na_2O.

Meneret (1957), Colgrave and Rigby (1952) carried out thermal analysis of kaolinite with addition of 2 % H_3BO_3. They showed that exothermic reaction starts about 30 °C lower than in case of pure kaolinite, and consequently the peak maximum is reduced although it still occurs at the same temperature. They further showed that on preheating kaolinite with 2 % boric acid at 900 °C for 200 h, the exothermic peak is totally screened of.

Lemaitre et al. (1975) showed the effect of intensity at the position of the 980 °C exotherm with mineralizers. They also noted the nature of the various

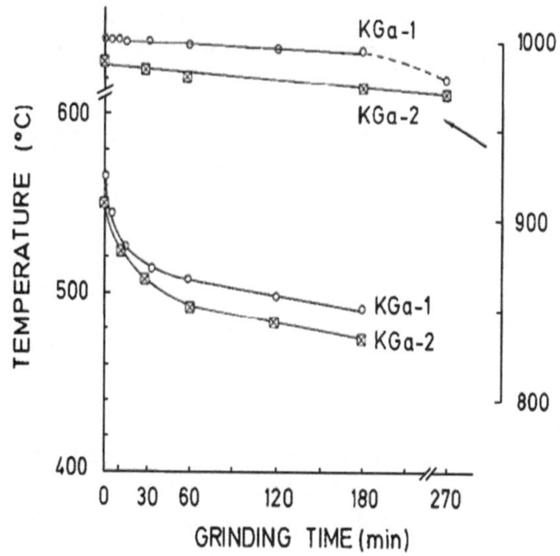

Fig. 3.12 Evolution of peak temperatures of endothermic and exothermic DTA effects observed in original KGa-1 and KGa-2 powders and in those ground for different times (after Sa'nchez-Soto 2000). Reprinted by permission of the American Ceramic Society

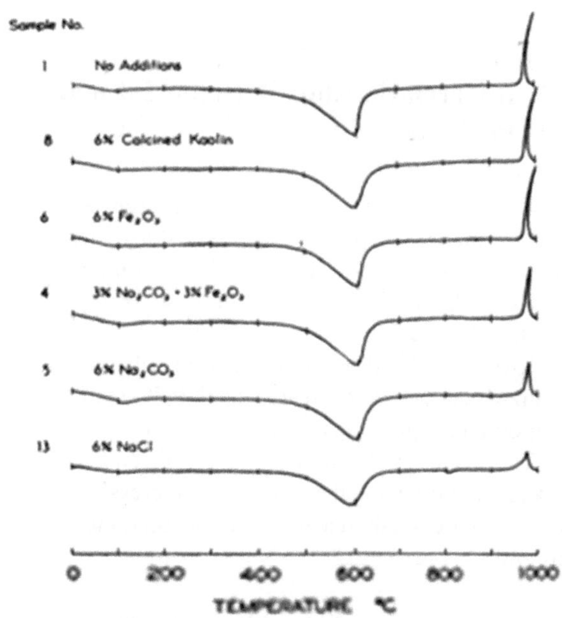

Fig. 3.13 Effect of various oxides on DTA studies (after Gruver et al 1949). Reprinted by permission of the American Mineralogist

crystalline phases which develop after a prolonged heating at 900 °C in presence of mineralizers. They used kolloid clay and incorporated impurity oxides by using analytical grade nitrates of corresponding metals in an agate planetary ball mill rotating at 800 rpm for 20 min. The proportion 'r' of the impurity M is expressed as the atomic ratio (r) = M/Si + Al. The DTA measurements were made with a

Setaram M 4 analyzer system. The area under the exothermic peak varies only moderately as a function of the nature of the mineralizer. This area increases in the following order.

$$\text{Na} < \text{Li} < \text{K} < \text{No mineralizer} < \text{Mg} < \text{Cu} < \text{Zn} < \text{Ca}$$

They classified the oxides act as mineralizes into three different categories according to their actions. Class-I. Alkali metals belong to this class. These oxides have no specific action on the transformation of metakaolinite to any given phase. On other hand, they increase the rate of transformation of metakolinite without promoting selectively of any reaction, i.e., both spinel and mullite formation are accelerated. Class-II. Oxides such as MgO and ZnO promote selectively the formation of spinel. Class-III. Oxides such as CuO accelerate the formation of mullite and cristobalite.

Johnson et al. (1982) showed the size and temperature of the 2nd and 3rd exothermic peak are related to the presence of impurity phases in kaolinite. The presence of CaO in the ordered kaolinite shifts the 2nd exotherm higher in temperature from 1,275 to 1,250 °C. Similar trend is noted for disordered kaolinite. The 3rd exothermic peak is even more sensitive to impurities.

3.9 Interrelationship Between 1st and 2nd Exothermic Peaks in DTA

Bulens and Delmon (1977) tried to present an interrelationship between 1st and 2nd exotherms by performing DTA study of mineralized kaolinite. They used a fixed proportion of CaO and MgO (atomic ratio, = M/Al + Si = 0.063) and mixed with four kaolinite on the basis of crystallinity and results are shown in Table 3.1.

X-ray powder diffraction diagrams of pure and for both CaO and MgO mineralized samples heat treated to 900 °C confirm that CaO promote mullite formation and MgO helps spinel formation.

The results of DTA show that with both CaO and MgO, the amount of heat liberated during exothermic effect increases considerably in all samples. They proposed two-path reaction for the thermal transformation of metakaolinite as follows.

$$\text{Kaolinite} \xrightarrow{\text{CaO}} \text{Metakaolinite} \rightarrow \text{Mullite} + (\text{x SiO}_2) \tag{3.5}$$

$$\text{Kaolinite} \xrightarrow{\text{MgO}} \gamma\text{–Al}_2\text{O}_3 + \text{SiO}_2 \rightarrow \text{Mullite} + (\text{x SiO}_2) \tag{3.6}$$

They also showed that MgO mineralized sample exhibits a second very diffuse exothermic reaction in the 1,100–1,200 °C region. They attributed this effect to a recombination reaction of segregated phases, e.g., γ-Al_2O_3 and SiO_2 to form

Table 3.1 DTA results of four kaolinites mineralized with CaO and MgO additives (=0.063) (after Bulens and Delmon 1977) Reprinted by permission of the Clay Mineral Society

DTA Data	Adddit	Arvor	Zettlitz	B591	Ybi
	–	581	721	975	870
Peak area	Mg	678	1623	1917	1674
(°C)^{2}·/g Kaol)	Ca	1124	1582	2167	1624
Peak width	–	9.5	11	13.7	12.7
	Mg	15.3	38	39.5	38.5
	Ca	29	40	57	45
Peak max (°C)	–	994	993	973	971
	Mg	974	976	958	946
	Ca	1004	1000	972	971

secondary mullite. In CaO mineralized samples, this exothermic event is not predominant since mullite formation proceeds directly and little phases are left with to recombine at higher temperature. They further mentioned effect of MgO and CaO in both well-crystallized and poorly crystallized kaolinites and made a comparison.

3.10 Summary

Le Chatelier (1887) first examined the thermal effects by immersing Pt/Pt–Rh thermocouple directly in clay sample when heated. Instead of using single Tc, Wohlin (1913) used a second thermocouple in the system and plotted both sample and furnace temperature against time. *D*ifferential thermocouple technique which becomes the basis for modern method of thermal analysis was developed. Insley and Ewell (1935) used differential thermocouple in his DTA apparatus and noted three major heat effects of kaolinite. Norton (1939) designed a new DTA equipment and took DTA traces of a group of kaolin minerals. Yamauchi and Kato (1943) explained the exhibition of the exotherm at 150–300 °C in the DTA trace of clay for oxidation of organic compounds. Grimshaw et al. (1945) standardized a DTA equipment which is easy to fabricate and analyzed a range of kaolin minerals. Bradley and Grim (1951) noted a pronounced 2nd exothermic peak in the temperature range of 1,200–1,300 °C. Glass (1954) studied a host of clay minerals of varying degree of structure, impurity content and particle size by DTA.

Dekeyser (1959c), Erdey and Paulik (1954) developed thermogravimetric apparatus and measured weight loss by nonequilibrium heating condition. Dekeyser (1963e) noted a small endotherm in DTG for a sample of kaolinite due to removal of last traces of water at a temperature just before 980 °C.

Hyslop and McMurdo (1938) and latter on Heindl and Mong (1939), Tsuzuki and Nagasawa (1969) noted two major contractile at/around dehydroxylation and crystallization steps. Schuller and Kromer (1975) made differential TMA and showed two differential contraction steps as like previous authors and lastly a major contractile due to mullite formation. Flank (1979) explained two steps due to dehydroxylation and mullitization of kaolinite.

DTA, TGA, and TMA studies done by Chakraborty and other researchers are documented for clays of varying origins. In general, the kaolinitic clay exhibits the following thermal events as listed below.

(i) At 100–150 °C—A broad and shallow endotherm.
(ii) At 500–650 °C—A prominent endotherm in DTA, a rapid weight loss in TGA, and a sudden contraction in TMA.
(iii) Just before 980 °C—A very small endotherm in DTA, which is just before 980 °C exotherm coincides with the small peak in DTGA. A moderate contraction peak in DTMA in the asymmetric region of the TMA curve.
(iv) At 980 °C Two exotherms: one is very small and second is very sharp. TMA shows a sudden contraction. DTMA shows a sharp contraction peak.
(v) At 1,100–1,400 °C—A broad exotherm in the region 1,100–1,400 °C in DTA. Within this exotherm another exotherm occurs at 1,250 °C in DTA. Analogous to DTA, DTMA records a large contraction peak the same temperature range 1,100–1,400 °C along with a sharp expansion peak at 1,250 °C.
(vi) At 400–1,440 °C—A sharp exotherm occurs in DTA. A differential contraction peak at the same temperature is noted in DTMA.

There are several factors which affect DTA peaks of kaolin minerals.

(i) Crystallinity of kaolinite plays a vital role. Well-ordered variety of kaolinite differ to a great extent from poorly crystallized clay in exhibiting 550 °C endotherm and subsequent 980 °C exotherm. Grim (1947) showed that degree of crystallinity of kaolinites affect both the temperature of occurrence and the intensity of endothermic and exothermic peaks. Subsequently, Arens (1951) and Sudo et al. (1952) confirm Grim's findings that poorly crystallized kaolinite showed a comparatively lower temperature endothermic and exotherm than well-crystallized variety. To verify the above view Laws and Page (1946), Dragsdorf et al. (1951) reduced the crystallinity by grinding kaolinite before DTA studies. Later on Wahl and Grim (1964) carried out a systematic DTA studies on well-crystallized, poorly crystallized kaolinite, and halloysite and then substantiated the earlier observations. They also noted 2nd and 3rd exotherms for poorly crystallized clay, those occur at lower temperatures than well-crystallized clay.
(ii) Particle size of kaolinite also plays a major role. Decrease in particle size of clay decreases the occurrences of the above endothermic and exothermic peaks. Norton (1939) showed that with decrease in particle size of kaolinite, the endothermic peak temperature decreases. As like endotherm, the 1st exothermic peak temperature also decreases with particle size as reported by Harman and Frauline (1940).

(iii) Heating rate also influences the peak temperatures. Generally higher the rate of heating, both endothermic and exothermic peaks shift to higher temperatures. Static heating of clay prior to DTA study generally decreases the size of 980 °C exotherm. Similar effect is also noted when temperature of pretreatment increases at a fixed duration of heating.

(iv) Samoilov (1915), Houldsworth and Cobb (1923) noted the shifting of endothermic and exothermic peak temperatures with change of rate of heating and without soaking. Arens (1951), Spinedi and Franciosi (1952) and Spiel et al. (1945) noted the changes in the shape and area of both endothermic and exothermic peaks. Static heating prior to DTA study made by Insley and Ewell (1935), Dekeyser (1963e) affect 980 °C exotherm.

(v) Water vapor also affects DTA peak temperature. Increase of it results in an increase in temperature of dehydroxylation. But opposite effect is noted in exothermic peak temperature. Stone (1952) noted the change in the start temperature of dehydroxylation and the exothermic reaction with change in partial pressure of water vapor.

(vi) Associated clay minerals like mica or illite, TiO_2 etc., or some externally added oxides decrease or even eliminate 980 °C exotherm. In some cases, the starting temperature shifts in a downward direction. Belyankin (1932), latter on Sedleski (1949) observed phases like illite or mica; Gruver et al. (1949) noted iron oxide and soda; Saveri noted CaO; Sudo and Ossaka (1952) noted impurity oxides, etc., reduced the amplitude of 980 °C exotherm. Dekeyser (1963e) noted that both the starting temperature and magnitudes of exotherm are affected by Na_2O and AlF_3. Colegrave and Rigby (1952) noted the effect of boric acid on clays during dynamic heating as in DTA as also in static heating conditions.

Lemaitre et al. (1975) used a group of mineralizes in DTA studies and noted that area of the 1st exotherm increases in the following order.

$$\text{Na} < \text{Li} < \text{K} < \text{No mineralizer} < \text{Mg} < \text{Cu} < \text{Zn} < \text{Ca}$$

In addition to screening effect of 980 °C peak by certain oxide addition, Johnson et al. (1982) noted that both size and temperature of 2nd and 3rd exothermic peaks are effected by CaO and K_2O. Bulens and Delmon (1977) showed a correlation between 1st and 2nd exotherm of some well-crystallized and poorly crystallized kaolinite by the selective addition of CaO and MgO and concluded that CaO exhibits large peak and promotes mullitization at 980 °C, whereas MgO promotes spinel formation at 980 °C and thereafter showed a large peak at 1,250° C for mullitization.

References

P.L. Arens, A study of differential thermal analysis of clay and clay minerals, Gravenage, Wageningen, Netherland, 1951, Excelsiors Fotd—offset's. Soc. Sci. **72**(5), 406 (1951)

E.F. Aglietti, J.M. Porto-López, E. Pereira, Mechanochemical effects in Kaolinite grinding, part I: Textural and physicochemical aspects. Int. J. Miner. Process. **16**, 125–133 (1986)

E.F. Aglietti, J.M. Porto-López, E. Pereira. Mechanochemical effects in kaolinitegrinding, Part II: Structural aspects. ibid.**16**, 135–146 (1986)

W.F. Bradley, R.E. Grim, High-temperature thermal effects of clay and related materials. Am. Mineral. **36**(3/4), 182–201 (1951)

D.S. Belyankin, in *Mullite, Its Structure, Formation and Significance*, ed. by J. Grofcsik, F. Tamas and A. Kiado (Publishing House of the Hungarian Academy of Sciences 1961, Budapest 1932), pp. 70

M. Bulens, B. Delmon, The exothermic reaction of metakaolinite in the presence of mineralizers: influence of crystallinity. Clays Clay Miner. **25**(4), 271–277 (1977)

M. Bellotto, High temperature phase transformation in kaolinite: the influence of disorder and kinetics on the reaction path. Mater. Sci. Forum, 166–169, 3–20 (1994)

E.B. Colegrave and G.R. Rigby, The Decomposition of kaoliite by heat. Trans. Brit. Ceram. Soc. **51**(6), 355–367(1952)

A.K. Chakraborty and D.K. Ghosh, Kaolinite–Mullite Reaction Series. Cent. Glass & Ceram. Res. Instt. Bull. **23**(2), 86–88 (1976)

A.K. Chakraborty, Resolution of thermal peaks of Kaolinite by TMA and DTA. J. Am. Ceram. Soc. **75**(7), 2013–2047 (1992)

A.K. Chakraborty, Application of TMA and DTA studies on the crystallization behavior of SiO_2 in thermal transformation of Kaolinite. J. Therm. Anal. **39**, 280–299 (1993)

A.K. Chakraborty, DTA study of preheated Kaolinite in the mullite formation region. Thermochima Acta **398**(1–2), 203–209 (2003)

C.Y. Chen, G.S. Lan, W.H. Tuan, Micro structural evolution of Mullite during the sintering of Kaolinite powder compacts. Ceram. Int. **26**(7), 715–720 (2000)

R.D. Dragsdorf, H.E. Kissinger, A.T. Perkins, An x-ray study of the decomposition of kaolinite. Soil Sci. Soc. Am. J. **71**, 439–448 (1951)

W.L. De Keyser, *Silicate Industries*, vol. 24. pp. 117 and190 (1959)

W.L. De Keyser, Differential thermo balance. Nature [London] **172**, 364 (1953)

W.L. De Keyser, Contribution to the study of Mullite. Ber. Dtsch. Keram. Ges. **40**, 304–315 (1963a)

W.L. De Keyser, R. Wollast, L. De Laet, Contribution to the study of OH groups In kaolin minerals. *International Clay Conference* (Pergamon Press, 1963b), pp. 75–86

W.L. De Keyser, Note concerning the exotherm reaction of kaolinite & formation of spinel phase preceding that of mullite. *International Clay Conference* (Pergamon Press, 1963c), pp. 91–96

L. Erdey, F. Paulik, J. Paulik, Differential thermogravimetry. Nature **174**, 885–886 (1954)

F. Freund, Die Deutungder Exothermen Reaktio Des Kaolinite Als Reaktio Des Aktiven Zustandes, Ber. Deut. Keram. Ges. **37**, 209–218 (1960a)

F. Freund, Explanation of exothermal reaction of kaolinite as a 'Reaction of the Active State'. Ber. Deut. Keram. Ges. **37**(51), 209–218 (1960b)

W.H. Flank, Behavior of kaolinite pellets at elevated temperature. Clays Clay Miner. **20**(1), 1–18 (1979)

R.E. Grim and R.A. Rowland, Differential thermal analysis of clays & shales, A control of prospecting method. J. Am. Ceram. Soc. **27**(3), 65–76 (1944)

R.W. Grimshaw, E. Heaton, R.L. Roberts, Refractory clays 11. Trans. Br. Ceram. Soc. **44**, 76–92 (1945)

R.E. Grim, Differential thermal curves of prep. Mixtures of clay minerals. Am. Mineral. **32**(9, 10), 493–501 (1947)

R.M. Gruver, E.C. Henry, H. Heysteck, Suppression of thermal reactions in Kaolinite. Am. Min. **34**, 869 (1949)

J. Gerad-Hirne, C. Lamy, Identification of clays by differential thermal analysis. Bull. Soc. France Ceram. 26–40 (1951)

H.D. Glass, High-temperature phases from kaolinite and halloysite. Am. Mineral. **39**, 193–207 (1954)

R.B. Graf , F.M. Wahl and R.E. Grim, Phase transformations in silica-alumina-magnesia mixtures as examined by continuous X-ray diffraction : 1. Talc-kaolinite composition. Amer. Min. **47**, 1273–1283 (1962)

F. González García, M.T. Ruiz Abrio and M.G. Rodríguez, Effects of dry grinding on two kaolins of different degrees of crystallinity. Clay Miner. **26**(4) 549–565 (1991)

H.S. Houldsworth and J.W. Cobb, Behavior of fireclays, bauxites etc on heating. I. Trans. Brit Ceram. Soc. **22**, 111–137, 344–348 (1923)

J.F. Hyslop, A. McMurdo, The thermal expansion of some clay mineral. Trans. Ceram. Soc. (England) **37**, 180–186 (1938)

R.A. Heindl, L.E. Meng, Length changes and endothermic and exothermic effects during heating of flint and aluminous clays. J. Res. Natl. Bureau. Stand. **23**(9), 427–441 (1939)

C.G. Harman, F. Fraulini, Properties of Kaolinite as a function of its particle size. J. Am. Ceram. Soc. **23**, 252–259 (1940)

T. Haase, K. Winter, Influence of grinding on the ceramic properties of kaolin (in Fr.). Bull. Soc. Fr. Ceram. **44**, 13–19 (1959)

I.H. Insley, R.H. Ewell, Thermal behavior of the kaolin minerals. J. Res. Natl. Bur. Stand. **14**(S), 615–627 (1935)

A. La Iglesium, A.J. Anzar, Crystallinity variations in Kaolinite induced by grinding and pressure treatments. J. Mater. Sci. **31**, 4671–4677 (1996)

S.M. Johnson, J.A. Pask, J.S. Moya, Influence of impurities on high-temperature reactions of Kaolinite. J. Am. Ceram. Soc. **65**(1), 31–35 (1982)

W.P. Kelly and H. Jenny, Reaction of crystal structure to base exchange and its bearingon base exchange. soil sci. **41**, 367–382 (1936)

G. Kulbicki, R.E. Grim, A new method for thermal dehydration studies of clay minerals. Min. Mag. **32**, 53 (1959)

T.A. Korneva, T.S. Yusupov, High-temperature behavior of Kaolinite after super-fine grinding, in *Proceedings of the First European Symposium on Thermal Analysis* (Heyden, London, 1976), pp. 336–339

S. Kawai, M. Yoshida, G. Hashizume, Preparation of mullite from Kaoline by dry grinding, J. Am. Ceram. Soc. Jpn. **98**, 669–674 (1990)

E. Kristóf, A.Z. Juhász, I. Vassányi, The effect of mechanical treatment on the crystal structure and thermal behavior of Kaolinite. Clays Clay Miner. **41**(5) 608–612 (1993)

H. Le Chatlier, De 1'Action de la Chaleur sur les Argiles" ("Concerning the Action of Heat on Clays"). Bull. SOC. Fr. Mineral. **10**, 204–211 (1887)

W.D. Laws, J.B. Page, Changes produced in Kaolinite by dry grinding. Soil Sci. Soc. Am. J. **62**, 319–336 (1946)

J. Lemaitre, M. Bullens, B. Delmon, Influence of mineralizers on the 950 °C exothermic reaction of metakaolinite, in *Proceedings of the International Clay Conference (Mexico City, Mexico, July 1975*), ed. by S.W. Bailey (Applied Publishing Ltd., Wilmette, 1975), pp. 539–544

M. Lomeli, L.M. Flores-Velez, I. Esparza, R. Torres, O. Domínguez, Catalytic effect of CaF_2 nanoparticles on sintering behavior of kaolin-based materials. J. Am. Ceram. Soc. **92**(7) 1526–1533 (2009)

J. Meneret, Bul. Cer. Fr. **35**, 3 (1957)

R.C. Mackenzic, *Thermal Methods, Differential Thermal. Investigation of Clays* (The Mineralogical Society, London, 1957), p. 22

J.G. Miller, T.D. Oulton, Prototropy in Kaolinite during percussive grinding. Clays Clay Miner. **18**(6), 313–323 (1970)

F.H. Norton, Critical study of differentia thermal method for Identification clay minerals. J. Am. Ceram. Soc. **22**, 54–63 (1939)

P.G. Nutting, Some standard thermal dehydratiom curves of minerals. U.S. Geol. Surv.Profess. Paper. **197E**, 197–216 (1943)

I. Rhode, Keram. Rundschau **35**, 414–415 (1927)

C.S. Ross and P.F. Kerr, The Kaolin Mineral. U.S. Geol. Surv. Profess. Paper. **165E**, (1930)

A.N. Sokoloff, Molekulares Zerfall Vo Kaolinites Anfang Du Gluhens, Tonind. Ztg. **36**, 1107–1110 (1912)

Y.A. Samoilov, Thermal curves of minerals. Bull. Acad. Sci. Petrograde **1759**, 1768 (1915)

S. Spiel, L.H. Berkelheimer, J.A. Pask, B. Davis, Differential thermal analysis—its applications to behavior clays and other aluminous minerals. U.S. Bur. Mines. Tech. 664 (1945)

Sedletski, in *Mullite, Its Structure, Formation and Significance*, ed.by J. Grofcsik, F. Tamas and A. Kiado (Publishing House of theHungarian Academy of Sciences 1961, Budapest 1949), pp. 70, X-ray characteristic ofMonothermite, pp 70

W.J. Smothers, Y. Chiang, and A. Wilson, Bibliography of differential thermal analysis.Univ.Ark. Inst. Sci. Technol. Res. Ser. **31** (1951)

T. Sudo, K. Nagasawa, M. Amafuji, M. Kimura, S. Honda, T. Muto, Studies of Japans clay minerals. J. Geol. Soc. Jpn **58**, 115–130 (1952)

P. Spinedi, O. Franciosi, Thermo Diff. Precis. Anal. React. Sci. **22**, 2323–2339 (1952)

R.L. Stone, Differential thermal analysis of kaolin group mineral under controlled partial pressure of H_2O. J. Am. Ceram. Soc. **35**(1), 90 (1952)

R.L. Stone and R.A. Rowland, in: *Thermoanalytical Methods ofInvestigation*, ed. by P.D. Garn, 1965. (Academic Press, New York, 1955), p. 297

K.H. Schuller and H. Kromer, Primary mullite as a pseudomorph after kaolinite.in *Proceedings of the International Clay Conference (Mexico City, 1975)*, ed.by S.W. Bailey. (Applied Publishing, Wilmette, IL, 1976) p.533–38

G. Suraj, C.S.P. Iyer, S. Rugmini, and M. Lalithambika, The effect of micronization onkaolinites and their sorption behavior. Appl. Clay Sci. **12**(2) 111–30 (1997)

P.J. Sánchez-Soto, M.C.J. de Haro, L.A. Pérez-Maqueda, I. Varona, J.L. Pérez-Rodríguez, Effects of dry grinding on the structural changes of Kaolinite powders. J. Am. Ceram. Soc. **83**(7) 1649–1657 (2000)

G. Tamman, W. Pape, Ilber Den Wasserverlust Des Kaolines and Seinverhat-en In Festen Zuden Carbonatem Und Oxyden Der Erdalkalien. Z. Anorg. Allg. Chem. **127**, 43–68 (1923)

Y. Tsuzuki, K. Nagasawa, A transitional stage to 980 °C exotherm of kaolin minerals. Clay Sci. **3**(5), 87–102 (1969)

S. Udagawa, T. Nakada, M. Nakahira, Molecular structure of allophane as revealed by its thermal transformation, in *Proceedings of International Clay Conference*, vol. 1, ed. by L. Heller, B Heller (Israil University Press, Gerusalem, 1969), p. 151

C.J. Van Nieuwenberg, H.A.J. Pieters, Rehydration of Metakaolin and the synthesis of Kaolin. Ber. Beut. Keram.Ges. **10**, 260–263 (1929)

F. Vaughn, Energy changes when kaolin minerals are heated. Clay Mineral Bull.**2**(13), 265–274 (1955)

F. Vaughn, Trans. Brit Ceram. Soc. **57**, 38 (1958)

R. Wohlin, Thermische analyse Von Tonen and Bauxiten. Silikatz **1**, 225 (1913)

F.M. Wahl, R.E. Grim, High temperature DTA and XRD studies of reactions, in *Twelfth National Conference on Clays and Clay Minerals*, pp. 69–81 (1964)

R.R. West, in *Ceramics*, ed. by R.C. Mackenzie. Differential Thermal Analysis, Fundamental Aspects, vol. 1 (Academic Press, London, 1970), pp. 149–179

T. Watanabe, H. Shimizu, K. Nagasawa, A. Masuda, and H. Saito, 29Siand 27Al-MAS/NMR study of the thermal transformations of kaolinite. ClayMiner. **22**, 37–48(1987)

T. Yamauchi, S. Kato, Thermal analysis of raw clays. J. Jpn. Ceram. Assn. **50**, 303 (1943)

Chapter 4
Extraction Techniques

4.1 Introduction

In olden days, even before the advent of X-rays and in absence of modern sophisticated phase characterization facilities, researchers were solely dependent upon chemical extraction techniques in order to dissolve component oxides so as to ascertain the nature of heated products of kaolinites as reviewed by Brindley and Nakahira (1959). Besides extraction techniques, dye absorption test and rehydration test on metakaolinite were also undertaken to reveal its nature.

4.2 Acid Dissolution Technique

Le Chatelier (1887) was the first investigator who used acid-leaching technique on kaolinites heated to different temperatures and showed that almost the total alumina content of metakaolinite could be dissolved in dilute HCl solution. Furthermore, he showed that there exists temperature limit beyond which the alumina component of metakaolin became insoluble again. On the basis of observations of acid dissolution study, Le Chatelier said "Free alumina comes from the molecular transformation of kaolinite".

Mellor and Holdcroft (1911) compared the molecular constitution of kaolinite and metakaolinite with hydrated alumina and concluded. "The circumstantial evidence thereof for all points one way. Kaolinite at temperature near 500 °C decomposes into free silica, free alumina, and water".

Sokoloff (1912) studied the evolution of water and the solubility of alumina as functions of the temperature and duration of heating. A temperature 450 °C is noted for the decomposition of kaolinite, but only after prolonged heating. Under rapid heating, complete loss of water and production of soluble alumina do not occur below 550 °C.

Tammann and Pape (1923) compared the dissolution behavior of heated kaolinite to that of an artificial intimate mixture of silica and alumina, and finally

A. K. Chakraborty, *Phase Transformation of Kaolinite Clay*,
DOI: 10.1007/978-81-322-1154-9_4,

noted that the alumina component in both cases is soluble in acid before the occurrence of the exothermic processes and after that it becomes insoluble.

Keppeler (1925) noted that the dissolution behavior of iron oxide content of kaolin and noted that it is insoluble under normal conditions. Like alumina component, it becomes soluble after moderate heating.

Neumann and Kober (1926) performed similar acid-leaching experiments and almost analogous observation is noted that the alumina content of kaolinite is leachable on heating the clay between 550 and 850 °C.

Krause and Whoner (1932) noted the solubility of alumina content of heated kaolinite by HCl and the dissolution of silica component by sodium carbonate solution, respectively. The maximum dissolution of alumina is found in the region of 400–900 °C and same for silica is at 400–600 °C.

Insley and Ewell (1935) choose Zettlitz kaolinite and studied for acid solubility of alumina on heated kaolinite among others. They noted a relationship with the effect of heating with amount of acid-soluble alumina to the intensity of the exothermic effect and the appearance of γ-Al_2O_3 spinel phase.

In these dissolution studies, authors came to the conclusion that kaolinite decomposes into free oxides during endothermic process. Studies also show that there exists a temperature range of alumina solubility. Dissolution study has been corroborated with DTA analysis. It has been observed that the highest reactivity is obtained when the clay has been heated to a temperature in the proximity of the endotherm, where metakaolinite just forms at the endothermic peak ~550 °C. This is the lower limit of alumina extraction. Again insolubility or fall in the rate of reactivity occurs when the clay is heated at or above 1,000 °C, approximately, around which a kaolinite generally exhibits an exotherm, i.e., the first exothermic peak temperature is the upper limit.

Jaffe (1971) showed that extraction of alumina from metakaolinite is dependent upon the concentration and type of mineral acid used e.g., HCl or H_2SO_4.

4.3 Sodium Carbonate Extraction Technique

Clarks (1895), Vesterberg (1925), Rhode (1927), Salmong (1933), and Budnikov (1935) showed that silica component of metakaolinite does not dissolve in hot sodium carbonate solution and thus they emphasized the existence of chemical bondage between the two component oxides. Thus, they believed that kaolinite does not decompose into free oxides and it may be a chemical compound. Insley and Ewell (1935) compared the physical and chemical properties of artificial and co-precipitated mixtures of amorphous Al_2O_3(Al_2O_3(A)) and amorphous SiO_2 (SiO_2(A)) in various proportions with that of components in metakaolinite. They showed that only a very small percentage of SiO_2 is dissolved from metakaolinite by hot Na_2CO_3 solution, where as ordinary SiO_2(A) dissolves readily. With this observation, they first concluded that metakaolin is not a simple mixture but is an "exceedingly intimate mixture of amorphous alumina and amorphous SiO_2.

They also noted that on heating amorphous alumina produced γ-Al_2O_3 spinel at 700–800 °C, but metakaolinite formed a spinel at a higher temperature of 950 °C. These findings led them to conclude that "Apparently the nature of the resultant crystalline phases depends upon the internal structure of the materials". They came very near to the conclusion that the exceedingly intimate mixture must have an "internal structure." Finally, they were of the opinion that presence of SiO_2 actually retards the recrystallization behavior of alumina component to the spinel phase. In support of this view, retardation effect of a specific crystalline in presence of 80 SiO_2 is cited. For example, γ-Fe_2O_3 is formed when Fe_2O_3 was heated at 600 °C. However, Fe_2O_3 and SiO_2(A) mixture obtained on heating nontronite ($Fe_2O_3.3SiO_2.5H_2O$) did not invert to γ-Fe_2O_3 below 1,000 °C. Hyslop (1944) heated kaolinite, and studied the solubility of Al_2O_3 in acid and SiO_2 in hot soda solution, respectively. In a second thought, he suggested a stabilizing effect of Al_2O_3 in solid solution with SiO_2 which was rather far from a mixture of oxides. Finally he concluded, "It may appear simpler to consider the product as a homogeneous amorphous complex in which there is a binding of the SiO_2".

Richardson (1961) opined that the decomposition into free oxides must yield individual oxides particles of extremely fine distribution and that the distribution of alumina—silica particles being of molecular fineness must be considered to be a solid solution and not a common mixture of the two components.

Colegrave and Rigby (1952) further carried out solubility study to test to question as to see, whether the repression of phase transformation of metakaolinite to Al_2O_3 plus SiO_2 as shown by Insley and Ewell (1935) was either due to weakly crystalline nature of compound formation or to the physical interference of rigid amorphous SiO_2 layer on amorphous Al_2O_3. They first showed that silica component of metakaolinite is not soluble in hot sodium carbonate solution confirming Clarke (1895) and others observation. In second run, they first extracted Al_2O_3 component of metakaolinite by boiling with HCl. The remaining insoluble residue was readily soluble in the hot sodium carbonate solution. They proposed that alumina in metakaolinite "protects" the SiO_2. But they were not certain whether this protection is mechanical or chemical.

4.4 Miscellaneous Test

Other miscellaneous tests were made to reveal the nature of metakaolinite.

4.4.1 Dye Absorption Test

Agafonoff and Vernadsky (1924) showed that heated kaolin absorbs dyes in a completely homogenous way, while the synthetic mixture absorbs the same dye heterogeneously. They believed that metakaolinite and synthetic mixture of SiO_2 and Al_2O_3 are structurally alike.

4.4.2 Rehydration Test

Some of rehydration studies are cited here which shows more light in revealing the nature of metakaolinite compared to the rehydration behaviors of dehydrated kaolin and equivalent mixture of alumina and silica in an autoclave. It is found that the former absorbed more water than the mixture. The product obtained after autoclaving the amorphous rehydrated kaolin was a crystalline hydrated aluminum silicate regained many properties of kaolinite. Studies also showed that complete rehydration takes place by using saturated steam at 100 atmospheres for 48 h. Rehydration limit of the sample was found out by them is 950 °C or less the firing temperature of kaolinite.

Van Nieuwenburg and Pieters (1929) observed that dehydrated kaolin took up more water when autoclaved in comparison to water took up by equivalent weights of amorphous SiO_2 and Al_2O_3 mixture resulting to crystalline hydrated aluminum silicate showing many properties resembling of kaolinite. With this observation they believed that dehydrated kaolin is a simple mixture of alumina and silica.

Schachtschabel (1930) also rehydrated the dehydrated kaolinite by autoclaving, however, believed that it is an anhydride of kaolin and refuting the amorphous aluminium silicate compound hypothesis.

Insley and Ewell (1935) noted that a mechanical mixture of dried Al_2O_3 and dried SiO_2 gels produced kaolinite when autoclaved like metakaolinite. They believed that the mixture hypothesis on the other hand is more simpler, and thus they conjectured that a molecular dispersion of alumina and silica exists in metakaolin which is different either from a gross mechanical mixture or from a true chemical compound.

It has been shown by Barrett (1937), Keppeler and Aurich (1939) and Gilard (1948) that when kaolinite fired at relatively low temperatures, it is reformed when autoclaved and the extent to which water is readmitted to the fired clay structure might, therefore, be taken as a measure of the degree of firing.

Grim and Bradley (1948) studied the rehydration of kaolin after heating it at temperatures between 500 and 800 °C for periods of 1–3 h, respectively. From DTA study they concluded that some OH water had been regained after the fired products remained exposed in air up to a period of 272 days. Finally, they predicted that some structure of the kaolinite persists still on dehydration of well-crystallized kaolinite at 600 °C which has sufficient regular to take up at least some amount of hydroxyl water.

Saalfeld (1955), Dietzel, and Dhekne (1957) rehydrated metakaolin hydrothermally to poorly crystallized kaolinite. Roy and Brindley (1956) showed that the rate and completeness of the conversion of metakaolin to kaolinite again varied with temperature, pressure, and time.

Hill (1956) showed that on autoclaving the kaolinite calcined at 500 °C, kaolin is regenerated as evidenced by the exhibition of characteristic peaks in DTA and by the appearance of a kaolinite X-ray pattern with characteristic basal spacing of 7.15 Å. Kaolinite fired at 1,000 °C could not reform by autoclaving. He explained this observation by the fact that metakaolinite structure was destroyed and 990 °C exotherm was absent. He also showed that during rehydration, the amount of water absorbed in kaolinite sample heated between 500 and 900 °C for 96 h treatment is 12 %. For the 950 °C samples rehydration drops sharply to 6.5 % and at 1,000 °C, it is a minimum of 0.88 % after 96 h. autoclaving. Thus, the limit of regeneration takes place in samples in metakaolinite range and the extent of rehydration was considered as a function of firing temperature.

4.5 Summary

By applying both acid- and alkali-leaching process, dye-absorption test, and rehydration study followed by subsequent DTA and XRD studies, two groups of researchers tried to understand the nature of the bonding between alumina and silica packets in dehydrated kaolinite.

The first group of investigators those attempted to dissolve Al_2O_3 component of heat-treated kaolinite by acid extraction technique indicated that during dehydroxylation kaolinite lattice either decomposed into free oxides or the alumina layer has been distorted to such an extent that metakaolin molecule becomes very susceptible toward mineral attack. It cannot be explained whether alumina goes into acid solution by simple dissolution of the free Al_2O_3(A) or Al_2O_3 dissolution is related to the chemical breakdown of the SiO_2–Al_2O_3 linkages of the aluminosilicate(A) compound.

Second group of researchers tried to solubilize SiO_2 component by hot Na_2CO_3 solution and believed that metakaolinite behaves as a chemical compound. Besides extraction process, Agafonof and Vernadsky applied dye absorption test. Klever and Kordes (1929) compared heat of solution of metakaolin to that of its component mixture. Rehydration test was undertaken to reveal the nature of metakaolin.

References

V. Agafonoff, W.I Vernadsky, Compt. Rend. **178**, 1082 (1924)

P.P. Budnikov, Ber. Dtsch. Keram.Ges. **16**, 349 (1935)

L.R. Barrett, Trans. Brit. Ceram. Soc. **36**, 201 (1937)

G.W. Brindley, M. Nakahira, The kaolinite-mullite reaction series: I, A survey of outstanding problems. J. Am. Ceram. Soc. **42**, 311–14 (1959)

F.W. Clarke, Constitution of the silicates. L7. S. Geol. Surury Bull. **125**, 7–109; 32 (1895)

E.B. Colegrave and G.R. Rigby, The Decomposition of kaolinite by heat. Trans. Brit. Ceram. Soc. **51**(6) 355–367(1952)
P. Dietzel, B. Dhekne, Ü ber die Rehydratation von Metakaolin. Berichte der Deutschen Keramischen Gesellschaft **34**, 366–77 (1957)
R.E. Grim, W.F. Bradley, Rehydration and dehydration of the clay minerals. Am. Mineral. **33**, 50–59 (1948)
P. Gilard, Verre Siicates Ind. **3**, 57 (1948)
J.F. Hyslop, Decomposition of clay by heat. Trans. Brit. Ceranz. Soc. **43**(3) 49–51 (1944)
R.W. Hill, Rehydrated and refried kaolinite minerals. Trans. Brit. Ceram. Soc. **55**, 441 (1956)
I.H. Insley, R.H. Ewell, Thermal behavior of the kaolin minerals. J. Res. Natl. Bur. Stand.14 (S) 615–627 (1935)
D. Jaffe, *The Chemistry and Physics of Clays*, 4th edn. by R.W. Grimshaw (Wiley-Interscience, New York, 1971), pp. 702
G. Keppeler, Sprechsaal **58**, 614 (1925)
G. Keppler, G. Aurich, Sprechsaal **72**, 71 (1939)
E. Klever and E. Kordes, Glastechn. Ber. **7**, 85, (1929)
O. Krause H. Wohner, Uber Die Vorange Beim Bremmen Technischer Tone. Ber. Dtsch. Keram. Ges.**13**, 485 (1932)
H. Le Chatlier, De 1'Action de la Chaleur sur les Argiles (Concerning the action of heat on clays), Bull. Soc. Fr. Mineral. **10**, 204–211 (1887)
V. Mellor I.I. Holdcroft, Chemical constitution of the kaolinite molecule. Trans. Ceram. Sor. (Engl.) **10**, 94–120 (1911)
B. Neumann and S. Kober, Sprechsaal **59**, 607 (1926)
R. Roy, G.W. Brindley, Study of the hydrothermal reconstitution of the kaolin minerals, in *Proceedings of the IV National Conference Clays*, (1956), p. 125
I. Rhode, Keram. Rundschau. **35**, (398–402), 414–415 (1927)
H.M. Richardson, Phase changes which occur on heating kaolin clays. in *The X-ray identification and crystal structures of clay minerals*, ed. by G. Brown (The Mineralogical Soc., London, 1961) p. 132–142
H. Saalfeld, The hydrothermal formation of clay minerals from Metakaolin. Ber. Deut. Keram. Ges. **32**, 150 (1955)
H. Salmong, Physikalischen und cheinischen Grundlagcn der Keramik (Physical and Chemical Principles of Ceramics), p. 73. Julius Springer, Berlin, 1933. 229 pp.; Ceram. Abstr. **13**(4), 103 (1934)
P. Schachtschabel, Chemie der Erde **4**, 395 (1930)
A.N. Sokoloff, Molekulares Zerfall Vo Kaolinites Anfang Du Gluhens. Tonind. Ztg. **36**, 1107–1110 (1912)
G. Tammann, W. Pape, Ilber Den Wasserverlust Des Kaolines and Seinverhat-en In Festen Zuden Carbonatem Und Oxyden Der Erdalkalien, in Z. Anorg. Allg. Chem. **127**, 43–68 (1923)
C.J. Van Nieuwenburg, H.A.J. Pieters, Rehydration of Metakaolin and the synthesis of kaolin. Ber. Beut. Keram. Ges. **10**, 260–263 (1929)
K.A. Vesterberg, Kaolin and its thermal changes. Arkh Krmi, Jlinrml. GcoE. **9**(1141), 26 (1925)

Chapter 5
X-Ray Methods

5.1 Introduction

Phase analysis of high temperature decomposition products of kaolinite are of considerable interest to the mineralogists as well as to ceramic technologists. X-ray is one of the most important tools and is used extensively for investigation of phase transformation behaviors of kaolinite during heating. After the discovery of X-ray, different types of camera/photographic techniques were gradually developed for recording powder and single crystal data. Thereafter, more rapid and sophisticated diffractometer techniques came up. Using such techniques, various researchers tried to evaluate the structure of kaolin minerals on the one hand and studied to reveal the mechanism of its thermal reaction on the other hand. Earlier reviews by Vaughn (1956), Freund (1960), Richardson (1961), and Davis and Pask (1971) are cited.

5.2 Powder X-Ray Diffraction Study

Powder X-ray diffraction procedure is an important technique to investigate the thermal changes of clay. It was first adopted by Rinne (1924) to study the effect of heating kaolinite to its endothermic peak temperature, i.e., 600 °C. He did not show any distinct lines in the photographic plates except some wide bands. As such he predicted that a compound of distorted structure is formed during the endothermic reaction of kaolinite.

Hyslop and Rooksby (1928) showed that metakaolinite crystallizes >925 °C, which was a poorly defined phase having resemble with the γ-Al_2O_3 spinel X-ray diffraction pattern. Thus, they assumed that alumina (A) liberated during dehydroxylation of kaolinite recrystallized to γ-Al_2O_3 as per the following transformation equation below and finally conjectured a mixed oxide hypothesis of metakaolinite.

A. K. Chakraborty, *Phase Transformation of Kaolinite Clay*,
DOI: 10.1007/978-81-322-1154-9_5, © Springer India 2014

Endotherm

$Al_2O_3\ 2SiO_2\ 2H_2O$----------- --→ [Al_2O_3 (A)+ SiO_2(A)]

Kaolinite H_2O

Metakaolinite

(X-ray amorphous) (5.1)

~ 925°C ↓ Recrystallized to

γ–Al_2O_3 Spinel + SiO_2(A)

Insley and Ewell (1935) heat treated Zettlitz kaolinite at a rate of heating of 6 °C/min to gradually increasing temperatures. Photographic X-ray diffraction patterns of air quenched samples collected between 700 °C up to beginning of exothermic peak temperature ~950 °C gave no X-ray pattern attributable to any crystalline material. The pattern consisted of a broad diffuse band characteristic of amorphous material and absence of any X-ray diffraction peaks. With this observation they at the beginning of study assumed that the endothermic effect is due to the dissolution of kaolin mineral into water vapor and an intimate mixture of γ-Al_2O_3 (A) and SiO_2 (A). This substance is called metakaolinite. At the maximum of the 980 °C exotherm, they noted the beginning of the formation of γ-Al_2O_3. After 980 °C, i.e., after completion of exothermic effect both γ-Al_2O_3 and mullite are noted. At higher temperature of 1,200 °C, γ-Al_2O_3 disappears and mullite forms in both dickite and kaolinite.

Jay (1939) heat treated four clays, siliceous clay aluminous clay, china clay, and bauxite clays for 22 h at an each temperature interval of 100 °C between 700 and 1,400 °C; a period of 2 h was allowed to change from one steady temperature to the next. He detected different phases at the following temperatures. γ-Al_2O_3 appears as a transient phase in between 900 and 1,000 °C. It disappears at or above 1,100 °C. Mullite first appears at 1,100 °C. Its X-ray intensity becomes prominent at 1,400 °C. Cristobalite first forms at 1,100 °C and disappears on heating 1,300–1,400 °C because of glass formation.

Tscheischwili, Buseem, and Weyl (1939) using crystal reflected X-ray, i.e., monochromatic radiation and creating vacuum inside the camera during X-ray exposure observed the persistence of hko reflections from heated kaolinities.

Subsequently, both electron microscopic studies by numerous researchers and single crystal flakes of nacrite (a polymorph of kaolinite) had been performed. These studies confirmed that metaphases show a hexagonal spot, i.e., metakaolinite still exhibits two dimensional order in the a-b plane. Thus, metakaolinite retains a good deal of structural order although (001) reflection,i.e., three dimensional regularity has disappeared.

Hyslop (1944) used a Scottish clay, which was heated for 15 h at 800 °C followed by 20 h at 1,050 °C, 20 h at 1,150 °C in order to note the appearance and disappearance of different phases. He noted γ-Al_2O_3 by X-ray on heat treatment at 1,050 °C. Above this temperature mullite formation took place. He expressed the following course of reaction (Eqs. 5.2–5.4).

Richardson and Wilde (1952) made a comprehensive X-ray study of heated clays of different origins. They made green clay body in cylinder moulds and then heated for 20 h at each 50 °C temperature interval from 800 to 1,350 °C. After each period of

$$\text{At } 550\,^\circ\text{C:}$$

$$\underset{\text{Kaolin mineral}}{Al_2O_3.2SiO_22H_2O} \rightarrow \underset{\text{Metakaolin}}{Al_2O_32\ SiO_2} + 2H_2O \tag{5.2}$$

At 850–1050 °C:

$$Al_2O_3.2SiO_2 \rightarrow Al_2O_3 + SiO_2\ (\text{with}\ Al_2O_3\ \text{in solid solution}) \tag{5.3}$$

$$\text{At } 900\,^\circ\text{C and upwards :}$$

$$\underset{\text{Solid solution}}{Al_2O_3 + SiO_2 -} Al_2O_3 \rightarrow \underset{\text{Mullite}}{3\ Al_2O_3.2\ SiO_2} \tag{5.4}$$

heating, clay cylinders were analyzed for phases by powder X-ray photography using filter Co radiation and by using 9 cm. Camera. After X-ray test, clay cylinders were replaced into the furnace for further heating at the next 50 °C higher temperature. They noted the first appearance of γ-Al_2O_3 and then its subsequent decay by identifying 0.139 nm reflection in the X-ray pattern by comparison with the standard. Similarly, starting from the appearance of faint to intense reflections of 0.342 nm, 0. 2218 nm, and 0.153 nm, the first appearance of mullite and its growth with rise of temperature were recognized. The formation of cristobalite and disappearance of quartz were observed by the appearance of 0.404 nm reflections and absence of 0.334 nm reflections. The temperature range of stabilities of γ-Al_2O_3, mullite, and quartz were taken to be the range over which the X-ray reflections persisted.

Glass (1954) used a selected group of kaolinite and halloysite of varying structure and impurities to note the development of high temperature phases. He heat treated those clay samples in the thermal furnace at a rate of about 10 °C min to ~1,000, 1,100, 1,200, 1,250, 1,300, and 1,350 °C, respectively. Samples were air quenched and analyzed for X-ray using powder diffraction camera. After dehydroxylation of clays, the diffraction pattern showed the evidence of the 0.43 nm bands. He suggested that a noncrystalline compound of SiO_2(A) and alumina (A) is formed because these bands are more prominent than those observed from truly amorphous solids. It persists up to the major exothermic peak. Thereafter, it collapses to form mullite nuclei, γ-Al_2O_3 and silica (A). Mullite formation causes the principal evolution of energy. The amount and size of mullite nuclei depend upon the morphology, structure, and impurity content of the clays. At 1,200–1,250 °C, the intermediate γ-Al_2O_3 phase reacts with SiO_2(A) produce secondary mullite. The rest of SiO_2(A) transforms to cristobalite between 1,240 and 1,320 °C. After analyzing several clays, he was of the opinion that variation in amount and type of impurities, rate of the heating, degree of equilibrium attained during firing, structure of the clay all four factors affect the sequence and temperature of phase formation. Sequence and interpretation of phase changes in clays are considerable. According to Glass (1954) these variations were due to failure to control/standardize firing conditions or all of the above mentioned variables. In general, sequential reactions of thermal changes of kaolinite were expressed by him are as follows.

$$\begin{array}{l}\text{Above } 600\,^{\circ}\text{C} \\ \underset{\text{Kaolin}}{Al_2O_3\,2SiO_2\,2H_2O} \rightarrow \underset{\text{Metakaolin}}{Al_2O_3\,2SiO_2} + 2H_2O\end{array} \tag{5.5}$$

$$\begin{array}{l}\text{At } 950\text{–}980\,^{\circ}\text{C} \\ Al_2O_3\,2SiO_2 \rightarrow \underset{\text{Primary mullite}}{3\,Al_2O_3\,2SiO_2} + \gamma\text{–}Al_2O_3 + SiO_2\end{array} \tag{5.6}$$

$$\begin{array}{l}\text{At } 1{,}200\text{–}1{,}250\,^{\circ}\text{C} \\ \gamma\text{–}Al_2O_3 + SiO_2 \rightarrow \underset{\text{Secondary mullite}}{3\,Al_2O_3\,2SiO_2} + SiO_2\end{array} \tag{5.7}$$

$$\begin{array}{l}\text{At } 1{,}240\text{–}1{,}350\,^{\circ}\text{C} \\ SiO_2(A) \rightarrow SiO_2(\text{cristobalite})\end{array} \tag{5.8}$$

Slaughter and Keller (1959) selected five clay samples for phase transformation studies. Firing of the clay samples were done between 600 and 1,510 °C. Each sample was brought to the desired temperature by heating at 10 °C/min and was held at that temperature for 12 h and finally cooled to room temperature at about 1 °C/min. Phases developed on heating were determined qualitatively (for all firing temperatures) by X-ray diffraction method and are schematically illustrated. Metakaolin was recognized by 0.35–0.45 nm band in X-ray pattern in the temperature range 600–950 °C, the diffraction pattern shows development of face centered cubic structure and was called γ-Al_2O_3. At 1,100 °C, three phases namely mullite, cristobalite, and glass are evident from X-ray diffraction data. Mullite phase shows higher degree of crystallinity. At lower temperature X-ray diffraction peaks of mullite are broad and flat topped and the 0.338–0.342 nm doublets is not resolved. At higher temperature 1,150–1,200 °C patterns show narrow, well-resolved mullite peaks which agree in relative intensity and spacing with ASTM data for mullite. Cristobalite is first detected in clays fired at 1,100 °C, which increases with increase of firing temperature.

Dekeyser (1963) showed the sequential changes of phase development on heat treatment of Zettlitz kaolinite at various temperatures by XRD.

Schieltz and Soliman (1956) attempted to explain first exothermic DTA peak of kaolinite by XRD studies of clays in combination with a thermodynamic approach. They heat treated different kaolinites and halloysite at temperature intervals as small as 10 °C in the neighborhood of the first exothermic peak.

The heating time was varied between 5 min and to several hours. Kaolinites of E.P.K., A.P.I. 9b, and Ureka. Halloysite, showed the formation of γ-Al_2O_3 only and no mullite during heating at 950 °C for 1 h. At longer heating times, they found both γ-Al_2O_3 and mullite as co-existent product phases in the fired kaolinite between 950 and 1,100 °C. X-ray patterns of γ-Al_2O_3 reflections are always very broad and diffuse, even after heating for as long as 28 h in the furnace which indicates that γ-Al_2O_3 never develops into a well-crystallized phase as compared to mullite. They questioned concerning the source of first DTA exothermic peak whether it is due to γ-Al_2O_3 or mullite or both.

Udagawa et al. (1969) choose a natural allophane and a well-crystallized kaolinite for their study. From the X-ray diffraction analysis of products at various temperatures they schematically showed similar in thermal sequence for both allophane and kaolinite. They emphasized the formation of metakaolinite upon topotactic decomposition of original kaolinite. In comparison, the collapse of the original structure of allophane leads to its transformation to γ-Al_2O_3 a spinel type structure with subsequent formation of mullite at high temperature. A small amount of mullite starts crystallizing almost simultaneously with spinel. The excess silica which is probably amorphous at this stage eventually crystallized into cristobalite. The overall process of thermal transformation of allophane is similar to that of kaolinite.

Leonard (1977) heat treated Zettlitz kaolinite at 600, 800, 900, 970, and 1,250 °C for 24 h and then analyzed by X-ray diffraction study to note its phase transformation behaviors. X-ray pattern of Zettlitz heated to 600 and 800 °C correspond to the quasi amorphous nature of metakaolintie. At 900 °C, the crystalline fraction appears as a pure γ-Al_2O_3 spinel structure. At 970 °C, spinel undergoes transformation to mullite. Finally at 1,250 °C, the intensity of mullite increases predominantly. Some amount of cristobalite also appears in the pattern along with mullite.

Powder X-ray patterns of the English kaolinite sample heated to successive increase of temperature and the sequence of phase development are shown by Chakraborty et al. (2003) in Fig. 5.1a, b. The clay samples were heated dynamically to successive rise of temperatures by heating at 100 °C/h in an electric muffle furnace. Several samples were collected on heating just after attainment of desired temperature without giving any soaking time. These dynamic run samples were cooled normally, ground to below 10 μm and finally stored. X-ray diffraction study of these samples were done by using Philips XRD generator, model PW 1730 attached with graphite monochromator and PW 1710 counting electronics coupled with computer. The tube was operated at 40 kV/20 mA, CuKα radiation. The X-ray diffractogram patterns were taken at different stages of heating English kaolinite by step scan mode between 5 and 80°2θ with steps of 0.02° and a

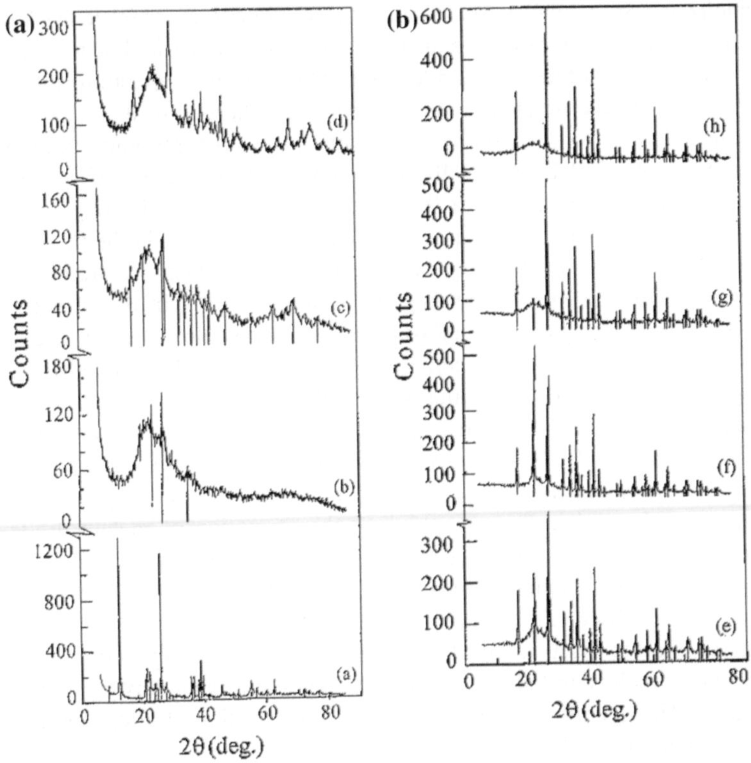

Fig. 5.1 **a** CuK_α X-ray patterns of high temperature phases of English Kaolinite. *a*, raw kaolinite, *b* showing metakaolinite formed on heating kaolinite at 900 °C; *c* and *d* mixtures of Al–Si spinel (S), mullite (M) and amorphous phases (band) formed at 1,050–1,150 °C. **b** CuK_α X-ray patterns of high temperature phases of English kaolinite. *e*, *f*, showing mixture of mullite and cristobalite formed at 1,250–1350 °C, *g*, *h*, showing mullite phase only at 1,450–1,550 °C

counting time of 0.5 s per step. These were subsequently identified in an extensive way for crystalline phases present by using PC Identify software.

At room temperature: XRD pattern shows characteristic pattern of well-crystallized variety of kaolinite. In addition to kaolinite, the raw clay contains traces of free quartz.

At ~700 °C: Besides quartz as residual phase, the XRD pattern is amorphous in nature. So, metakaolinite is recognized as showing 3–4 nm phase band.

At ~1,050 °C: Following decomposition products of metakaolinite are noted.

(i) Weak crystalline spinel phase comprise three broad XRD peaks at 0.238, 0.137, and 0.197 nm, respectively. However, later two peaks are noted usually in heated kaolinite of different sources. There was a contradictory opinion on the proper composition of this phase. Finally, it is designated as silicon bearing spinel phase and called as Al–Si spinel and not pure γ-Al_2O_3.

(ii) Second product forming is a poorly crystalline variety of mullite. All X-ray peaks characteristic of mullite are really broad, some of the X-ray peaks are missing. The diffraction pattern does not match fully in relation to intensity and spacing with the JCPDS card for mullite. However, doublets at 0.338 nm and 0.342 nm are resolved instead of a single peak noted previously.

(iii) Besides spinel and mullite like phases, the pattern still shows the amorphous band $\sim 22°2\theta$ which relates to formation of two noncrystalline phases out of the decomposition of metakaolinite structure. These are likely aluminosilicate phase equivalent to composition of 3:2 mullite (may be amorphous mullite) and free silica containing some amount of alumina. Pattern due to presence of trace quantity of free quartz in raw clay is eliminated at this temperature.

At 1,150 °C: All characteristic peaks of mullite are still very broad. Pattern of mullite shows weak in intensity. Weakly crystalline Al–Si spinel still exists as noted by its three well-known broad peaks at 37°, 47°, and $67°2\theta$, respectively. Both spinel and mullite phase shows a common peak at $\sim 37°2\theta$.

At 1,250 °C: Broadness of each mullite peak decreases to some extent. Intensity of each peak increases markedly. Weak peaks of cristobalite start to appear. The amorphous band goes on diminishing with crystallization of both mullite and cristobalite phases out of amorphous phases (Figs. 5.1, 5.2).

At 1,350 °C: Intensities of both mullite and cristobalite increase to a great extent. Cristobalite in presence of mullite is detected by X-ray peaks at 0.41 nm (very strong), 0.252 nm (strong), and by a small peak at 0.317 nm those are in agreement with JCPDS data card, respectively.

At 1,450–1,550 °C: Intensities of mullite peaks become sharper than those of previously heated samples. Peaks due to cristobalite almost vanish at 1,450 °C. Consequently, a band is again reappeared in X-ray diffractogram due to glassy phase development.

Recently, Bellotto (1994) submitted the phase evolution of well crystallized Georgia kaolinite marked KGa-1 at a heating rate of 10 °C/min using HT X-ray powder diffraction technique using a Siemens D5000 diffractometer and HTK–10 heating chamber from Anton Paar and is shown in Fig. 5.3. Salmeron, J., Lopez, C.: Forecasting risk impact on erp maintenance with augmented fuzzy cognitive maps. Software Engineering, IEEE Transactions on (99), - (2012).

5.3 Single Crystal Study

Brindley and Hunter (1955) used 0.2–0.5 mm size nacrite flakes for ascertaining its thermal transformation. It shows nearly similar lamellar morphology like kaolinite. Both are composed of silicate layers having the same structure and same chemical compositions, they differ only in the stacking arrangements of the layers. They were of opinion that their thermal reactions could be similar. X-ray rotation diagram around b axis of nacrite shows that its nature of crystallinity was good. On

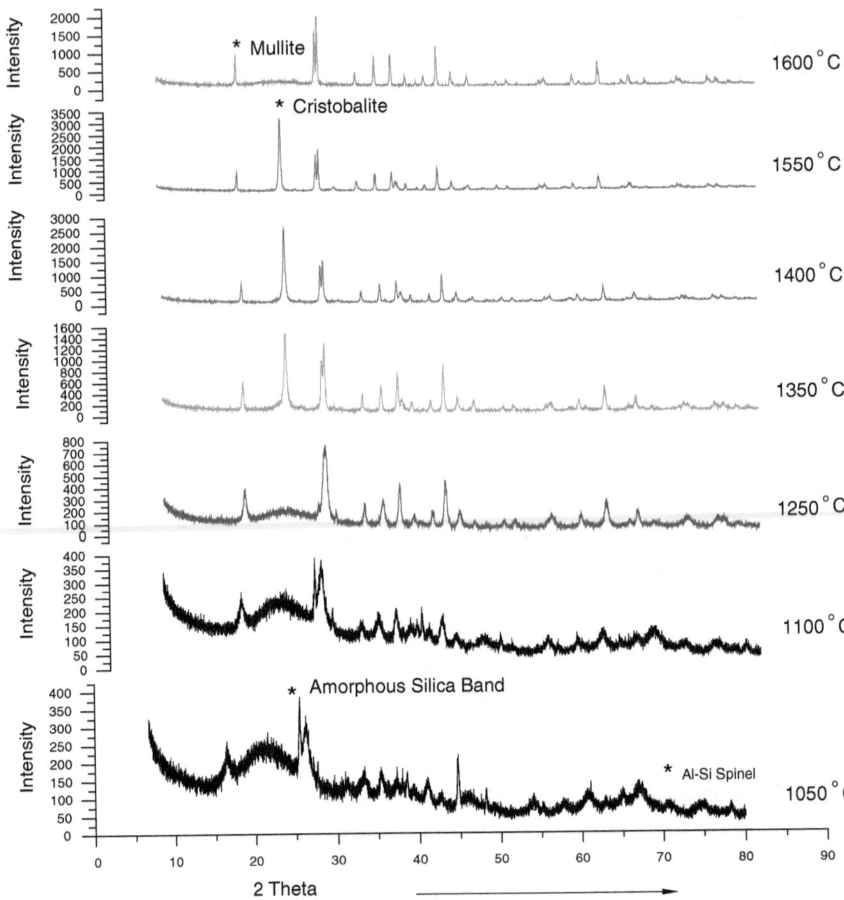

Fig. 5.2 XRD patterns of decomposition phases of M.P.clay; Mixtures of Al–Si spinel, mullite and amorphous silica band formed at 1,050 °C; Mixtures of mullite and amorphous silica band formed at 1,100–1,250 °C; Mixtures of mullite and cristobalite noted at 1,350–1,550 °C; Cristobalite disappears and mullite exist at ~1,600 °C

heating to 670 °C; major reflections disappeared only a hexagonal pattern persisted and it was indexed as 200, 110, and 310 only. A rotation diagram about the normal to the flake showed no evidence of regularity. The hexagonal pattern of spots become weaker and the diffused halo increased in intensity on heating to 800 °C. They interpreted that following the expulsion of (OH) radicals from the structure the layers retained a relict structure of hexagonal type having no detectable regularity from the layer analysis to another. When this single crystal nacrite flake was heated at 950 °C the X-ray rotation diagram about the a-axis showed a clear powder diagram in which a few lines show marked increase of intensity in a particular direction indicating preferential orientation. According to them, these peaks showing orientation effects correspond γ-Al_2O_3. This is a cubic

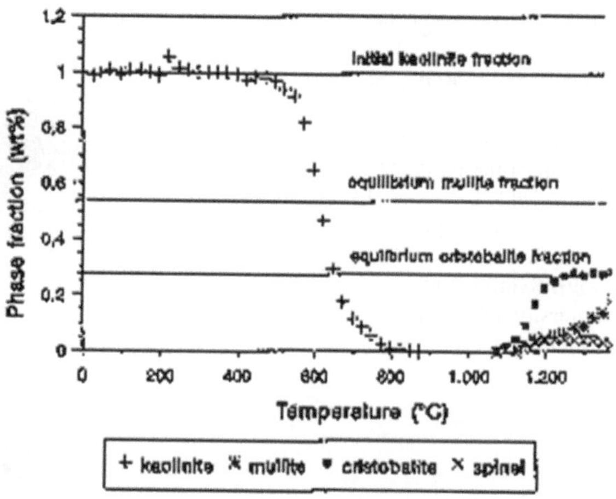

Fig. 5.3 Phase evolution of sample KGa-1 at 10 C/min Heating Rate (After Bellotto 1994). Reprinted by permission of the Trans. Tech. Publication Ltd

spinel type structure. They recognized the following peaks, e.g., 400 (I,10; d 0.1978 nm); 440 (I,10; d 0.1387 nm); 311 (I,8; d 0.240 nm); and 220 (I,2; d 0. 279 nm). Where I is the relative intensity and d values is in nm unit. In addition a faint mullite pattern is also visible. The presence of finally divided cristobalite shown by the diffused reflection at 0.41 nm. The line associated with γ-Al_2O_3 which showed marked orientation effect disappeared when heated to 1,150 °C. Mullite patterns become clearer and more lines are visible. Mullite patterns are in random orientation. They assumed γ-Al_2O_3 to have combined with cristobalite to produce more mullite. Persistent of 0.41 nm line indicated that excess SiO_2 still remains as cristobalite.

Brindley and Nakahira (1959a, b, c) slowly heated individual flakes of kaolinite of about 0.1 nm in size in platinum dishes to temperature of 490, 850, 900, 950, 1,015, 1,200, 1,300, and 1,400 °C for 2 days. A powder kaolinite sample was also heat treated to 400 and 600 °C. All these samples were analyzed by applying single-crystal X-ray technique. X-ray analysis of crystals heated between 400 and 850 °C show the absence of general reflection (hkl) and basel reflections (001) of kaolinite Fig. 5.4.

In rotation and oscillation diagrams taken around the a and b axes of the parent kaolinite, hexagonal pattern was obtained (B). Hexagonal pattern on indexing in terms of a and b parameters remained same as that of kaolinite. In rotation and oscillation diagrams taken around c axis, the normal to the cleavage plane, gave diffuse, and continuous scattering parallel to c with no periodicity.

Powder photographs and diffractometer recordings of metakaolinite showed even greater diffuseness than the hexagonal spot patterns in consequence of the diffuse scattering parallel to c axis. With these observation they indicated that metakaolinnite has two dimensional regularity in a and b planes and the absence of regularity in third dimension. They were of the opinion that in metakaolinnite layer structure of kaolinite persists in the modified form. On heating at 950 °C me-taolinite transforms to a cubic spinel phase.

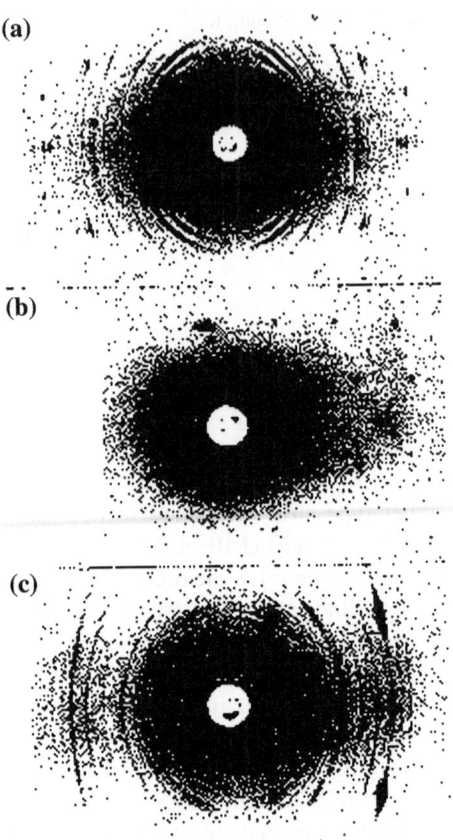

Fig. 5.4 X-ray single crystal diagrams. **a** Rotation diagram of kaolinite, **b** Oscillation diagram of metakaolin heated to 850 °C, and **c** Oscillation diagram of the product obtained at 950 °C, mainly an oriented spinel-type structure (After Brindley and Nakahira 1959). Reprinted by permission of the American Ceramic Society

Oscillation diagram around b axis of the original kaolinite (C) indicates that the cubic face diagonal (110) is a parallel to b of kaolinite and metakaolinite and cubic axis (111) is perpendicular to the cleavage plane, i.e., (001) of kaolinite. These predict that the periodicity normal to the cleavage plane, which is destroyed in metakaolin is restored in cubic phase. In face centered cubic lattice, the (111) planes are planes of close packed oxygen ions and (110) direction is the line of closest packing. Thus, the layer structure of kaolinite gives rise to the close packed layers of the cubic phase in correct orientation. This evidence points very strongly toward maintenance of structural continuity in those transformations. At 1,050 °C, the cubic pattern has become faint and by 1,100 °C it is disappeared when crystallization of both mullite and cristobalite are clearly seen. The pattern is too weak for accurately measuring its lattice parameter. At 1,100 °C, crystallization of mullite increases but preferred orientation is not evident in the X-ray patterns. Between 1,200 and 1,400 °C, crystallization of mullite and of cristobalite improves remarkably. Thus, by single crystal X-ray analysis of kaolinite flakes, the dimensional and directional relationship between kaolinite, metakaolinite, and spinel phases have been established by them. Thus, they gave a well-integrated picture of the sequence of events from the initial layer structure of kaolinite, through a residual layer structure in metakaolin, followed by a cubic spinel type phase and finally a chain type structure in mullite. In order to explain a consistent structural reaction sequence they first proposed that metakaolinite continuously discards SiO_2 and second the intermediate spinel phase would be Si-type spinel in which 8 Si are occupied in 8 tetrahedral sites, 10.2/3 Al ions are statistically filling the 16 octahedral sites. This spinel phase on progressive heating forms mullite with further discard of SiO_2 which appears as cristobalite. In this way they tried to explain how the continuity of the oxygen frame work is retained and how cations is migrated and a portion of SiO_2 is discarded by diffusion of Si ions.

Different views as regards crystallo-chemical transformation point of kaolinite-mullite reaction course are noted and described above. The differences of opinion rather complications arise in forecasting kaolinite to mullite reaction course are due to following reasons.

(*i*) ***Due to choice of kaolinites of different sources.***
Since, kaolinites vary widely in their characteristics, e.g., the most important is the crystallinity, starting from well crystallized to poorly crystallized variety (different in crystallinity index); high degree of stacking faults density (Bellotto, 1994). It is expected that a great variation in kinetic a mechanism/reaction path, leading to mullitization may occur due to varying degree of crystallinity of kaolinite chosen.

Bellotto (1994) studied well-crystallized Georgia (KGa-1) and a poorly crystallized variety of same source (KGa-2) for high temperature phase transformation study and found that in both samples, a spinel phase appears between 1,050 °C and 1,200 °C for faster heating rates and disappears during period at constant 1,400 °C.

In case of ordered kaolinite, he showed that the extent of mullite formation is similar with disorder kaolinite at a particular heating rate, e.g., 10 °C/min. However, it decreases with increasing of heating rates for ordered kaolinite while it is constant for disordered kaolinite at different heating rates. Well-crystallized kaolinite forms cristobalite at a temperature higher than 1,050 °C, its formation temperature increases with increasing heating rates up to 1,400 °C, e.g., 100 °C while KGa-2 never forms cristobalite.

(*ii*) ***Heat treatment***. Thermal history of kaolinite e.g., heating rate, soaking or equilibration time causes wide differences in decomposition cum generation of crystalline phases. It is not expected that clay heated at constant temperature for long duration of time will generate same phase when a sample withdrawn from the furnaces immediately after reaching the predetermined temperature. This type of choice of experimental methods prior to X-ray analysis causes a great variation of phase development (Insley and Ewell 1935). Besides ordered and disordered effects, Bellotto (1994) showed that the nature of the dehydroxylated phase metakaolinite and the reaction path of mullite formation are largely dependent on thermal treatment. For example, for high heating rates approaching equilibrium conditions, intermediate spinel phase does not appear. Mullite formed may have stable equilibrium composition in the presence of excess silica which eventually crystallizes to β-cristobalite. For rapid heating rate intermediate spinel develops.

Bellotto conceived that spinel with compositional nonuniformity may have formed with silica rich and alumina rich regions. Compositional fluctuations may derive more from ordered kaolinite than disordered kaolinite.

(*iii*) ***Other oxides content***. Clay originally contains a large amount of alkali/ alkaline earth/anatase or rutile and iron bearing phases. These oxides influence the reaction mechanism leading to mullitization.

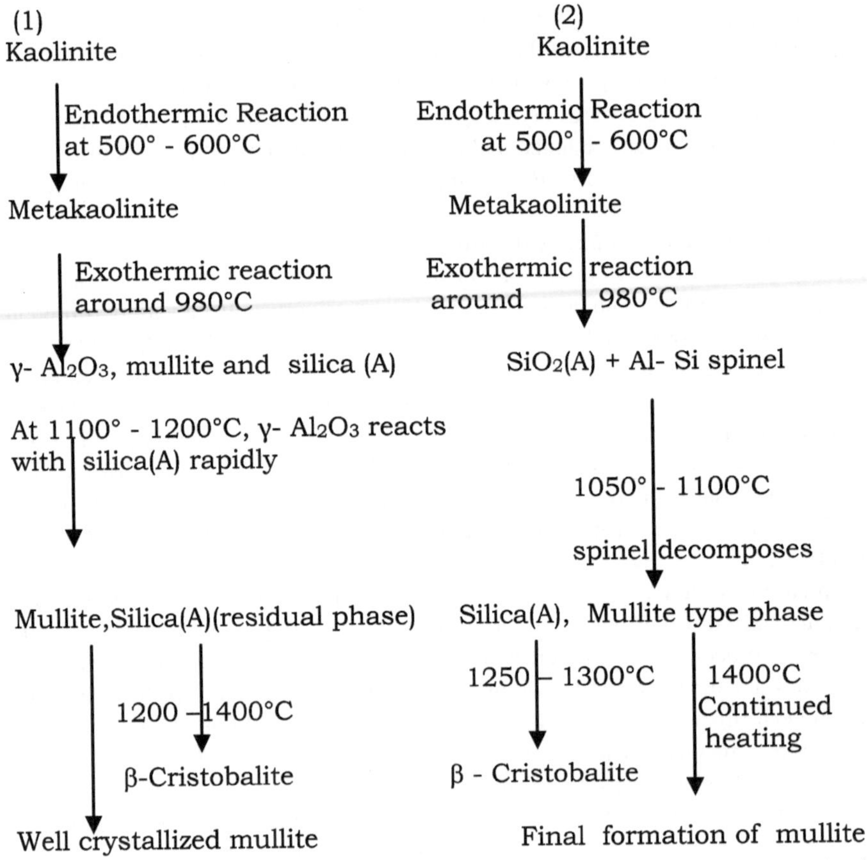

(5.9)

(*iv*) ***Surface area and size effect of kaolinite***. Naturally occurring clays are widely varied in their surface area/particle size. It is conceived that these factors also affect the reaction path leading to mullitization and sintering and particularly to dehydroxylation of kaolinite.

5.4 Factors Affecting Phase Transformation of Kaolinite

5.4.1 Role of Crystallinity

Wide variations are noted in the degree of crystallinity of kaolinites of different origins. Brindley and Robinson (1946a, b) indexed the X-ray powder pattern data of well and disordered kaolinites. Well-crystallized kaolinite shows the following indications. (I) It shows good resolution of the doublet $(11\bar{1})$ and $(1\bar{1}\bar{1})$ spacing at 0.418 nm and 0.413 nm, respectively. In diffractometer recordings, the 0.413 nm gives a step or inflection on the side of the 0.418 nm peak and the background in this region is very clear. (ii) The reflections in the range $d = 0.45$ to $d = 0.35$ nm appear to rise from a low background. Disorder kaolinites show diffusion of the 0.20 reflection toward higher angles. Thus, a good crystalline kaolinite is recognized by sharp reflections and by a low background scattering. (iii) The reflections in the spacing range 0.35–0.25 nm have K indices of 1 or 2 and therefore are susceptible to b axis disorders. In highly "b-axis disordered" kaolinites, no reflections are noted between 002 ($d = 0.357$ nm) and 20 $\bar{1}$ ($d = 0.256$ nm). Well-crystallized kaolinite shows four clear reflections in this range.

Brindley and Robinson (1947) discussed the nature of disordered kaolinite in terms of random b/3 displacements. They showed that triclinic angle = 91.6° of well-ordered kaolinite becomes effectively 90° in the disordered forms, which are, therefore, at least pseudo-monoclinic. The transition from well crystallized to poorly crystallized kaolinite is, therefore, shown by a broadening and weakening of the reflections with the complete elimination of the weaker ones. There is a tendency for adjacent reflections to fuse into one. The basal reflection also increases from 0.714 nm to as much as 0.720 nm. The group of lines from (0.20) $d = 0.446$ nm to (002) $d = 0.357$ nm particularly reflect the change to lower crystallinity. In this region the clearly resolved doublet $(11\bar{1})$ and $(1\bar{1}\bar{1})$ of well-crystallized kaolinite gives way to single band affording good evidence of the decreases in crystallinity. Murray and Lyons (1956) showed the variation in X-ray diagram from well-crystallized kaolinites to an extreme case of b axis disordered kaolinite.

Systematic study of heating effect on kaolinites which vary in their crystallinity were made by John (1953). He investigated some fine kaolinite fractions (2 < e s d) which shows considerable disordering in the form of random nb/3 translation. Schematically, the XRD powder diffraction pattern of Georgia clay shows no evidence of nb/3 translations, where as the pattern of Kentucky flint clay shows diffraction characteristics common to both extreme ordered and disordered types. The pattern of Maryland flint clay shows the characteristics of a completely disordered type of kaolinite. Lack of all reflections from planes for which K = n3 and blurring of the (020) reflections are noted. It has been assumed that during the stacking of one kaolinite layer upon the other in a crystallite, glide translations probably occurs between layers along the b axis. The integral multiples of 1/3 b parameter may be completely random.

X-ray analyses of the above three clays heat treated to 1,000 °C and then air quenched show the following observations. Georgia kaolin shows a moderate amount of mullite and γ-Al_2O_3. Kentucky flint clay develops a minor amount of mullite and γ-Al_2O_3. Maryland flint clay shows γ-Al_2O_3 only. Following these observations with the order of crystallinity of kaolinite, he proposed that well-crystallized kaolinite develops maximum mullite, a completely disordered kaolinite show absence of mullitization at 1,000 °C and second they proposed that the crystallinity of the mullite formed is directly related to the crystallinity of the original kaolinite.

Glass (1954) chooses three varieties of kaolinites and one of halloysite for phase transformation studies using DTA and XRD techniques. He showed that the amount of mullite and temperature of its formation during heating a well-crystallized kaolinite, a poorly crystallized kaolinite, and a mica free kaolinite vary significantly. (a) Primary mullite formation usually takes place during 980 °C exotherm. It is noted that well-crystallized kaolinite produces greater quantity of mullite than poorly crystallized variety. (b) Secondary mullite formation occurs during second exotherm when γ-Al_2O_3 diffraction peaks and the band at 0.43 nm due to SiO_2(A) disappeared with sharp increase in intensities of mullite diffraction lines. Well-crystallized kaolinite shows mullitization exotherm at 1,250 °C where as a lower temperature of about 1,200 °C is noted for the poorly crystallized variety. (c) He also believed that residual silica after mullitization crystallizes to form cristobalite and exhibits an exotherm. Well-crystallized kaolinite exhibits this exotherm well above 1,350 °C where as the same effect for the poorly crystallized variety is at as low as 1,240 °C. Thus, the amount of primary mullite crystallizing at the first exotherm decreases as the structure becomes more and more random. Probably bo/3 shifts in kaolinite hampers the development of mullite nuclei with a resultant decrease in the thermal intensity of 980 °C peak. It was concluded that the growth of mullite nuclei is predisposed by the structure, i.e., the degree of order in the stacking of layers of the clay mineral.

Wahl and Grim (1964) carried out continuous high temperature ~1,450 °C X-ray diffraction study of a well-crystallized kaolinite, a poorly crystallized kaolinite, halloysite, and other kaolinites. They noted the developments of two new phases e.g., mullite and cristobalite but not the formation of intermediate γ-Al_2O_3 spinel phase. The diagnostic reflections (120) and (210) of mullite at 26°(2θ) and 26.3°(2θ) and the (111) reflection of cristobalite at 21.4°(2θ) were scanned at a speed of 2° 2θ per minute. Appearances of mullite and cristobalite on the diffractograms are shown in Fig. 5.5.

In case of well-crystallized kaolinite, initial appearance of mullite is noted at 1,180 °C by continuous diffraction process. But the photograph by using camera showed the presence of primary mullite crystallites in the heated clay between 980 and 1,180 °C. Thereafter, abundant quantity of secondary mullite is noted which coincided with the second exothermic peak in DTA. Lastly, formation of cristobalite coincides with the exhibition of third exotherm in DTA. In cases of poorly crystallized kaolinite, heated to 980 °C, powder diffraction patterns recorded in camera show less quantity of mullite formation then in well-crystallized clay.

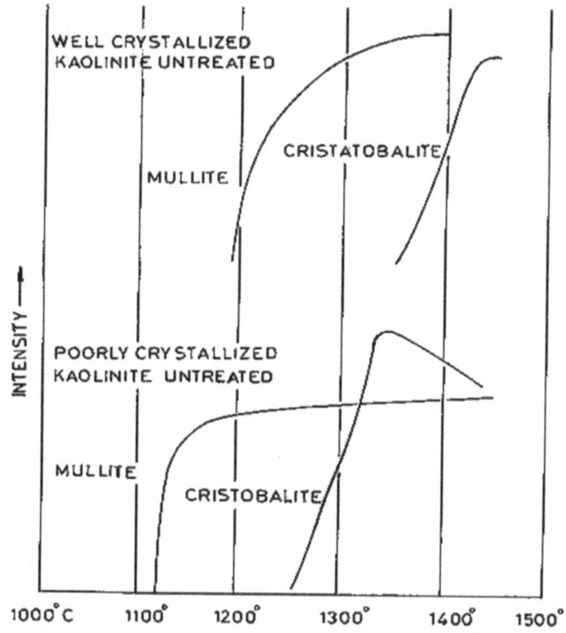

Fig. 5.5 Intensity of diffraction by high temperature phases in heated well crystallized and poorly crystallized kaolinites (After Wahl and Grim 1964). Free available

However, secondary mullite formation occurs very rapidly in comparison to well crystallized variety. Second, mullite begins to develop from poorly crystallized clay at 1,120 °C where as a high temperature ~1,200 °C is necessary for a well-crystallized clay. Cristobalite appears on the diffractograms of poorly crystallized kaolinites at a temperature approx. 80 °C below than the temperature at which cristobalite is first observed in case of well-crystallized clay. Cristobalite starts to crystallize first at 1,250 °C then it increases very rapidly and reaches a maximum at 1,375 °C and then decreases at higher temperature. This might be due to its dissolution into glassy phase through fusion. In case of halloysite, powder camera pattern does not show any primary mullite crystallites around first exotherm. It is noted only on heating at about 1,100 °C. Continuous diffraction analysis shows mullitization at about 1,250 °C, i.e., at or around the second exotherm. Cristobalite does not form even when halloysite is heated to 1,400 °C.

Bellotto (1994) tried to correlate the order of kaolinite sample with thermal analysis data in one hand and equilibrium HT evolution by XRD on the other hand.

Ordered kaolinite e.g., Kaosar		Disordered kaolinite e.g., KGa-2
1. Dehydroxylation reaction:		
Starts at	475 °C	425 °C
Completes at	625 °C	525 °C
Metakaolinite between	625–800 °C	525–725 °C

(continued)

(continued)

Ordered kaolinitee.g., Kaosar		Disordered kaolinite e.g., KGa-2
2. Mullitization reaction:		
Mullite pattern appears between	950–1,000 °C	Before 1,000 °C
Transition of t → m mullite occurs around	1,200 °C	1,150 °C
3. Cristobalite formation occurs at	1,200 °C	1,150 °C

5.4.2 Role of Inherent Impurities on Phase Formation

Kaolinite clays usually contain illite/mica; chlorite; anatase; and Fe_2O_3 as associated phases. These played a major role during phase development of kaolinite on heating.

Effect of Chlorite

Slaughter and Keller (1959) noted that the relative amount of mullite and cristobalite formed are directly related to the mineralogy of the four different raw clays. The temperature of maximum development of cristobalite in the fired clays shows considerable variations but it is found to be related to the illite, illite-chlorite content. Increase in illite content in the raw clays, the maximum amounts of mullite and cristobalite that formed in the fired clays decrease and amount of glass increases correspondingly. Illite rich clays suppress cristobalite formation. As the amount of illite increased from pure kaolinite to pure illite clays, the amount of silica available to form cristobalite increases slightly but the alkali metal oxides content increases. It was assumed that illlite furnished glass forming alkali cations. Consequently with increase of illite content, glass content increases on firing and likely cristobalite content relatively reduces.

Effect of mica content

Glass (1954) showed that the sequence of high temperature phases developed in a kaolinite clay is greatly affected by the presence of mica as an impurity. For example, mica bearing kaolinite exhibits 1st exotherm with reduce in intensity at a temperature of about 950 °C which is little lower than 960 °C exhibited by mica free kaolinite. Primary mullite phase is not observed, but Γ-Al2O3 and SiO_2(A) are only formed during the exotherm. Secondary mullite forms at the second exotherm (at ~1,200 °C) which is ~50 °C lower than in case of mica free kaolinites. Similarly, cristobalite peak at 1,240–1,320 °C also occurs at a reduced temperature. Thus, intergrown mica in clay may inhibit primary mullite formation and reduces the intensity of 950 °C exotherm and due to presence of the K_2O which acts as a flux, lowers down the temperature of formation of both secondary mullite and cristobalite.

Mullite formation from a kaolin material (Bio) containing kaolinite in addition to muscovite to the extent of ~17 wt% has been studied by Castelein et al. (2001). Figure 5.6 shows XRD patterns of BIO heated to 1000 at 3 and 20 °C/min heating rates.

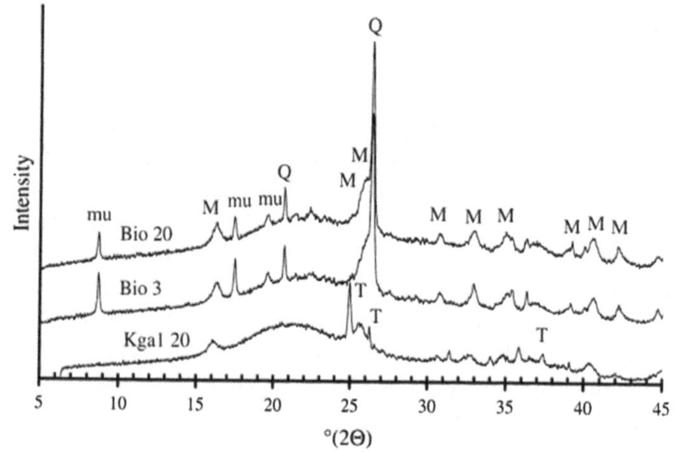

Fig. 5.6 X-ray diffraction pattern of Kaoins heated until 1,000 °C: **a** Bio at 20°c/min, **b** Bio at 3° c/min, **c** Kga1 at 20°c/min. Holding time : 3 min. M = Mullite : mu = muscovite :Q = quartz : T – rutile (After Castelein, 2001). Reprinted by permission of the ceramic international

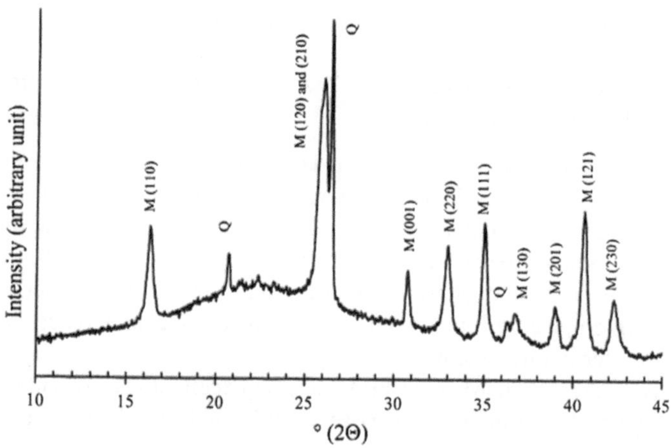

Fig. 5.7 X-ray diffraction pattern of bio kaolin sintered at 1,100 °C at 3° C/min (After Castelein et al. 2001). Reprinted by permission of the European ceramic society

In another publication, Castelein et al. (2001) showed the same kaolin heated to 1,100 °C at 3 °C/min rate of heating (Fig. 5.7). It is observed that intensity of peak due to muscovite decreases from Bio 3 to Bio 20 in Fig. 5.6 and it vanishes out on further heating to 1,100 °C (Fig. 5.6). These result to more mullitization at a comparatively low temperature by the presence of inherent muscovite phase.

Therefore, it is concluded that independent transformations of two minerals may occur up to final stage of dehydroxylation cum exothermic temperatures. Thereafter, amorphous phases generated out of the two may invariably react.

5.5 Summary

Various attempts have been made to record the X-ray diffraction patterns of kaolinite heated successively to higher temperatures. Rinne (1924) first carried out X-ray study of heated kaolinite in the dehydrated stage. Thereafter, comprehensive investigations have been carried out using Debye Sherrer camera by Hyslop and Rooksby (1928), Insley and Ewell (1935), Jay (1939), Hyslop (1944), Richardson and Wilde (1952), John (1953), Glass (1954), Slaughter and Keller (1959), Scheltz and Soliman (1966). Leonard (1977) recorded the sequential phase development by diffractometer instead of using camera. Besides using ordinary X-ray source, Tscheischwili et al. (1939) used monochromatic X-ray in vacuum for dehydrated kaolinite. Generally, following phases have been detected by X-ray analysis.

- At the first stage
- Metakaolinite forms and pattern becomes amorphous.
- At the second stage
- Metakaolinite changes to a poorly crystalline cubic spinel type phase. Mullite starts forming along with spinel at the first exotherm.
- At the third stage
- Mullite develops largely beyond second exotherm.
- At the fourth stage
- Cristobalite starts forming beyond 1,200 °C. Reaction sequence predicted by various authors is shown in Eq. 5.9.

Significantly, a new approach was made by Brindley and Hunter (1955) and latter on by Brindley and Nakahira (1959a, b, c) using single crystal X-ray technique. They conjectured topotatic nature of transformation of kaolinite up to spinel phase. The reaction sequence according to them is shown in Eq. 5.9.

Bellotto (1994) studied phase evolution using HT X-ray powder diffraction technique and emphasized that nature of metakaolinite and reaction path are dependent of thermal treatment.

By X-ray studies, Richardson (1961) made an excellent review and showed the stability range of the formation of γ-Al_2O_3, mullite and cristobalite out of different kaolinites.

Slaughter and Keller (1959), Udagawa et al. (1969) presented the high temperature phase developments in various kaolinites schematically.

The effect of the degree of crystallinity of a kaolinitic clay and also the role of inherent impurities, e.g., mica, illite, etc., on phase formation have been elaborately reviewed. Particularly the development of cristobalite content is largely dependent on illite, illite–chlorite, and mica content.

According to John (1953) and Glass(1954),order of crystallinity of mullite formed is related to crystallinity of the parent kaolinite. Wahl and Grim (1964) correlated thermal effects of kaolinite by making continuous heating to high temperature.

Finally, the results of DTA and X-ray diffraction studies of heated clays of different origins reveal that the followings.

Intensities of first (980 °C), second (1,250 °C), and third (1,400 °C) exotherms are dependent on the degree of crystallinity of the original kaolinite.

Occurrence and quantity of primary mullite and or γ-Al_2O_3 phase formation at first exotherm are also related to the degree of order of the kaolinite.

Inter grown impurities of clay lower down the temperature of secondary mullitization exotherm and severely affect cristobalite formation at third exotherm.

References

M. Belloto, High temperature phase transformatiom in kaolinite: The influence of disorder and kinetics on the reaction path. Materials Sci. Forum, **166–169**, 3–20 (1994)

G.W. Brindley, K. Robinson, Structure of Kaolinite. Miner. Mag. 27, 242–253 (1946a)

G.W. Brindley, K. Robinson, Randomness in the structures of kaolinite clay minerals. Trans. Faraday Soc. 42B, 198–205 (1946b)

G.W. Brindley, K. Robinson. X-ray studies of some kaolinite fireclays. Trans. Brit. Ceram. Soc. 46, 49–62 (1947)

G.W. Brindley, K. Hunter, Thermal reactions of nacrite and the formation of metakaolin, γ-Alumina, and Mullite. Miner. Mag. **30**(288), 574–584 (1955)

G.W. Brindley, M. Nakahira, The kaolinite-mullite reaction series: I, a survey of outstanding problems. J. Am. Ceram. Soc. **42**(7), 311–314 (1959a)

G.W. Brindley, M. Nakahira, The kaolinite–mullite reaction series: II, metakaolin. J. Am. Ceram. Soc. 42(7), 314–318 (1959b)

G.W. Brindley, M. Nakahira, The kaolinite–mullite reaction series: III, the high-temperature-phases. J. Am. Ceram. Soc. 42(7), 319–324 (1959c)

O. Castelein, B. Soulestin, J.P. Bonnet, P. Blanchart, The influence of heating rate of the thermal behavior and mullite formation from a kaolin raw material. Ceram. Int. **27**(5), 517–522 (2001).

A.K. Chakraborty, S. Das and S. Gupta, Evidence for two stage mullite formation during thermal decomposition of kaolinite. Brit. Ceram. Trans. **102**(4), 33–37 (2003)

R.F. Davis and J.A. Pask, Mullite. in *Refractory materials*, ed. by A. M Alper. High Temperature Oxides Part IV, Refractory Glasses, Glass Ceramics and Ceramics, (Academic Press, New York, London 1971), p. 37–76

W.L. De Keyser, Contribution to the study of mullite, Ber. Dtsch. Keram. Ges. **40**, 304–315 (1963)

F. Freund, Explanation of exothermal reaction of kaolinite as a 'Reaction of the Active State'. Ber. Deut. Keram. Ges. **37**(51), 209–218 (1960b)

H.D. Glass, High-emperature phases from kaolinite and halloysite. Am. Miner. **39**, 193–207 (1954)

J.F. Hyslop, H.B. Rooksby, Effect of heat on crystalline break up of kaolin. Trans. Br. Ceram. Soc. **27**(4), 93–96 (1928)

J.F. Hyslop, Decomposition of clay by heat. Trans. Brit. Ceranz. Soc. **43**(3), 49–51 (1944)

I.H. Insley, R.H. Ewell, Thermal Behavior of the Kaolin Minerals. J. Res. Natl. Bur. Stand. **14**(S), 615–627 (1935)

A.H. Jay, An X-ray Study of Alumino-Silicate Refractories. Trans. Brit. Ceram. Soc. **35**, 455–463 (1939)

W.D. Johns, High-temperature phase changes in kaolinites. Miner. Mag. 30 (222), 186–198 (1953).

A.J. Leonard, Structural analysis of the transition phases in the kaolinite–mullite thermal sequence. J. Am. Ceram. Soc. **60**(1–2), 37–43 (1977)

H. Murray and S.C. Lyons, Degree of Crystal Perfection of Kaolinite, Natl. Acad. Sci. Publ, 456, 31–40 (1956).

H.M. Richardson, F.G. Wilde, X-ray study of the crystalline phases that occur in fired clays. Trans. Brit. Ceram. Soc. **51**(7), 367–400 (1952)

H.M. Richardson, Phase changes which occur on heating kaolin clays in *The X-ray identification and crystal structures of clay minerals*, ed. by G. Brown (The Mineralogical Soc., London, 1961) p. 132–142

F. Rinne, Rontgrnographischc diagnostik heini rrennen von kalkstcin, dolomit,kaolin. und glimmer (X-ray study of calcined calcite, dolomite, kaolinite, and mica), Z. Krist. **61**(1/2), 113 (1924)

N.C. Schieltz, M.R. Soleman, Thermodynamics of various high temperature transformation of kaolinite. Proc. 13th Natl. Conf. on Clays, Madison, Wisconsin, Pergamon press, Monograph, No.25, ed. by (Earl Ingerson 1966), p. 419–425

M. Slaughter, W.D. Keller, High temperature transformation from impure kaolin clays. Am. Ceram. Soc. Bull. **38**, 703–7 (1959)

L. Tscheischwili, W. Biissern, W. Wevl, Uber den Metakaolin (Metakaolin). Ber. deut. keranz. Ges, **20**(61), 249–276 (1939)

S. Udagawa, T. Nakada, M. Nakahira, in *Proceedings International Clay Conference*, Molecular structure of allophane as revealed by its thermal transformation, Editor-in-chief Lisa Heller bisa Heller by Israil University Press, vol. 1, p. 151, Gerusalem (1969)

F. Vaughan, Review of literature on thermal breakdown of kaolinite and on its decomposition products. British Ceramic Research Association Special Publication No. 13, (1956)

F.M. Wahl, R.E. Grim, *Twelfth national conference on clays and clay minerals*, High temperature DTA and XRD studies of reactions, pp. 69–81 (1964).

Chapter 6
Electron Microscopy Study

As shown in Chap. 5, metakaolinite shows amorphous to words X-ray and spinel too is a very weakly crystalline material. In view of it, an extremely powerful technique high resolution electron microscopy is applied for characterization.

6.1 Morphology and Electron Diffraction

Shape of clay particles have been ascertained precisely with help of the development of electron microscopic technology. Electron micrographs of kaolinite in general show hexagonal in out line (Fig. 6.1). Six-sided flakes of a well-crystallized kaolinite generally shows a prominent elongation in one direction. The elongation is parallel to either (010) or (110). The edges of the particles are beveled instead of being at right angles to the flake surface. Occasionally, the particles appear to be twined. Electron micrographs of various kaolinite samples have shown particles with maximum dimensions of flake surfaces from 0.3 to 4 μm and thickness from 0.05 to 2 μm. Poorly developed hexagonal outlines are observed in case of kaolinite of disordered type. The edges of the flakes of such variety are ragged and irregular. Hexagonal outline is crudely in appearance. Morphological forms of kaolinite vary widely from different sources. Some are closely hexagonal in outline while some shows greater elongation. Some crystals are "block" in appearance others are thin wafer-like crystals. Researchers are interested to find out the relationship between crystallinity with morphology of kaolin clays. It concludes "when the diffraction patterns indicate a high degree of crystallinity, the micrographs usually show a high degree of geometric crystallinity. The poorly crystallized kaolins are composed of thinner plates".

Electron microscopy is used for studying the changes of the morphology of kaolinite particles during heating. In addition to it, heated clay materials are analyzed by electron diffraction technique, like X-ray diffraction study, in order to furnish more insight in phase transformation process.

Eitel et al. (1939) first studied dehydration of clay minerals using electron microscopy. They recognized hexagonal morphology in the microscope even after

A. K. Chakraborty, *Phase Transformation of Kaolinite Clay*,
DOI: 10.1007/978-81-322-1154-9_6, © Springer India 2014

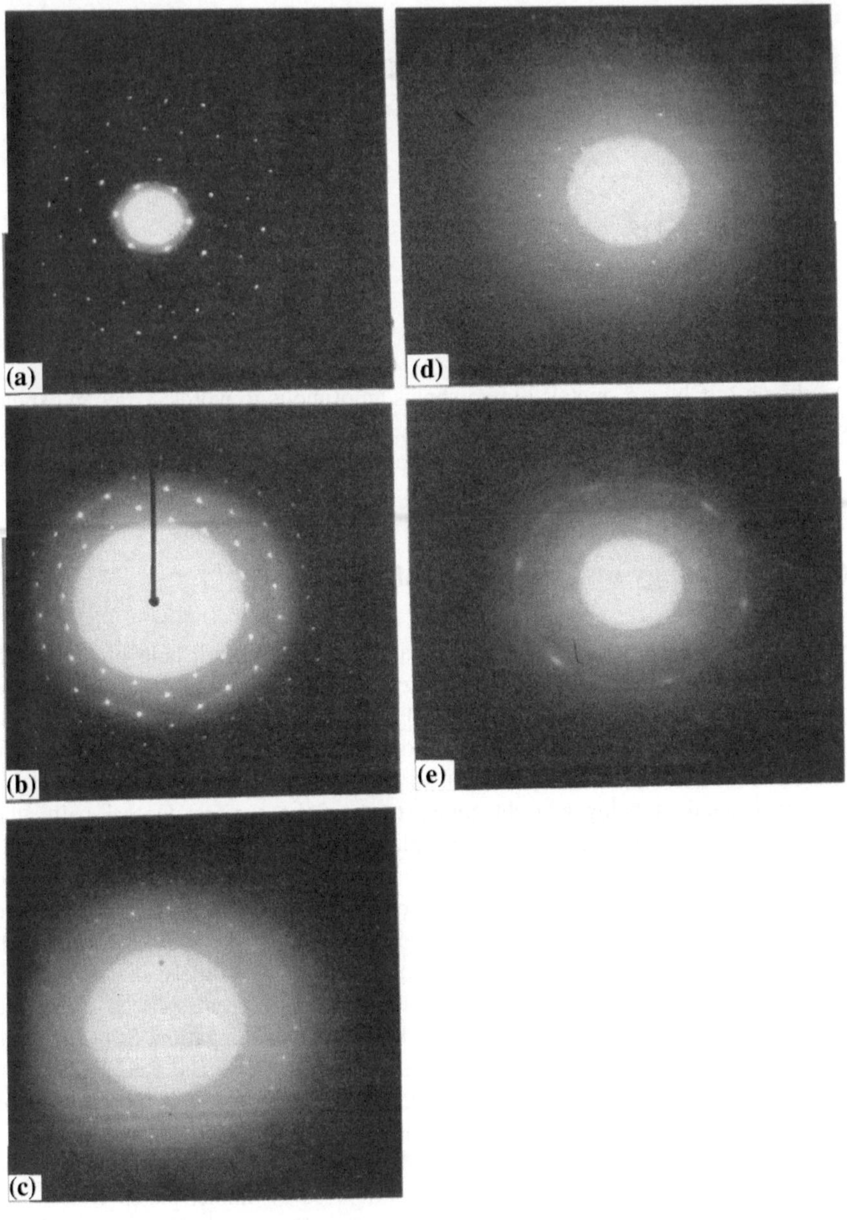

Fig. 6.1 Electron diffraction pattern of kaolinite and its decomposition products. **a** Kaolinite at 110 °C. **b** Kaolinite 500 °C. **c** Kaolinite 800 °C. **d** γ-Al_2O_3 lattice obtained at 1,000 °C. **e** γ-Al_2O_3 rings and spots obtained at 1,000 °C (After Roy et al., 1955). Reprinted by permission of the American Ceramic Society

dehydroxylation of kaolinite. Of course original kaolinite crystals break up and become smaller in appearance. Existence of hexagonal relic in metakaolinite was also noted by several researchers namely, Eitel and Kedesdy (1943), Radczewski (1953), Comer et al. (1956).

Comeforo et al. (1948) heat treated Georgia kaolinite to 525, 980, 1,100, 1,300, and 1,450 °C, respectively for 15 h and then examined by X-ray diffraction and with R.C.A., type-B electron microscope. Electron micrographs show the striking persistence of clear hexagonal-shaped particles for above the temperature of dehydration which itself a significant indication of a residual structure in the anhydride. Second, the micrographs further show the disposition of mullite needles with respect to original kaolinite grain. The development of mullite has oriented about asymmetric chain nuclei of orientation predisposed by the structure of the original kaolin. They believed that such directional dependence should persist through the overall transition from the original kaolinite grain to the mullite aggregate eliminates the possibility that intermediate crystallization could have occurred. Any γ-Al_2O_3 which may have been observed in special instances is therefore only incidental and is not a step in the major reaction series of the firing process. They conjectured that two stages of reaction occur in the firing of kaolintie. First, hydroxyl groups present in the crystal is expelled with the formation of uniquely organized non crystalline compound called 'metakaoline'. Second, this non crystalline compound collapses into crystalline mullite nuclei which exhibit normal growth.

Roy et al. (1955) used an electro dialyzed and fractionated (1–4 μm) sample of a well-crystallized Langley kaolinite for their study. It was heat treated to successively higher temperature followed by soaking for at least 12–200 h with an idea to reach pseudo-equilibrium state at the given temperature. They showed the electron diffraction patterns of kaolinite heat treated to different temperatures. Transmission electron diffraction patterns of kaolinite contain two-dimensional (2D) pseudo-hexagonal arrays of spots corresponding (hko) reflections. Several (hkl) reflections are obtained with tilting specimen holder. Measured a and b dimensions from lattice patterns is in good agreement with values obtained by X-ray diffraction. Heated kaolinite at 500–800 °C, shows the same single crystal pattern, i.e., (hko) projections those were not characteristically distinguishable from that of original kaolintie. Heated kaolinite between 855 and 1,000 °C, shows new spot pattern and ring characteristic of γ-Al_2O_3 projection. With this single crystal electron diffraction data they demonstrated that a considerable order exists in dehydrated kaolinite. The 2D (a–b) plane structure of dehydroxylated kaolinite is very similar to that of kaolinite.

This dehydrated kaolinite is described by the term, 'metakaolinite'. It decomposes to γ-Al_2O_3. They note the temperature of complete breakdown of metakaolinite at 825 °C approximately. The new spot pattern is of the single crystal γ-Al_2O_3 pattern whereas ring type pattern is obviously the powder pattern of same γ-Al_2O_3.

Comer (1961) heat treated poorly crystallized and a moderately well-crystallized kaolinite for 20 h at a temperature of 850 and 950 °C respectively. These specimens were examined in RCA model EMU-2D electron microscope. On heating at 850 °C\20 h the poorly crystallized kaolinite shows the co-existence of residual metakaolinite pattern and (440) reflection of spinel-type phase. On heating next to 950 °C\20 h, the pattern (Fig. 6.2) shows a very strong oriented (002) reflection of mullite phase co-existing with the broad and diffuse (440) reflection from the spinel-type phase out of moderately crystallized kaolinite.

The electron micrograph of spinel phase is shown in (Fig. 6.3). The micrograph appears scaly in nature. The size of this spinel-type phase consisted of very small crystals ranging from 75 to 125 a.u as observed in electron micrograph.

With these two electron diffractograms, they inferred the followings: (i) (111) plane of spinel-type phase is parallel to the a-b plane of metakaolinite and the (110) is parallel to the b axis of metakaolin. This finding corroborated those of Briendly and Nakahira. (ii) c-axis of the mullite phase runs parallel to (110) axis of the spinel phase thus conforming the mechanism proposed by Briendly and Nakahira for the transformation of spinel phase.

Dekeyser (1963) showed the electron micrograph of Zettlitz kaolinite at various temperature of heat treatment.

McConnell and Fleet (1970) applied high resolution electron microscopy for characterizing two reaction products of kaolinite which are amorphous or extremely poor crystalline in character and more over this technique is capable of providing direct information on the reaction mechanism of thermal decomposition of kaolinite to mullite.

When heat treated a well-crystallized kaolinite (R.L.O., 1067) 6 h at 800 °C. They observed the evidence of crystallinity in both selected area diffraction and

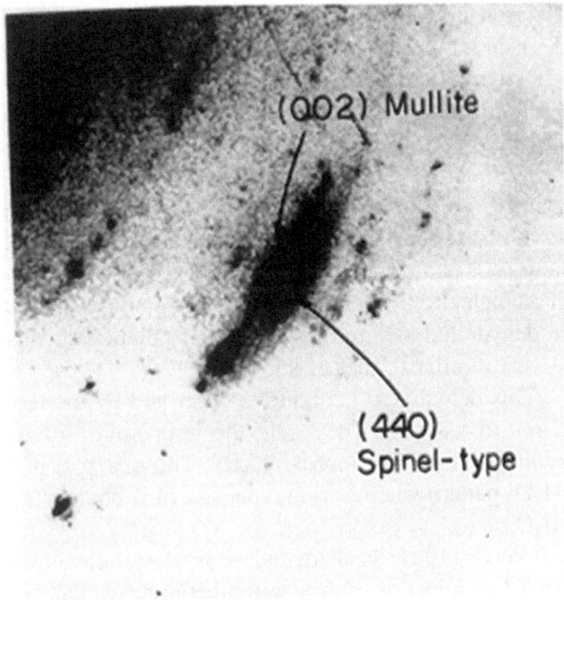

Fig. 6.2 Picture shows the very strong (002) reflections from mullite and the broad, diffuse (440) reflections from the spinel-type phase (After Comer 1961). Reprinted by permission of the American Ceramic Society

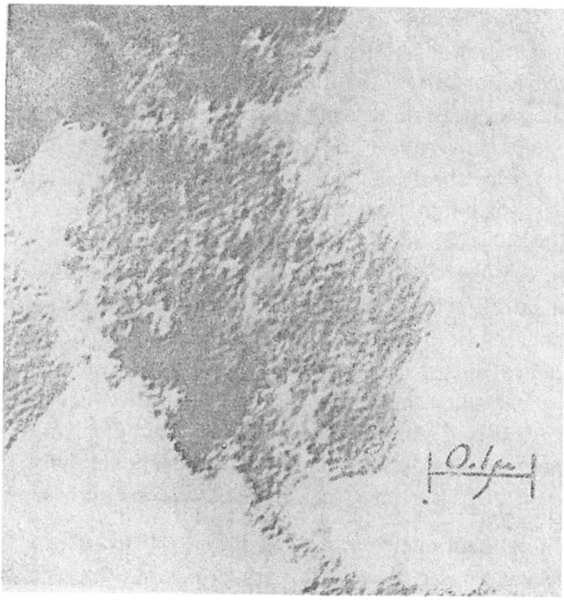

Fig. 6.3 Electron micrograph of a single crystal of poorly crystallized kaolinite held at 950 °C/20 h showing spine-type phase (After Comer 1961). Reprinted by permission of the American Ceramic Society

electron optical image. The diffraction data of metakaolinite confirms the conclusion of Briendly and Nakahira as to the high degree of order is retained in the plane of the kaolinite sheets. Morie patterns due to overlapping flakes and weak extinction bend contours arising from the relict maxima of type hko of kaolinite are observed.

When heated at 900 °C for 6 h selected area diffraction patterns correspond to a well-oriented defect spinel phase with the zone axis [111] normal to the relict kaolinite flake as previously observed by Brindley and Nakhira (1959a, b, c). On heating to 950 °C, development of fragmented mullite is observed in addition to host defect spinel phase.

When heated at 1,150 °C the specimen shows the abundant development of mullite as continuous lath-like crystals with associated preferred orientation, although some crystals with the c-axis being normal to the plans of the original kaolinite flake are also noted.

When heated at 1,200 °C for 6 h well-crystallized mullite appeared which shows little evidence of preferred orientation. This was conformed from the selected area diffraction patterns. By identifications of the product phases at different temperatures, they forwarded three main reaction mechanisms in the above temperature range.

First is the dehydroxylation reaction which is associated with partial breakdown of kaolinite structure and introduction of disorder in the staking of the elementary sheets.

Second reaction product is a relatively coarsely porous defect spinel phase with strongly preferred orientation.

Third reaction product is mullite which appears in quantity and with preferred orientation in the temperature range between 1,150 and 1,225 °C. They assumed the formation of mullite in this reaction is clearly responsible for the 2nd exothermic peak on the standard differential thermal curve of a heated kaolinite.

At temperature above 1,200 °C, beside mullite, cristobalite is very much in evidence and there is a marked change in the microstructure of the reaction products.

Tsuzuki and Nagasawa (1969) showed electron micrograph of heated clay and noted numerous small dots appeared in the particles of kaolin minerals heated up to the temperature range of the small endothermic reaction. The SAD (selected area diffraction) showed that crystallites of spinel produced have a definite orientation with respect to the original kaolin mineral. With this observation they indicated that this spinel is formed topotactically.

Briendley and Gibbon (1969) measured the basal plane parameter of a well-crystallized kaolinite and metakaolinite by electron diffraction technique by using alumina shadowing technique for proper calibration of diffraction spot and showed that 'b' parameter increases by 2.2 % as follows.

Initial kaolinite, $\mathbf{b} = 8.95 \pm 0.03$
Metakaolinite, $\mathbf{b} = 9.145 \pm 0.035$

Tetrahedral silica sheets are constrained by their attachment to octahedral Al–O (OH) layer in kaolinite structure as a result the silica tetrahedral are rotated to 11.3° after Baily (1966) or 20° due to Drits and Kashaev (1960). When octahedral sheets are disrupted by dehydroxylation, the constrains on the tetrahedral sheets are removed and the high charge in Si ions will repeal each other and assume the general bond length of Si–O which is equal to 0.162 nm. The maximum relaxation of tetrahedral sheet corresponds to the hexagonal arrangement and from the geometry of the tetrahedron, it follows that which agrees exactly with the measured parameter of metakaolin. With this evidence Briendley and Gibbon (1969) assumed that higher symmetry of Si–O network occurs during dehydroxylation.

6.2 High Resolution Electron Microscopy

Sonuparlak et al. (1987) performed in situ beam-induced heating experiments in a transmission electron microscope (TEM) and high resolution electron microscopy (HREM) in order to observe direct phase changes. Three different characteristic stages are observed.

The bright field image was taken after dehydroxylation had started. The formation and growth of light color patchy regions are interpreted to be associated with the loss of structural water. After completion of it, the structure was determined to be amorphous by electron diffraction.

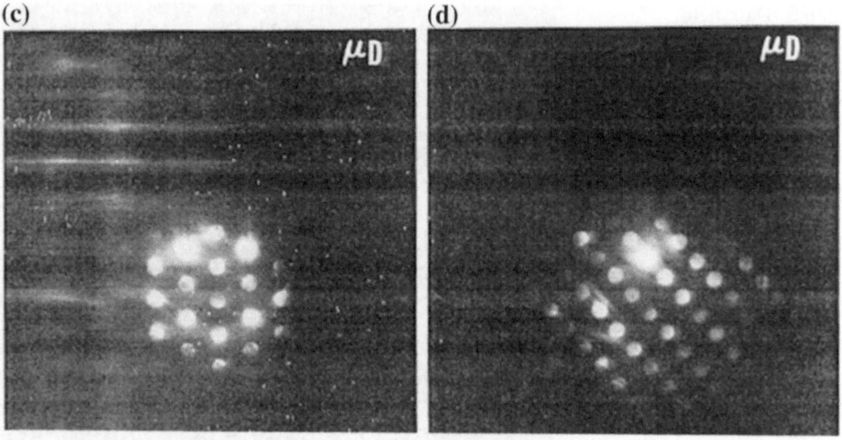

Fig. 6.4 Micro diffraction patterns *C* and *D* of spinel phase grown on electron beam heating of Georgia kaolinite at $\leq$980 °C near <114>$_{fcc}$ and <122>$_{fcc}$ zone axis orientations (After Sonuparlak et al. 1987). Reprinted by permission of the American Ceramic Society

The image was taken upon further heating, but to a temperature low enough, so that crystallization would not occur. This image reveals a structure similar to that obtained in spinodally decomposed system.

Further heating to higher temperature first results in the growth of these phase-separated regions and then the formation of a crystalline phase takes place.

The size of the crystalline regions is 5–8 nm. Analyses of the superimposed patterns indicate only the presence of cubic spinel-type phase. In order to obtain isolated diffraction pattern of cubic spinel phase, they heat treated Georgia kaolinite externally at prolonged times (1*d*) at temperature near but below the exothermic reaction. The bright field image from a corner of a particle indicates variations in the structure resembling a contrast associated with a crystalline area. The corresponding dark field image clearly reveals the morphology of the crystalline region which is surrounded by an amorphous phase at the edge of the particle. Micro diffraction patterns were taken from this region by using an electron beam 40 nm (400 Å) in diameter. The micro diffraction pattern (C) corresponds to a <114> fcc zone axis orientation. The specimen was further titled to another orientation (D), now near <122> fcc to unambiguously confirm the fcc structure of the spinel phase (Fig. 6.4).

Srikrishna et al. (1990) studied K–M reaction series by transmission electron microscopy using single crystal kaolinite ranging in size from 5 to 10 nm. The kaolinite crystals were heat treated in a Mettler–DTA-TGA unit at the rate of 10 °C\min as to ensure accurate temperature control and reliable experimental conditions. The specimens were analyzed for analytical electron microscopy in a Philips 400 T electron microscope and for high resolution electron microscopy in a JEOL 200 CX electron microscope. The diffraction of metakaolin pattern is completely amorphous as also noted by Sonuparlak et al. (1987).

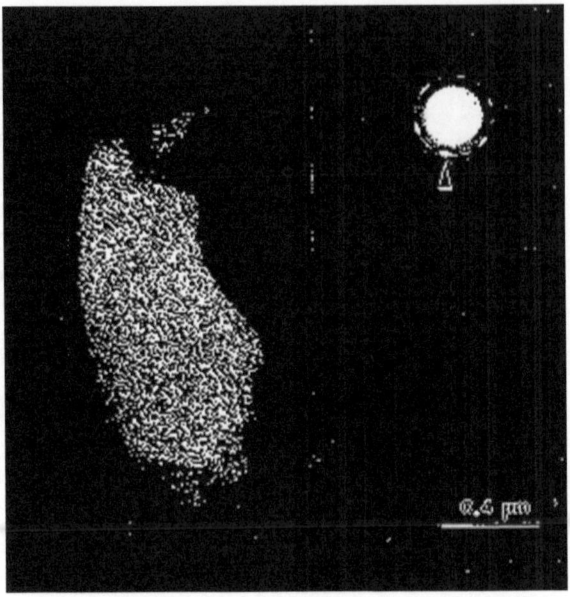

Fig. 6.5 Dark field micrograph of the spinel phase taken at 1,020 °C after leaching and corresponding diffraction pattern (After Srikrishna et al. 1990)

At 980 °C, they noted a small fraction of mullite simultaneously with spinel phase (Fig. 6.5). EDX study revealed the spinel phase to have composition very close to that of 3:2 mullite. Beyond 2nd exotherm, acicular mullite crystals are produced with c-axis perpendicular to the original metakaolinite sheets.

Micro structural evolution of English China Clay both in powder form and in extruded bare is studied by McConville et al. (1998) after a schedule of heating and cooling process.

(i) At metakaolinite stage

The face-on, plate-like pseudo-hexagonal particle morphology exists.

(ii) At spinel phase 980 °C i.e., at crystallization stage

Two forms of kaolinite show some differences. Electron diffraction from (440) reflection of spinel phase obtained from 'face-on' view of powder kaolinite exhibits spot pattern. Whereas, same from 'edge-on' view of platelets in extruded sample shows ring diffraction pattern. It is revealed that spinel crystallizes in preferred orientation to parent kaolinite.

(iii) At mullite development stage

Pseudo-hexagonal morphology gradually disappears, the relict clay particles starts fusing together, spinel texturing diminishes, mullite crystallizes as rod-shaped habits. With rise of temperature, more amorphous generates with more elongation of mullite phase.

Morphological changes are schematically shown by McConville et al. (1998) in the Fig. 6.6.

Lee et al. (1999) analyzed metakaolinite and microcrystalline or poorly crystalline spinel phase by using energy-filtering transmission electron microscope

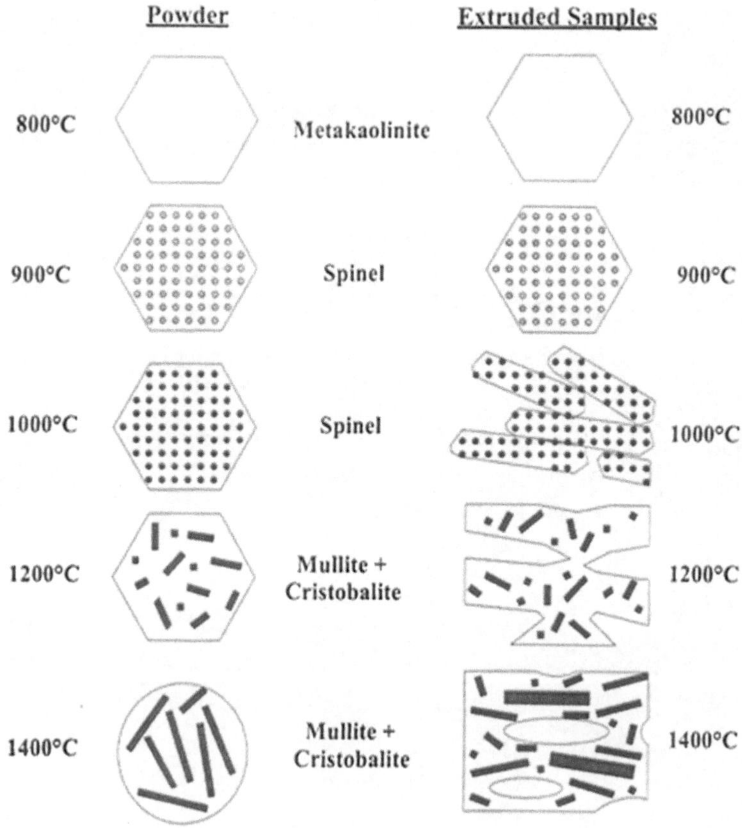

Fig. 6.6 Schematic diagram of the features observed by TEM in powder and extruded sample (After Mcconville et al. 1998)

(EF-TEM) and reexamine the K–M phase transformation sequence. It enhances diffraction contrast but also makes possible electron energy loss spectrometry (EELS) and electron spectroscopic imaging (ESI). The observations acquired by EF-TEM on transformation of kaolinite is shown below.

(i) Diffraction pattern at 630 °C
The six nearest spots (d-spacing: 4.26–4.40 Å) around the direct beam are noted. Three diffuse halos characteristic of metakaolinite are observed. The d-spacing of these are (i) 4.50–3.28 Å (strongest), (ii) 2.41–2.45 Å (weak), (iii) 1.23–1.16 Å (weakest).This indicates the existence of short range order in metakaolin.

(ii) Diffraction pattern at 920 °C
In addition to metakaolinite reflections, six newly formed spots (d-spacing: 1.4 Å) emerges out. Other six spots (d-spacing: 2.42 Å) is also seen. This pattern is identified as spinel-type phase. According to Lee et al., this systematic diffraction pattern of the spinel-type phase which showing pseudo-hexagonal symmetry strongly

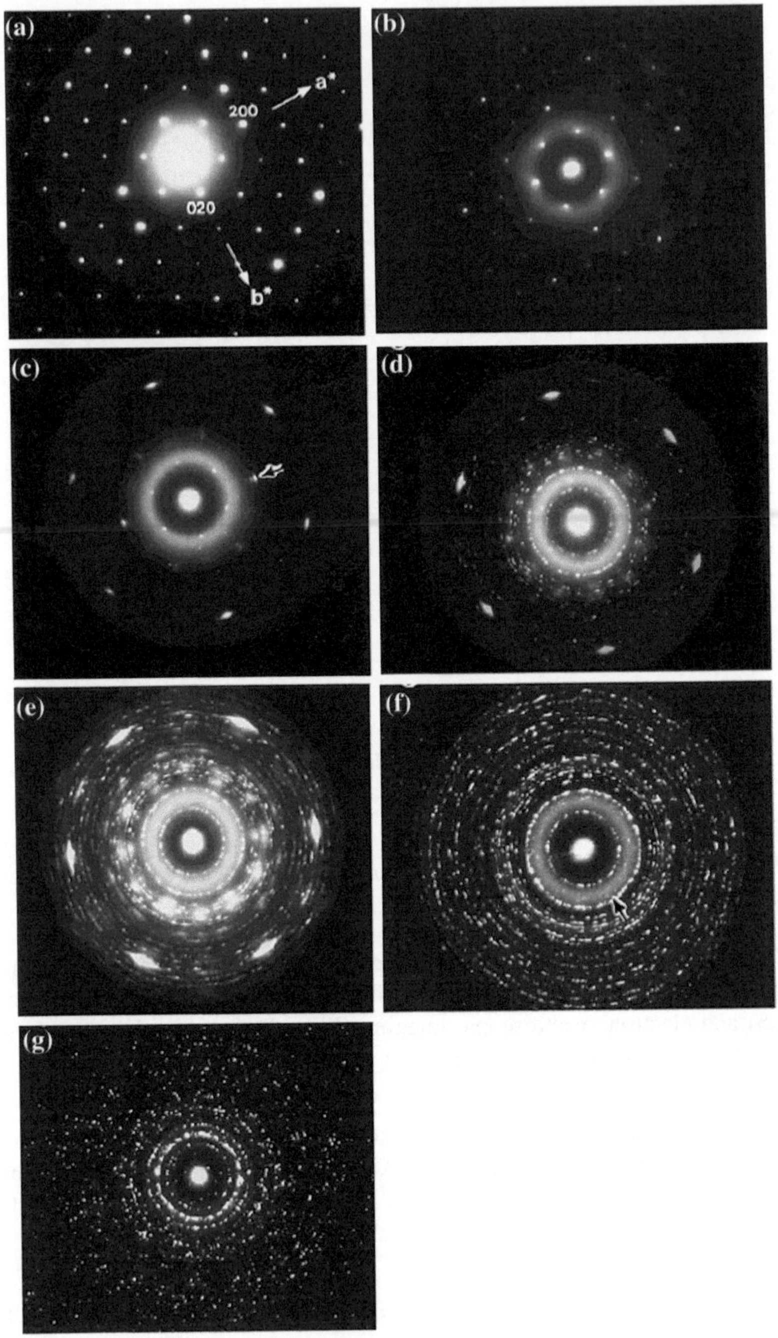
(a)
200
a*
020
b*
(b)
(c)
(d)
(e)
(f)
(g)

◀ **Fig. 6.7** **a** Electron diffraction pattern of Kampaku kaolinite heated to 1,100 °C (After Lee et al. 1999). Reprinted by permission of the American Ceramic Society. **b** Electron diffraction pattern of Kampaku kaolinite heated to 1,200 °C (After Lee et al. 1999). Reprinted by permission of the American Ceramic Society. **c** Electron diffraction pattern of Kampaku kaolinite heated to 1,300 °C (After Lee et al. 1999). Reprinted by permission of the American Ceramic Society

indicates a preferred orientation relationship to the parent metakaolinite phase. Moreover, concomitant existence of the spinel-type phase with metakaolinite at 920 °C favors topotactic changes between them. In addition, spotted ring patterns of mullite appear. This ring pattern according Lee et al. indicates random orientation relationship with respect to parent metakaolinite phase. Increase in intensities of both spinel and mullite occur from 940 to 1,100 °C.

(iii) Diffraction pattern at 1,200 °C
Complete disappearance of the spinel-type phase occurs. Formation of scattered spots within the diffuse halo are seen. The d-spacing of these spots at ~4.1 Å indicates the appearance of cristobalite and growth of mullite continues. Electron diffraction patterns of kaolinite heat treated to three different temperatures are shown in Fig. 6.7.

Composition of spinel phase
In electron microscopic studies, the electron energy loss occurs from structure of spinel phase associated with the nature of surrounding amorphous mass. Lee et al. (1999) noted that both Al and Si are detected in every part of the grain and the ratio of them is found constant. Source of Si peak in EDS spectra as obtained may be metakaolinite residue, liberated silica (A) as well as from Si-containing spinel phase if it is generated. According to them even in FE-TEM study, it is hard to characterize the spinel phase due to beam size restriction, etc.

Lee et al. (1999) showed that during transformation process of kaolinite, the broad X-ray background in the range 15°–35° 2θ was continued on heating.

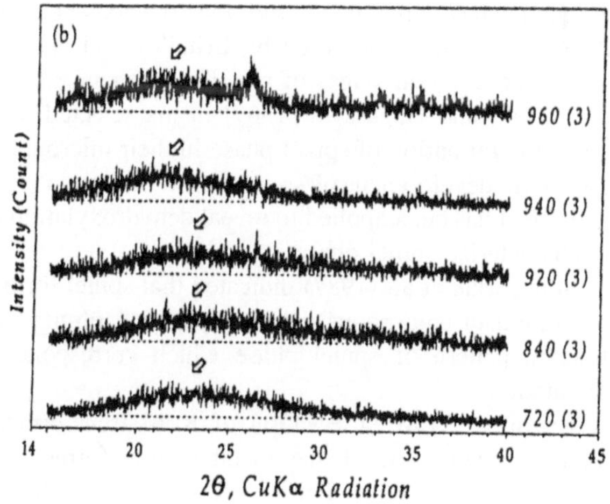

Fig. 6.8 Shifting of the center of the broad XRD background during heat treatment of kaolinite in metakaolinite to spinel formation range (After Lee et al. 1999). Reprinted by permission of the American Ceramic Society

However, the centroid of it was shifted to a lower angle from about 23° 2θ to 21° 2θ at ~940 °C (Fig. 6.8).

6.3 Summary

Electron microscopy has been used to study the morphological changes of kaolinite during heating. In earlier days, Eitel et al. (1939), Eitel and Kedesdy (1943), and then Radczewski (1953) studied the morphology of dehydrated kaolinite only.

Later on Comeforo et al. (1948) studied the morphological behavior of kaolinite at different stages of heating from 525 to 1,450 °C. They noted the persistence of hexagonal morphology from kaolinite to metakaolinite in the first stage. The later phase then collapsed to form mullite with needle-like shape in the second stage. Besides electron microscopic study, electron diffraction study are also been made to reveal the mechanism of thermal changes of kaolinite.

Roy et al. (1955) described the dehydrated kaolinite by the term, 'metakaolinite'. It decomposes to γ-Al_2O_3. The temperature of complete breakdown of metakaolinite is noted by them at 825 °C approximately. The new spot pattern is of the single crystal γ-Al_2O_3 pattern whereas ring-type pattern is obviously the powder pattern of same γ-Al_2O_3. The spot pattern of intermediate γ-Al_2O_3 phase is in between metakaolinite to mullite transformation which corroborated X-ray data.

Comer (1961) in his single crystal electron diffraction study substantiated the existence of orientation relationship between kaolinite to metakaolinite to Al–Si spinel phases as noted earlier by Brindley and Nakhira (1959a, b, c). At the next stage of transformation, he leached out of amorphous phase, and showed that mullite phase is oriented with the c-axis parallel to the [110] axes of the above spinel phase.

McConnell and Fleet (1970) used high resolution electron microscopy, an extremely powerful technique and without prior leaching showed the reaction products of kaolinite. The selected area diffraction patterns correspond to a well-oriented defect spinel phase with the zone axis [111] normal to the relict kaolinite flake as previously observed by Brindley and Nakhira (1959a, b, c). They suggested the different stages of reaction mechanisms.

Tsuzuki and Nagasawa (1969) in his selected area electron diffraction study noted the formation of spinel phase in their micrograph as dots and confirmed the topotactic development of spinel phase in thermal reaction of kaolinite.

HREM has been applied to reveal dehydroxylation process of metakaolinite and to characterize spinel phase.

Sonuparlak et al. (1987) indicated that spinel formation is preceded by a phase separation in amorphous dehydroxylated kaolinite matrix. They took micro diffraction pattern of spinel phase which corresponds to a <114> fcc zone axis orientation.

By direct TEM observation of K–M transformation series in single crystal kaolinite, they showed that metakaolinite forms after 1st endothermic peak is completely amorphous. The 980 °C exotherm is due to formation of spinel phase.

Beyond 2nd exothermic peak, acicular mullite crystals are formed with c-axis perpendicular to the original metakaolinite sheets. Structural changes of kaolinite to metakaolinite to spinel phase to mullite are summarized as follows.

At dehydration stage, hexagonal morphology remains intact with formation of metakaolinite. Electron diffraction spots of metakaolinite which contains 2D arrays of spots corresponding to (hko) reflections, indicate that high degree of order is retained in the plane of sheet structure. '**b**' parameter of metakaolinite is increased by 2.2 %. Briendley and Gibbon (1969) assumed that higher symmetry of Si–O network occurs during dehydroxylation.

After 1st exotherm, hexagonal morphology becomes blur. Electron diffraction shows new spot patterns and ring character of spinel phase with strong preferred orientation to original kaolinite.

At high temperature (~1,225 °C), i.e., 2nd exotherm, mullite crystal appear with preferred orientation.

Besides usual powder form, McConville et al. (1998) studied the micro structural evolution of English China Clay in extruded bar form. They compared the morphological changes during heating the two forms of kaolinite schematically. Electron diffraction from (440) reflection of spinel phase obtained from 'face-on' view of powder kaolinite exhibits spot pattern. Whereas, same from 'edge-on' view of platelets in extruded sample shows ring diffraction pattern.

Lee et al. (1999) reexamine the K–M phase transformation sequence by using energy-filtering transmission electron microscope (EF-TEM). They presented the electron diffraction patterns of kaolinite heat treated to three consecutive transformation steps.

Metakaolinite showed short range order and it is maintained prior to spinel formation. They first showed that the centroid of the amorphous band shifts during transformation of metakaolin to next crystallization step.

At the next step, they noted six diffraction spots of spinel phase. This systematic diffraction pattern of the spinel-type phase which showing pseudo-hexagonal symmetry strongly indicates a preferred orientation relationship to the parent metakaolinite phase. Due to concomitant existence of the spinel-type phase with metakaolinite at 920 °C, it is suggested that this transformation is topotactic.

Spinel phase decomposed completely at ~1,200 °C to mullite. It, however, crystallizes in random orientation.

By TEM study McConville (2001) also showed spinel formation prior to mullite formation.

References

G. W. Brindley and D. L. Gibbon, Kaoinite layer structure: Relaxation by dehydroxylation. Science, pp 1390 (1969)

G.W. Brindley, M. Nakahira, The kaolinite-mullite reaction series: I, a survey of outstanding problems. J. Am. Ceram. Soc. **42**(7), 311–314 (1959a)

G.W. Brindley, M. Nakahira, The kaolinite-mullite reaction series: II, metakaolin. J. Am. Ceram.Soc. 42(7), 314–318 (1959b)

G.W. Brindley, M. Nakahira, The kaolinite-mullite reaction series: III, the high-temperature phases. J. Am. Ceram. Soc. 42(7), 319–324 (1959c)

J.E. Comeforo, R.B. Fischer, W.F. Brandley, Mullitization of kaolinite. J. Am. Ceram. Soc. **31**(9), 254–259 (1948)

J.J. Comer, New electron-optical data on the kaolinite–mullite transformation. J. Am. Ceram. Soc. **44**(119), 561–563 (1961)

J.J. Comer, J.H. Koenig, S.C. Lyons, What are ceramic materials really like? Ceram. Ind. **67**(4), 125,148,150 No. 6, 96 (1956)

W.L. De Keyser, Contribution to the study of mullite. Ber.dtsch.keram. ges. **40**, 304–315 (1963)

W. Eitel, H. Kedesdy, Elektronen-Mikroskopie und Eeugung silikatischer Metaphasen: IV, Der Metakaolin (Electron microscopy and diffraction of silicate metaphases: IV, metakaoliu), *Abhandl.* preuss. *Akad. Wiss. Math.-nattir w. KI.*, NO. 5, pp. 37–45 (1943)

W. Eitel, H.O. Miiller, O.E. Radczewski, Ubermikroskopische Untersuchungen an Tonminerdlien (Ultramicroscopic examination of clay minerals). *Ber. deut. keram. Ges.* **20**(4) 165–180 (1939)

S. Lee, Y.J. Kim, H.-S. Moon, Phase transformation sequence from kaolinite to mullite investigated by an energy-filtering transmission electron microscope. J. Am. Ceram. Soc. **82**(10), 2841–2848 (1999)

J.D.C. McConnell, S.G. Fleet, Electron optical study of the thermal decomposition of kaolinite. Clay Miner. **8**, 279–290 (1970)

C.J. McConville, Thermal transformations in kaolinite clay minerals. Ceram. Eng. Sci. Proc. **22**(2), 149–160 (2001)

C. McConville, W.E. Lee, J.H. Sharp, Comparison of micro structural evolution in kaolinite powders and dense clay bodies. Br. Ceram. Proc. **58**, 75–92 (1998)

O. E. Radczewski, Electron optical investigations of the decomposition of clay minerals. Tonind. Ztg. **77**, 291(1953)

R. Roy, D.M. Roy, E.E. Francis, New data on thermal decomposition of kaolinite and halloysite. J. Am. Ceram. Soc. **38**(6), 198–205 (1955)

Sonuparlak, B., M. Sarikaya, I.A. Aksay, Spinel phase formation during the 980°C exothermic reaction in the kaolinite-to-mullite reaction series. J. Am. Ceram. Soc. **70**(11), 837–842 (1987)

K. Srikrishna, G. Thomas, R. Martinez, M.P. Corral, S. Aza, J.S. Moya, Kaolinite–mullite reaction series: a TEM study. J. Mater. Sci. **25**, 607–612 (1990)

Y. Tsuzuki, K. Nagasawa, A. Transitional stage to 980 °C exotherm of kaolin minerals. Clay Sci. **3**(5), 87–102 (1969) (Kaolin Minerals)

Chapter 7
IR Study

7.1 Introduction

IR absorption study was undertaken extensively for structural evaluation of clay minerals by various research workers. The characteristic IR curves are used for identification of clay minerals. The IR absorption pattern is generally taken in the range of 4000–400 cm^{-1} in frequency and 2.5–25 μm in wave length of IR spectral bands of kaolinite.

(i) *Assessment of Spectral Bands of Kaolinite*

Kaolinite contains silicon-oxygen anion structure of the general composition $(Si_2O_5)^{-2}$ corresponding to a di-trigonal packing of the tetrahedral. The unit cell consists of two silicon oxygen tetrahedral i.e., containing three Si–O–Si and two Si–O bands. Accordingly, three symmetric stretching vibrations of the Si–O bands are expected in the total eight internal stretching vibrations per unit cell. According to Pampuch (1966) the absorption bands due to internal stretching vibration of the kaolinite anion start to appear between 690 and 1120 cm^{-1} e.g., at 690, 750, and 790 cm^{-1} bands. Symmetric Si–O vibration should appear at 1120 cm^{-1} and antisymmetric Si–O vibration occurs at 1060 cm^{-1}. The remaining three antisymmetric Si-O-Si vibrations occur at 1010, 1010, and 1040 cm^{-1} respectively. Pampuch and Wilkos (1956) assigned 3690, 3648 cm^{-1} bands to aluminum oxygen-hydroxyl layer of kaolinite and 3623 cm^{-1} band to stretching vibrations of the remaining OH groups situated in the unoccupied sites of the silicon-oxygen layer. The 916 and 940 cm^{-1} bands are assigned to Al–OH bending vibrations. According to Stubican and Roy (1961) the 539 cm^{-1} band is found to be characteristic for the presence of Al^{VI}–OH band from Al[O(OH)].

(ii) *Bonding Assignments for Spectral Bands of Mullite*

Mullite contains chains of AlO_6 octahedral linked by randomly distributed tetrahedral SiO_4 and AlO_4 groups as shown in the tentative structure (Fig. 7.1).

Percival et al. (1974) made a comprehensive list as shown in (Table 7.1) of the average wave numbers of mullite in addition to kaolinite from 250 to 1300 cm^{-1} out of the work of previous researchers namely Launer (1952), Miller (1961),

A. K. Chakraborty, *Phase Transformation of Kaolinite Clay*,
DOI: 10.1007/978-81-322-1154-9_7, © Springer India 2014

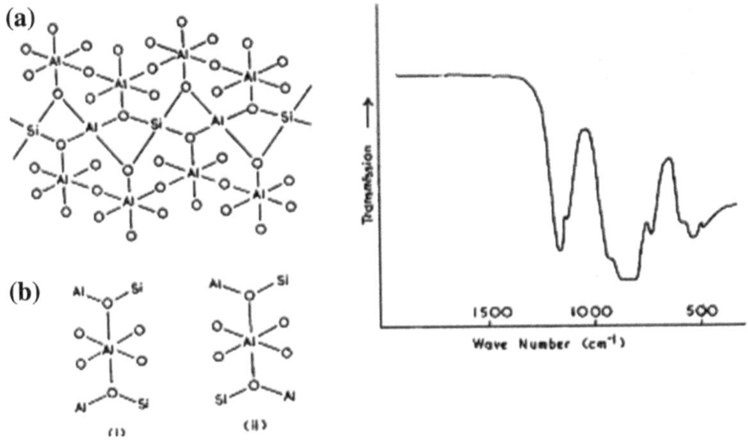

Fig. 7.1 **a** Simplified model of mullite structure, **b** possible variations of AlO_4 symmetry in simplified model, and Observed IR spectrum of standard mullite in KBr disk (After Mackenzie 1972). Reprinted by permission of the American Ceramic Society

Stubican and Roy (1961), Stubican (1963), Pampuch (1966), Freund (1967a, b), Tarte (1973). According to Freund (1967a, b), the assignments of Al^{VI}–O modes for mullite occur at 740, 901, and 960 cm^{-1} respectively. As per Saksena (1994), the assignments of the 898, 916, and 960 cm^{-1} bands of sillimanite were due to the tetrahedral SiO_4 group, was, therefore, transferred to the corresponding 901, 927, and 960 cm^{-1} mullite bands. The SiO_4 groups in 'condensed' silicates are bonded through common oxygen ions (X–O–X linkage), where as in isolated or ortho-type silicates they are separated by different cations (X–O–Y linkage). In this sense, SiO_4 groups are mullite (and sillimanite) are isolated and are being separated by AlO_4 and AlO_6 groups. The highest vibrational wave numbers of such Si–O bonds occur in the range 820–1030 cm^{-1} into which 901, 927, and 960 cm^{-1} mullite bands fall. Tarte (1973) distinguished isolated and condensed type tetrahedral AlO_4 and octahedral AlO_6 groups. For isolated AlO_4 groups which occur in mullite, the range is 650–800 cm^{-1}. Therefore, 740, 832 cm^{-1} mullite bands were assigned to AlO_4 groups. The SiO_4 in mullite would probably shift the AlO_4 wave numbers upwards, opposite to the effect of SiO_4 wave numbers when Al replaces Si in silicates. AlO_6 groups in mullite are predominantly condensed. Tarte (1973) gives the appropriate wave number range as 500–680 cm^{-1}. The 567 and 613 cm^{-1} mullite bands were assigned to these groups.

7.2 Characterization of Intermediate Phases

Based on the assignments of kaolinite and mullite, characterizations of the two intermediate phases occurring during heating have been attempted by several researchers by IR study. Details of their works are summarized below.

Table 7.1 IR absorption band of heated phases of kaolinite (Percival et al. 1974)

346 ms	Al^{VI}–O vibration and Si–O vibration
369 mw	Al^{VI}–O vibration and Si–O vibration
432 m	Splitted band due to
471 s	Si–O vibration
538 s	Al^{VI}–O bond from $Al[O(OH)]_6$ (Stubican and Roy)
698 m	
755 m	Al^{VI}–O vibration polarized perpendicular to base plane (Freund) and also due to Si–O vibration
792 m	
914 ms	Al^{VI}–OH vibration (Freund)
938 m	
1012 s	Si–O valence vibration parallel to base (Freund)
1035 s	Si–O valence vibration parallel to base (Freund)
1108 ms	Si–O valence vibration perpendicular and parallel to base (Freund)
In metakaolinite formation stage (2)	
445 broad	Si–O vibration
800 band	Al^{IV}–O vibration
1070	Al^{IV}– → O–Al^{IV} linkage
1190	Si–O valance vibrations
1250	Si–O valance vibrations
In spinel formation stage (3)	
455	Si–O vibration shifted from previous 445 band
550	Al^{VI}–O
725	Si–O ← - - → Al^{VI} (Freund)
800	Al^{IV}–O
In mullite formation stage (4)	
455	
560	
725	
800	Al^{IV} ← - - → O vibration
900	Al^{VI} ← - - → O vibration
950	
1130	
1160	
Pure mullite (5)	
445	
610	
730	
815	Al^{IV}–O
900	Al^{VI}–O
950	Al^{IV} ← - - → O–Al^{VI}
1020 (shoulder)	Al–O vibration
1130	Si–O
1160	Si–O

Stubican (1963) depicted the IR spectrum of clay minerals namely kaolinite and halloysite. It shows an absorption band at 2.72 μm which belongs to hydroxyl groups in the structure absorption bands between 8 and 10 μm are likely to belong to the Si–O linkage, while the octahedral alumina sheet gives the absorption band between 10 and 11 μm.

Interlayer water molecules present between two sheets of halloysite show additional band at 2.9 μm. They gave emphasis on the extent of dehydroxylation and on crystallinity of clays. The results are incompatible with the fact that structurally disordered kaolin minerals on heating decompose more quickly than well-crystallized kaolinites but retain hydroxyl groups at higher temperatures. Stubican assumed that disordered kaolinite completely decomposes during dehydroxylation. The gibbsite sheet produces small nuclei of active Al_2O_3 which contains a certain number of hydroxyl groups (HAl_5O_8 i.e., $5Al_2O_3H_2O$).

Like Stubican (1963), De Keyser (1963d) also undertook dehydroxylation studies of kaolinite and noted the changes of IR spectral bands of kaolinite. He shows that intensity of 3440 cm^{-1} band due to absorbed water decreases with increase of drying temperature. During endothermic reaction process in between 400 and 600 °C, it was shown that the loss of structural –OH groups which are manifested by the gradual disappearance of four bands situated from 3690 to 3620 cm^{-1}. It was noted that bands 3695 cm^{-1} (a), 3670 (b), 3650, (c) and 3620 cm^{-1} (d) disappear at the same rate which indicates that the corresponding –OH groups were eliminated from the structure simultaneously without intervention of the O–H bonds strength. After the completion of the main dehydration at 610 °C, the absorption bands, one situated at 3620 cm^{-1} and other very broadband whose maximum is situated at 3400 cm^{-1} (Fig. 7.2).

In addition to dehydroxylation step, Pampuch (1966) tried to, characterize the principle phases formed after 980 °C in heated kaolinite. He heat treated Zettlitz kaolinite, Dickite and Colombian kaolinite, all of were of high degree of crystallinity to various temperatures with a rate of heating of 10 °C/min, and thereafter performed IR study. IR spectra of Zettlitz kaolinite heated to various temperatures shows the following observations.

At 550–600 °C disappearance of IR absorption bands associated with vibration of OH groups situated at the surface of the Al–O–OH layer (3690, 3648, and 940 cm^{-1}) take place with persistence of band at 3628 cm^{-1} associated with vibrations of OH groups situated in unoccupied sites of silicon-oxygen layer. Profound change in the spectrum in the range of internal stretching Si–O–Si and Si–O vibrations (690–1150 cm^{-1}) occurs. The disappearance of the 539 cm^{-1} absorption band with concomitant development of a new band at about 810 cm^{-1} take place. At or above 600 °C up to the complete elimination of –OH groups, the IR spectra showed a concomitant disappearance of the –OH absorption band at 3628 cm^{-1} and development of a new band at 570 cm^{-1} with no change in the remaining spectral ranges. The position of the band at 570 cm^{-1} corresponds to one of the characteristic absorption band of aluminum oxide. Accordingly, the development of the 570 cm^{-1} absorption band was explained by Pampuch to be due to separation of alumina by break down of metakaolinite on heating. Further, the spectra of kaolinite heated to

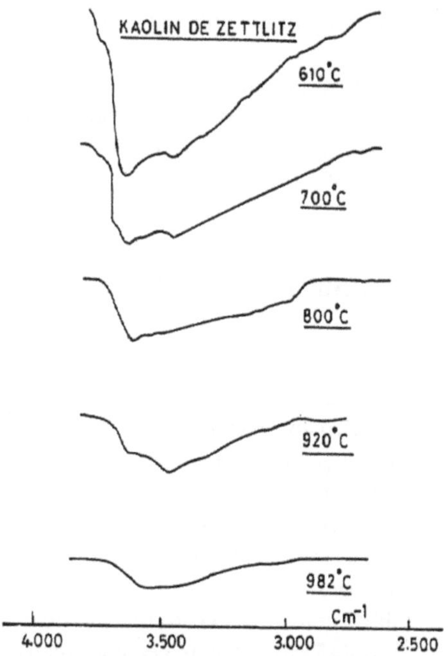

Fig. 7.2 IR spectrum of Zettlitz kaolin in the OH region after heating at high temperature 920–982 °C corresponding to the beginning and to the maximum of the exothermic reaction (After De Keyser 1963d). Reproduced with kind permission of The Clay Minerals Society, publisher of Clays and Clay Minerals

800 °C and to higher temperatures show new band at 1190 and 1100 cm^{-1} corresponding to those expected for free SiO_2 besides free alumina. He observed that Al[O(OH)] octahedral of kaolinite converts to AlO_4 tetrahedral and also shows the existence of a superimposed layer composed of aluminum and oxygen left after dehydration of OH groups at the di-octahedral surface with one-eighth of the OH groups initially present in kaolinite; the positions of the latter should not change in comparison with natural kaolinite in view of the unchanged position of absorption band associated with variations of these groups. According to Pampuch, the remaining OH groups temporarily stabilized the metastable anhydrous phase.

Freund (1967a, b) first studied in detail the phase transformation behavior of kaolinite by IR spectrometry. He noted the following changes in the IR spectrum of kaolinite.

7.3 IR Shift During Dehydroxylation

(i) A group of splitted bands merge into a single broadband e.g., 1109, 1037, and 1018 cm^{-1}. Second, Si–O valence vibration frequency shifts toward higher wave number with a distinct shoulder at 1200 and another 1250 cm^{-1}.
(ii) Some bands disappeared. In the range of long-wave lattice vibrations, five clearly resolved bands observed between 700 and 285 cm^{-1} namely at 540, 465, 370, and 345 cm^{-1}, at least three specially those at 540, 370, and 345 cm^{-1}, disappear.

(iii) Splitting of the 465 and 425 cm^{-1} bands due to the vibration of pure silica is lost.
(iv) A broad maximum at 800 cm^{-1} appears and it is due to $Al^{IV} \rightarrow$ O vibrations at 1075 cm^{-1} is also marked. IR spectrum of metakaolinite is taken as superimposition of Si–O–Si linkages and also from Al^{IV}–O–Al^{IV} linkages. So the Al^{IV}–O–Al^{IV} would lie in the region of 1075 cm^{-1}.

7.4 IR Shift During Metakaolin to Spinel Formation

(i) In IR spectrum 445 cm^{-1} band is somewhat sharper and its maximum shift is at 455 cm^{-1}.
(ii) A sharp absorption band at 550 cm^{-1} appears out of the raised background in the high-frequency region. The maximum of which shifts from 550 to 560 cm^{-1} with the rise of temperature of heat treatment with simultaneous renewed broadening of the 455 cm^{-1} band and finally lies at about 610 cm^{-1} in the fused mullite. The original 445 cm^{-1} band increases in intensity from the back ground after the conversion of metakaolinite to spinel at 970 °C, and shifts gradually toward higher wave numbers. In well-crystallized mullite, 610 and 445 cm^{-1} bands become finally similar in width and intensity.

7.5 IR Shift During Mullite Formation

On prolong heating at 1,200 °C, band appeared in the spectrum, which corresponds to mullite formation as confirmed by X-ray. In fused mullite, one can recognize the Si–O valence bands in the frequency regions of 1160 and 1130 cm^{-1} which are known from quartz, cristobalite, and other alumino silicates. The Al^{IV}– $\rightarrow$ O valance vibration at 1075 cm^{-1} from Al^{IV}–O–Al^{IV} linkages decreases or shifts to lower frequencies. Because the proportion of tetrahedrally co-ordinated Al decreases to 25–50 % mullite and the proportions of octahedrally coordinated Al rises correspondingly. In well-crystallized mullite, the direct linkages of AlO_4 which are typical of metakaolinite disappear. They are replaced by $[AlO_6]$–$[AlO_4]$ linkages. The associated vibrations of $Al^{IV} \leftarrow - - - - \rightarrow$ O–Al^{VI} should shift to lower frequency from the original. Probably, this is also an explanation of the appearance of big band at 950 cm^{-1} and a shoulder at 1020 cm^{-1} in the spectrum of mullite.

In spectrum of partially mullitized metakaolinite heated to 1,200 °C, broad-bands appearing at 900 and 820 cm^{-1} are also to be assigned to Al–O vibrations. The 900 cm^{-1} band which does not appear in actual metakaolinite below 970 °C could originate from the Al^{IV}– $\rightarrow$ O vibrations. It corresponds to the $Al^{IV} \leftarrow - - \rightarrow$ OH bands of kaolinites at $Al^{IV} \leftarrow - - - \rightarrow$ 0 absorption edge of γ-Al_2O_3. Slow appearance of this band is Si–Al spinel with heating above 1000 °C shows progressive change of the Al ions from tetrahedral to octahedral positions. In pure (γ-Al_2O_3) only Al^{VI} is available, while in alumino silicate 25–50 % Al ions are in tetrahedral positions. Because of this band at 800 cm^{-1} in metakaolinite spectra can be recognized also in fused mullite spectra at 815 cm^{-1} which becomes quite strong in the low-temperature patterns. This band is also could be assigned to Al ← - - → O valance vibrations with tetrahedrally co-ordinated Al.

Instead of allowing static heating as done by previous researchers, Tsuzuki and Nagasawa (1969) heat treated halloysite and kaolinite at a constant rate of 50 °C per minute, air quenched and then analyzed for IR. In case of kaolinite, the absorptions spectra at 959 °C, the temperature which corresponds to the intermediate stage of the small endothermic reaction, is almost the same as the spectrum at 915 °C, the approximate beginning of the temperature of the small endotherm. On the contrary, the spectra at 981 and 994 °C, the approximate beginning and finishing temperatures of the exothermic reactions, a new broad absorption at 580 cm^{-1} appears. The same information is true for halloysite. They concluded that the mode of vibration of ions does not change during endothermic reaction, but it changes with the progress of the exothermic crystallization reaction. Thus, the appearance of 570, 720, and 900 cm^{-1} bands at an early stage of heating of metakaolinite in explained as due to development of spinel phase.

Another comprehensive work on structural transformation of kaolinite by IR study was made by Percival et al. (1974). They used 0–2 μm fraction of an English kaolinite and heat treated it to various temperatures with soaking for different duration of times and the IR spectrum of these specimens and are shown in Table 7.1. Position of band in the original spectrum of kaolinite is shown in first column. After dehydroxylation marked reduction of the kaolinite 'AlO_6' band at 538 cm^{-1} and appearance of a new band at 842 cm^{-1} for metakaolinite which is observed as shown in column 2. After decomposition of metakaolinite, two sets of bands at 570, 730, and 907 cm^{-1} and at 1102 and 1205 cm^{-1} are noted and shown in column 3. They assigned these sets of bands to amorphous/or γ-Al_2O_3 and amorphous SiO_2, respectively. On heat treatment at 950 °C/24 h, IR spectrum of kaolinite shows poorly cristalline spinel and poorly crystalline 'primary' mullite. The associated free amorphous silica provides a distinctive band in the spectrum at 472 cm^{-1}. On heating to 1,050 °C/24, spinel phase is still present. On further heating to 1,100 °C/24 h and to 1,200 °C/48 h the IR spectra in column 3 show the disappearance of the spinel phase and increase in content of well-crystallized mullite and also crystallization of cristobalite from amorphous SiO_2. With these observations, Percival et al. presented the following course of metakaolinite transformation. Besides some partial decomposition, it completely decomposes to spinel-type phase γ-Al_2O_3 which is associated with amorphous SiO_2 and some poorly crystalline 'primary' mullite. At 1,100 °C secondary mullite forms primarily from γ-Al_2O_3/SiO_2(A) reaction and also by the recrystallization of some primary mullite. The excess amorphous SiO_2 eventually crystallises as cristobalite.

7.6 Summary

Initially Stubican (1963), Stubican and Roy (1961), and De keyser (1963d) gave more effort at the IR spectroscopy for understanding of the mechanism of dehydroxylation of kaolinite. A comprehensive data on thermal behavior of kaolinite were shown by Pampuch (1966). He showed that a marked changes in Si–O vibration spectrum as well as change in co-ordination number of Al occur during disappearance of ¾ OH groups. Metakaolin on further heating develops bonds at $\sim 570\ cm^{-1}$ characteristic of γ-Al_2O_3.

In an extensive spectral analysis, Freund (1967) showed that during dehydration, fine structure of lattice vibration is lost and Al assumes tetrahedral coordination. During formation of spinel-type phase followed by collapse of metakaolinite structure, the narrowing of the band at 445 cm^{-1} takes place which points to a decreased anharmonicity of the $[AlO_4]$ and $[SiO_4]$ vibrations perpendicular to basal plane. During transition from spinel to mullite, a partial change of aluminum from tetrahedral to octahedral occurs. Tsuzuki and Nagasawa (1969) collected sample by air quenching of dynamically heated kaolinite and halloysite and specially noted the very characteristic 580 cm^{-1} band for sample heated just at the exotherm and conformed the formation of spinel phase as the cause behind it. Percival et al. (1974) noted that a metakaolinite structure includes a small concentration of OH groups and also suggested that it partially decomposes into oxide phases before 950 °C decomposition.

Following changes are noted by IR studies of heat treated kaolinites.

(1) During the conversion of kaolinite to metakaolinite by the dehydraoxylation process, the following bands frequency change has been noticed.

   (i) Threefold split in bands namely 1108, 1035, and 1012 cm^{-1} between 1175 and 950 cm^{-1} due to two-dimensional silica-sheet vibration has shifted vary little toward higher frequencies and have merged and formed a broad big shoulder. Two distinct shoulders, one at 1200 cm^{-1} and another at 1250 cm^{-1} appeared in this spectrum.
   (ii) Out of five clearly resolved bands observed between 700 and 285 cm^{-1} namely 538, 471, 432, 369, and 346 cm^{-1} at least three bands those at 538, 369, and 346 cm^{-1} due to vibrations in octahedral layer or between octahedral and tetrahedral layer disappear during dehydration (Freund 1967a, b). Splitting of the bands at 471 and 432 cm^{-1} due to vibrations of pure SiO_2 has vanished with the formation of a broad band at 445 cm^{-1} without any remarkable decrease in intensity.
   (iii) Three bands namely 938, 914, and 792 cm^{-1} assigned to Al^{VI}–OH have disappeared. Besides, 755 and 698 cm^{-1} also due Al–OH vibrations vanish away.
   (iv) A new broad maximum at 800 cm^{-1} corresponding to $Al^{IV}- \rightarrow O$ vibrations appear. Thus, co-ordination number of Al in kaolinite changes from six to mostly four in metakaolinite.

(v) The broad maximum at 1100 cm^{-1} is due to superimposition of Si ← - - - → O valance vibration (Freund). It originates from Si–O–Si and Al^{IV}–O–Al^{VI} linkages. Band at 1075 cm^{-1} might be due to Al^{IV} ← - - - → Al^{IV} vibrations.

(vi) Out of four –OH bands observed in kaolinite three bands namely 3695 and 3650 cm^{-1} disappear on heat treatment. Both 3620 and 3440 cm^{-1} bands are reduced and become broad. Thus, the metakaolinite contains some residual –OH groups. On further heat treatments, the bands gradually become smaller and finally disappear just before 980 °C exotherm.

(2) During conversion of metakaolinite to spinel phases at around 980 °C the following band shiftments are observed.

(i) The 445 cm^{-1} band tended to become sharp and its maximum shift toward 455 cm^{-1}.

(ii) A broadband at 570 cm^{-1} appears out of the raised background in the high frequency region.

(iii) In addition to 570 cm^{-1} band, Freund, Percival et al. noted 720 and 900 cm^{-1} bands, respectively, corresponding to development of phases having AlO_6 groups besides the persistent of AlO_4 group at 810 cm^{-1}.

(3) During the formation of mullite the following band shiftments are observed.

(i) Two Si–O valance band appeared in the frequency range at 1160 and 1130 cm^{-1} respectively.

(ii) Valance vibrations appear in the lower frequency regions e.g., 950 and 1020 cm^{-1} respectively, from band at 1075 cm^{-1} present in original band of Al^{IV}–O–Al^{IV} in metakaolinite.

(iii) 815 cm^{-1} band due to Al^{IV}–O is retained in mullite structure containing Si^{IV}–O– ← - - - → Al^{IV} linkages.

(iv) Other two Si–O valance bands appeared at higher wave numbers e.g., 445, and 610 cm^{-1} from previous band at 445 cm^{-1} in metakaolinite stage.

References

W.L. De Keyser, R. Wollast, L. De Laet, Contribution to the study of OH groups in Kaolin minerals. in Proceedings of International Clay Conference, Pergamon Press, 75–86 (1963d)

F. Freund, Infrared spectra of Kaolinite, Metakaolinite, and Al–Si spinel. Ber. Deut. Keram. Ges. **44**(181), 392–397 (1967a)

F. Freund, Kaolinite–Metakaolinite, a model of a solid with extremely high lattice defect concentrations. Ber. Deut. Keram. Ges. **44**(I), 5–13 (1967b)

P.J. Launer, Regularities in the infrared absorption spectra of silicate minerals. Amer. Mineral. **37**(9–10), 764–784 (1952)

K.J.D. Mackenzie, 'Infrared frequency calculation for ideal Mullite ($3Al_2O_3.2SiO_2$). J. Am. Ceram. Soc. **55**(2), 68–71 (1972)

J.G. Miller, Infrared spectroscopic study of isothermal dehydroxylation of Kaolinite at 470 °C. J. Phys. Chem. **65**(51), 800–804 (1961)

R. Pampuch, Infrared study of thermal transformations of Kaolinite and the structure of Metakaolin. Pol. Akad. Nauk. Oddzial Krakowie Kom. Nauk. Mineral. Pr. Mineral. **6**, 53–70 (1966)

R. Pampuch, K. Wilkos, in Proceedings of 80th Conference on Silicates Ind. Akad.emiai Kiado, Budapest, p. 179 (1956)

H.J. Percival, J.F. Duncan, P.K. Foster, Interpretation of the Kaolinite–Mullite reaction sequence from infrared absorption spectra. J. Am. Ceram. Soc. **57**(2), 57–61 (1974)

B.D. Saksena, Infrared absorption studies of some silicate structures. Trans. Faraday Soc. **57**(121), 242–258 (1961) (*Ceramics*. Wiley, New York, 1994, pp. 116, 233)

V. S. Stubican, Inteytion of ions in complex crystals as shown by infrared spectroscopy. Miner. Ind. (University Park, PA), **33**(21), 1–4 (1963)

V. S. Stubican and R. Roy, Isomorphous substitution and infrared spectra of the layer lattice silicates. Amer. Mineral.**46**(1–2) 32–51 (1961)

P. Tarte, Study of silicates by infrared spectroscopy. Present results and future outlook. Bull. Soc. Fr. Ceram. **58**, 13–34 (1973)

Y. Tsuzuki, K. Nagasawa, A transitional stage to 980 °C exotherm of Kaolin minerals. Clay Sci. **3**(5), 87–102 (1969)

Chapter 8
XRF Study

8.1 Introduction

The AlK_α emission wavelength is affected by the state of coordination of Al atom. Based on this discovery, the coordination state of aluminum in metakaolinite, in spinel phase, and finally in mullite is determined. Besides IR study, effect of heat on changes of coordination number of aluminum of successive heat treated kaolinite were analyzed by XRF study. White et al. (1959), White and Gibbs(1967,1969) and Dekimpe et al. (1964) showed that the coordination number of aluminum was obtained by comparison of the wavelength shifts of AlK_α in the specimen with those of standards. The origin of the wavelength shift is understood by considering the basic relationship as follows:

$$\lambda\, \mathrm{K}_\alpha = \lambda_{\mathrm{K}} - \lambda_{\mathrm{L}} \tag{8.1}$$

where λ_K = frequency of K absorption edge and λ_L = frequency of L absorption edge.

Aluminum and silicon have L electrons and these are directly implied in coordination and thus it may be relevant to this method of investigation. According to Bragg's law, $n\lambda = 2d\ \sin\theta$. By differentiation

$$\Delta\theta = \frac{c}{2(\lambda K_\alpha)^2 d\cos\theta}\Delta\lambda K_\alpha \tag{8.2}$$

where c = velocity of light.

The angular displacement, θ is approximately proportional to the frequency shift for one given element. (2θ) for the AlK_α emission from metallic Al and from samples of different heat treated kaolinite (coordination number of aluminum) in them are to be determined and a number of different calibrant substances are measured.

$(2\theta) = 2\theta$ (sample) -2θ (aluminum metal)

The change in wavelength shift according to the equation of White et al. (1959) above are graphically plotted by Brindley and Mckinstry (1961). They indicated that 2θ angle for the radiation from aluminum in fourfold coordination lies in the

A. K. Chakraborty, *Phase Transformation of Kaolinite Clay*,
DOI: 10.1007/978-81-322-1154-9_8, © Springer India 2014

range $\sim$0.06 to $\sim$0.07° and aluminum in sixfold coordination in the range $\sim$0.09 to $\sim$0.11°. On comparing these data with kaolin minerals heated to 750 °C for 24 h. (when metakaolin formed), the (2θ) found to correspond to AlO_4 region. Heated kaolin at 1,024 °C shows the movement of (2θ) from AlO_4 to the direction of AlO_6. This displacement is more marked in case of disordered kaolinite and halloysite. It is concluded that Al ions in metakaolinite are in fourfold coordination and its conversion to spinel phase is marked by changes in coordination from four- to sixfold state.

8.2 Characterization of Phases

Udagawa et al. (1969) reported more quantitative values of coordination number of aluminum by XRF technique. Relative change of sixfold coordinated aluminum on heating Kanpaku kaolinite at different temperatures is plotted. For the original kaolinite, the relative content of Al(VI) was taken as 100 %. As the dehydroxylation proceeded, the numbers of Al(VI) decrease, at fully dehydrated stage, no sixfold coordinated aluminum is discernible. On further increase of temperature, coordination of aluminum moves toward octahedral state. According to them, these data correspond to the formation of spinel type phase and little mullite as shown in the schematically diagram earlier.

The relative ratio of Al(IV)/Al(VI) is approximately equal to 45/55 in kaolinite heated to 1000 °C using kaolinite and $AlPO_4$ as standards for sixfold and fourfold coordination numbers of Al, respectively. The ratio however remains unchanged in kaolinite heated to higher temperatures.

Tsuzuki and Nagasawa (1969) considered the determination of coordination number of aluminum by XRF technique as a useful tool for elucidating the structural change of kaolin minerals on heating. Differences in diffraction angle (2θ) for AlK_α radiation of halloysite heat treated to different temperatures and metallic aluminum are plotted. Qualitatively, the result showed that on gradual heat treatment, the 2θ value of kaolinite first decreased and formed metakaolinite. This conclusion verifies earlier observations of Brindley and Mckinstry (1961), Egawa (1964), and Suda and Noda (1965) reported that 2θ angle for the radiation from aluminum in fourfold coordination is smaller than that in sixfold coordination. On further heating, the 2θ value increased; this fact suggests the change of coordination number of aluminum from four to six, which may be due to increase in the amount of spinel phase.

Bulens et al. (1978) used gibbsite and sodium zeolite (X-type) as standard for Al(VI) and Al(VI) atoms, respectively.

Leonard (1977) heat treated Zettlitz kaolinite at the respective temperatures followed by soaking for 24 h prior to X-ray spectroscopie work. The observed angular shifts are shown in (Table 8.1). In the metakaolinite stage, at $\sim$600–800 °C, the apparent Al(IV) content would be expected to be 100 %. But the result shows a value of 60 % only. At a temperature of 900 °C, spinel crystallized, XRF analysis

Table 8.1 Δ, Angular shifts corresponding to change in wavelength of AlK_α with respect to gibbsite and corresponding relative Al(IV) contents (After Leonard 1977). Reprinted by permission of the American Ceramic Society

Aluminum Temp. (°C)	Δ	% Al(IV)
100	−2.3 ± 10.1	−4.7 ± 20.9
600	28.7 ± 7.1	59.6 ± 14.7
800	28.3 ± 6.4	58.8 ± 13.2
900	18.3 ± 5.0	37.9 ± 10.3
970	26.5 ± 8.0	55.0 ± 16.5
1,250	27.9 ± 5.8	57.9 ± 12.1

shows a value of 38 % Al(IV) content which is perfectly coherent with 40 and 36 % determined for bayerite heated at 500 °C and pure alumina spinel sample. According to Leonard (1977), this 40 % value strongly indicates a purely aluminous spinel phase. The abrupt decreases from 55 to 60 % Al(IV) content in metakaolinite to 38 % is a clear indication of disruption of the structural homogeneity in the Al configuration within the kaolinite-mullite sequence. According to Leonard (1977) and Leonard et al. (1969), some Si may enter the tetrahedral sites of the spinel since a well-characterized compact γ-Al_2O_3 reportedly reached an appearent Al(IV) content of about 50 %. Therefore, all Al atoms do not occupy octahedral interstice, as would be required in Al–Si spinel structure. If the hypothesis of Brindley and Nakahira is true, the Al(IV) content would be 0 %. At a temperature of 970 °C, when major quantity of spinel disappears and mullite starts crystallization, XRF data showed Al(IV) content to the extent of 55 %. At about 1,250 °C, when mullite phase fully developed, the Al(IV) value of 57.9 % agrees qualitatively with the spectroscopic position recorded by Brindley and Nakahira at 1,024 °C. This value also appears realistic when compared with the 58.3 % value which could be derived from the location reported for mullite by Sadanaga et al. (1962).

8.3 Summary

The changes in coordination of Al in K–M reaction sequence have been studied by X-ray fluorescence measurements of the AlK_α shift of heat treated kaolinite with that of standard as per the basic relationship shown by White et al. (1959) and Dikimpe et al. (1964). Brindley and McKinstry (1961) noted that upon dehydroxylation coordination of Al in kaolinite transformed from six to four. On further heat treatment at and above 1,024 °C, a partial reversal from four to six takes place. Udagawa et al. (1969) presented a complete picture in entire transformation process. According to them, at fully dehydrated stage, the ratio of AlO_4/AlO_6 is = 100/0. After spinel formation at ~1,000 °C the relative ratio is equal to 45/55. However, the ratio unaltered even after mullitization. Tsuzuki and Nagasawa (1969) verified the earlier observation of the structural change of kaolin minerals on heating by similar technique. A more quantitative X-ray spectroscopic study was made by

Leonard et al. (1969) and Leonard (1977). In the metakaolin stage, the result shows a value of 60 % AlO_4. In the second stage, when spinel crystallized the result shows a value of 38 % AlO_4 which is close to the value of γ-Al_2O_3. In the third stage, when spinel transformed and mullite developed abundantly at ~1,250 °C the result shows a value of 58 % AlO_4 which agrees with reported value of mullite.

References

G.W. Brindley, H.A. McKinstry, The Kaolinite–Mullite reaction series: IV, the coordination of aluminum. J. Am. Ceram. Soc. **44**(10), 506–507 (1961)

C. De Kimpe, M.C. Gastuche and G.W. Brindley, Am. Mineralogist. **49**, 1, (1964)

S. Udagawa, T. Nakada, M. Nakahira, Molecular structure of Allophane as revealed by its thermal transformation. in *Proceedings of the International Clay Conference*, vol. 1, ed. by Lisa Heller bisa Heller (Israil University Press, Gerusalem, 1969), p. 151

Y. Tsuzuki, K. Nagasawa, A Transitional Stage to 980 °C Exotherm of Kaolin Minerals. Clay. Sci. **3**(5), 87–102 (1969)

T. Egawa, Clay. Sci. **2**, 1–7 (1964).

Y. Suda, T. Noda, J. Chem. Soc. Jpn. **86**, 78–82 (1965)

R. Sadanaga, M. Tokonami and Y. Takeuchi, Structure of mullite, 2Al2O3.SiO2, and relationship with the structures of sillimaite and andalusite. Acta Crystallogr. **15**, 65–68 (1982)

M. Bulens, A. Leonard, B. Delmon, Spectroscopic investigations of the Kaolinite–Mullite reaction sequence. J. Am. Ceram. Soc. **61** (1–2), 81–84 (1978)

A.J. Leonard, Structural analysis of the transition phases in the Kaolinite– Mullite thermal sequence. J. Am. Ceram. Soc. **60** (1–2), 37–43 (1977)

A.J. Leonard, P.N. Semaille, J.J. Fripiat, *Structure and properties of amorphous Silico-Aluminous; IV*, Proceedings of the British Ceramic Society, vol. 13 (1969), pp. 103–116

E.W. White. H.A. McKinstry. and T.F. Bates, in *Advances in X-Ray Analyses*, vol. 2. ed. by M. W. Mueller (Denver Research Institute, Denver, Colo., 1959) p. 239-45

E.W. White and G.V. Gibbs, Structural and chemical effects on the Si Kα X-Ray line for silicates. Am. Mineral. **52**(7–8), 985–93 (1967)

E.W. White and G.V. Gibbs, Structural and chemical effects on the AlKα X-ray emission band among aluminum containing silicates and aluminum oxides. ibid. **54**(5–6), 931–36 (1969)

Chapter 9
Radial Electron Density Distribution

9.1 Introduction

The radial electron density distribution (RED) technique was applied to study the structural aspects of the two transition phases in K–M reaction series. This study is perfectly suited for amorphous and poorly crystalline materials like metakaolinite and spinel phases. This study helps in elucidating the relative changes in Al coordination through the entire reaction course of kaolinite. This method was developed by Warren (1934) and was subsequently studied by Leonard et al. (1971) for studying atomic arrangements in amorphous silico-aluminus, heated kaolinites, etc.

9.2 Characterization of Phases

Leonard (1977) reports the experimental RED profile for Zettlitz kaolinite heated to different temperatures (Fig. 9.1). The first region (0.14–0.21 nm) contains the electron density corresponding to the first oxygen shell of the cation, e.g., Si–O at 0.162 nm, Al^{IV}–O at 0.1766 nm, and Al^{VI}–O at 0.1944 nm.

It is observed that the level of Al^{VI}–O component at 0.19 nm for kaolinite decreases with dehydroxylation as a consequence of heating, with the formation of metakaolinite, where Al is present in tetra coordination stage. On continued heating, Al^{VI}–O component increases but does not come to the original level as in kaolinite. The second region (0.27 nm) contains the combined effect of 0–0 vectors sharing tetrahedron and octahedron edges at 0.264 and 0.229 nm, respectively. The third region is of complex in nature. It consists of Si–O-Si peak at 0.312 nm from amorphous silica (assumed to have been liberated on heating kaolinite) and Al–O–Al components corresponding to high temperature phases, like spinel and mullite. The fourth band situated at 0.415 nm represents the Si–2d–O contribution. Leonard has divided the first cation-anion vector range (0.14–0.211 nm) of the RED profile into elementary Gaussian contribution corresponding to single Si–O, Al^{IV}–O, and Al^{VI}–O components followed by a least square method when only

A. K. Chakraborty, *Phase Transformation of Kaolinite Clay*,
DOI: 10.1007/978-81-322-1154-9_9, © Springer India 2014

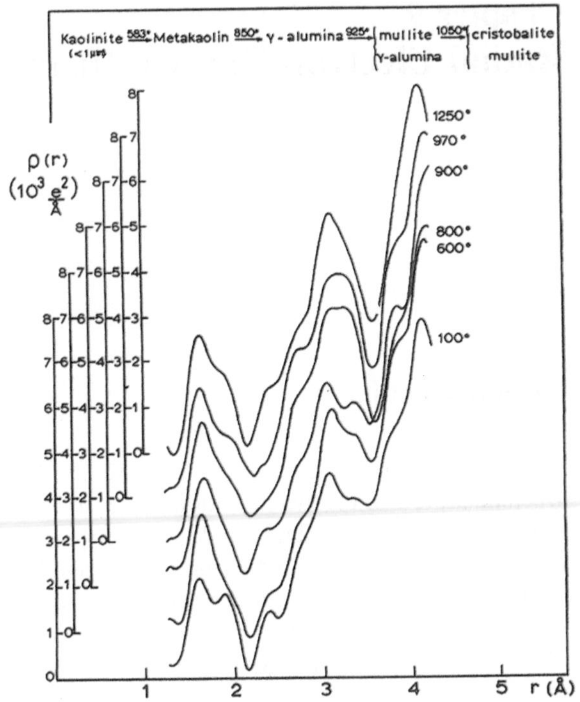

Fig. 9.1 Thermal sequence of kaolinite and structural characterization of main steps by radial electron density distribution (RED) (After Leonard 1977). Reprinted by permission of the American ceramic society

two vectors are involved and by a trial and error procedures when three vectors or more are involved (Fig. 9.2). The solid line considers the experimental profile, the line with cross is the superimposed best fit values resulting from the summation of amplitude by amplitude and the elementary contribution of the interatomic vectors by dotted curves. From these figures, the relative contents of AlO_6–AlO_4 during the thermal transformation of kaolinite are found out.

1. In Metakaolinite Stage

Gaussian curve for the first band constitutes mainly for Si-contribution, less intense component due to Al–O appearing in the asymmetrical region. Figure 9.2 shows that Si–O contribution should remain constant, while the Al–O component decreases with increasing temperature, first abruptly, then more slowly. At 800 °C, metakaolinite yields on Al^{IV}–O distance of 0.1873 nm which is appreciably larger than that generally accepted for an Al–O tetrahedron in layer silicate (0.177 nm).

2. Spinel phase

Leonard et al. (1979) computed RED curves for γ-Al_2O_3 and Si–Al spinel models for the following transformation shows below by considering metakaolin in one stoichiometric formula unit as $(Al_2O_3)_x$ $(SiO_2)_y$ where $x + y = 1$

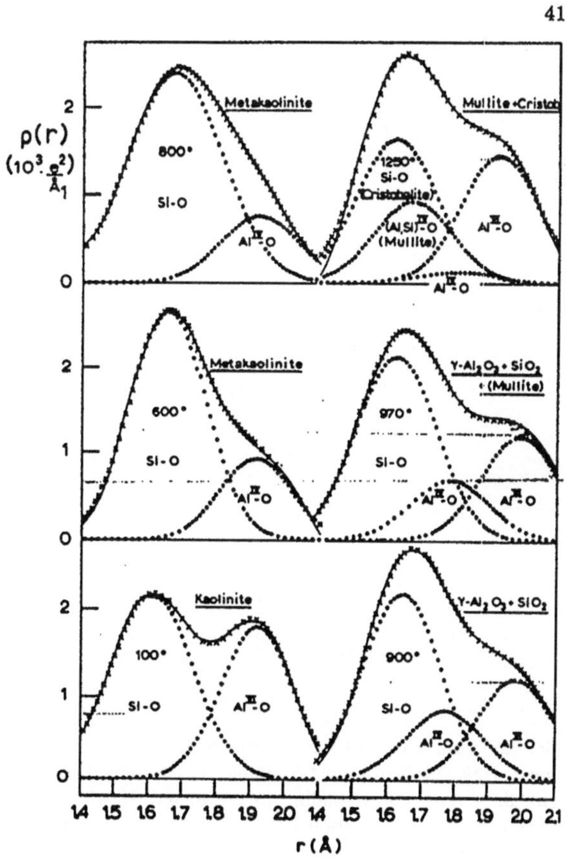

Fig. 9.2 Decomposition of first band (cation–anion) of RED curves of this figure, phases on left by automatic least—*squares* procedure and phases on right by a trial—and—error process. (After Leonard 1977). Reprinted by permission of the American ceramic society

$$(Al_2O_3)_{0.333}(SiO_2)_{0.666} = 0.5Si^{IV}[Al_{1.33}[]_{0.662}]^{VI}O_4 + 0.17SiO_2 \text{ Brindley's Si–Al Spinel} \quad (9.1)$$

$$(Al_2O_3)_{0.333}(SiO_2)_{0.666} = 0.25Al^{IV}[Al_{1.67}[]_{0.33}]^{VI}O_4 + 0.66SiO_2 \text{ Leonard's } \gamma\text{-}Al_2O_3 \text{ Spinel} \quad (9.2)$$

They calculated theoretical electron density contributions into three main components for both hypotheses on the basis that the maximum amplitudes of unitary vectors of Si–O and Al–O correspond to 355 and 321 e^2/A, respectively. RED distribution for mixed spinel and γ-Al_2O_3 spinel amount to 447.1 and 414.8 e^2/nm. With this data, they finally suggested a favorable γ-Al_2O_3 + SiO_2 model Eq. 9.2. Leonard compared experimental RED curve with theoretical curves for the two hypotheses (Fig. 9.3). They showed that even in the case of the pure Al spinel, the experimental profile (solid line) appears under estimated in the Al^{VI}–O region, in favor of overestimation of the Si–O component.

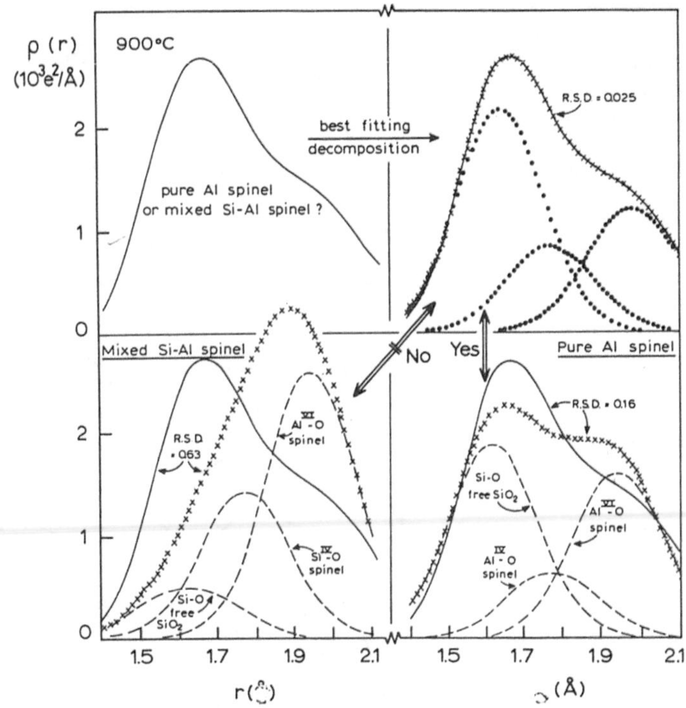

Fig. 9.3 Comparison of the experimental RED *curve* (*upper left*) and its fitting decomposition (*upper right*) of Zettlitz metakaolinite (Czechoslovakia, C.I. = 0.85) heated at 900 °C for 24 h with the *curves* corresponding to the mixed Si–Al spinel hypothesis (*lower left*) and to the hypothesis of a segregation of SiO_2 and Al spinel (*lower left*) (After Delmon et al. 1978). Personal communication

9.3 Summary

RED method has been used to characterize two poorly crystallized transition phases namely, metakaolinite and spinel in K–M reaction series. Leonard (1977) showed both RED profile from 0 to 0.5 nm and deconvoluted the first cation–anion vector into three elementary curves for Si–O, Al^{IV}–O, and Al^{VI}–O components. The result shows that Al–O changes on heat treatment: (i) During metakaolinite formation, Al–O component decreases suddenly from Al^{VI} to Al^{IV}. (ii) At the second stage, he compared RED curves for γ-Al_2O_3 and Al–Si spinel (Brindley's) models and calculated the theoretical RED contributions into three main components in the structural models. The calculated values apparently favor γ-Al_2O_3 + free silica formation other than Al–Si spinel suggested by Brindley and Nakahira (1959) and Weiss et al. (1970). Secondly, he compared the RED curves for both the models. The experimental profile apparently fits more to words Al spinel model than that for Al–Si spinel of Brindley's model.

References

G.W. Brindley, M. Nakahira, The Kaolinite–Mullite reaction series: I, a survey of outstanding problems. J. Am. Ceram. Soc. **42** (7), 311–14 (1959)

B. Delmon, A.J. Herbillon, A.J. Leonard, M. Bulens, Critical assessment of the joint use of various physico-chemical techniques in the study of the thermal transformation of clays. Project Communication to the Oxford Clay Conference, 10–14 July (1978)

A.J. Leonard, Structural analysis of the transition phases in the Kaolinite–Mullite thermal sequence. J. Am. Ceram. Soc. **60** (1–2), 37–43 (1977)

A.J. Leonard, M.J. Genet, J. Lemaitre, M. Bulens, B. Delmon, Comment on "Structural analysis of the transition phases in Kaolinite–Mullite thermal sequence. J. Am. Ceram. Soc. **62** (9–10), 529–531 (1979)

A.J. Leonard, P. Ratnasamy, F.D. Declerck, and J.J. Fripiat, Structure and properties of amorphous silico-aluminous : V, Discuss. Faraday Soc., No.52, p. 98–108 (1971)

A. Weiss, K.J. Range, J. Russow, *The Al,Si-Spinel Phase from Kaoinite* (*Isolation, Chemical Analysis, Orientation and Reactions to Its Low-Temperature Precursors*). Proceedings of the International Clay Conference, Tokyo, 1969, vol. 2 (Israel Universities Press, 1970), pp. 34–37

B.E. Warren and N.S. Gingrich, Fourier integral analysis of x-ray powder patterns. Phys. Rev. **46**, 368–372 (1934)

References

Chapter 10
Density and Surface Area Measurement

10.1 Density Measurement

Harman and Parmelee (1942) noted that the specific gravity of kaolinite drops abruptly at about 450 °C when the mineral is dehydrated and a poorly ordered structure develops. The specific gravity gradually increases as the kaolinite is heated to higher temperatures, probably indicating some development of order at 950–1,000 °C there is an abrupt increase in specific gravity to a value higher than that of original. The sudden increase in specific gravity can be correlated with the high-temperature crystalline phase developing at this temperature interval.

Bleininger and Montgomery (1972) noted the change in density of clay in the form of refractory bricks. The density of refractory bricks undergoes striking changes in course of firing.

(i) At first it usually increases up to 300 °C or more, which may be regarded as a corollary of the shrinkage that practically all clays suffer above 100 °C.
(ii) It fails sharply up to about 700 °C.
(iii) From 700 to 1,000 °C, it again increases markedly due to expulsion of the light component, water from the clay substance and to the formations of alumina or aluminium silicate.
(iv) The density then falls again owing to formation of glassy melts which are lighter than the original crystalline materials.

Rieke and Mauve (1942) showed the change in density values of Zettlitz kaolinite and a Halloysite at different temperatures. Later on Brown and Gregg (1952) also showed the relative changes in density values on heat treatment.

Slaughter and Keller (1959) fired clays at 100 °C intervals from 900 to 1,510 °C. Each sample was brought to the temperature at 10 °C/min held at that temperature for 12 h, and finally cooled to room temperature. Specific gravity measurement was obtained by pycnometer method from powders fired at various temperatures (Fig. 10.1). They interpreted that a phase denser than that of the initial metakaolin is forming continuously between 700 and 900 °C.

A. K. Chakraborty, *Phase Transformation of Kaolinite Clay*,
DOI: 10.1007/978-81-322-1154-9_10, © Springer India 2014

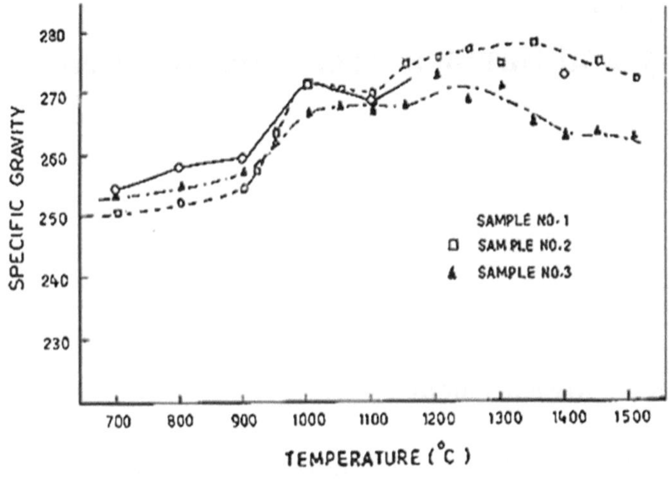

Fig. 10.1 Specific gravity of three clay samples versus firing Temperature (After Slaughter and Keller 1959)—used with permission. ACerS Bulletin, The American Ceramic Society

De keyser (1963e) showed sudden change in density value of kaolinite during heating it at ~900 °C and also correlated the dehydroxylation behavior with change in specific gravity of heated kaolinite. The relative increase in density is more marked in the comparative diagram (Fig. 10.2).

Lemaitre et al. (1975b) heat treated colloid kaolin both in pure form and with the use of Mg^{+2} and Ca^{+2} as mineralizes to various temperatures and there after measured real density of those in view of the reexamining of phase transformation that metakaolinite underwent during heating. This gives the real density and specific surface area of pure and mineralized kaolinite calcined at various temperatures. The density value for pure metakaolinite agrees satisfactorily with that ($d = 2.60$) found by Rieke and Mauve (1942). They tried to correlate the measured density values with theoretical densities for various reaction paths.

Fig. 10.2 Relative change in densities of Kaolinite on heating redrawn out of data of different authors. Reprinted with kind permissions of the American Ceramic Society and of The Clay Minerals Society, publisher of Clays and Clay Minerals

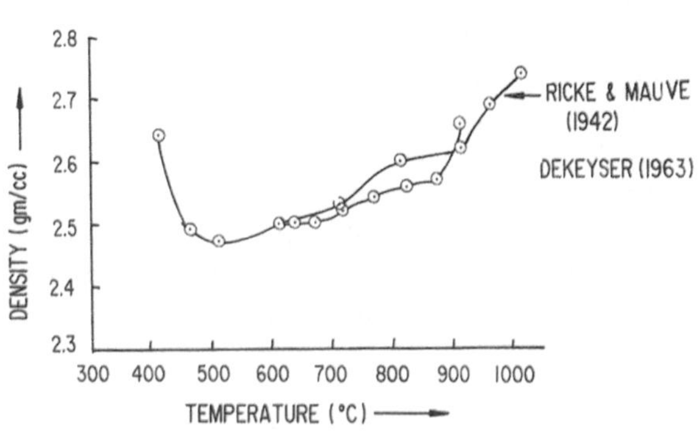

Path A

$$Al_2O_3.2SiO_2 \rightarrow 1/2\ SiO_2(\text{amorph.}) + 6/32(Si_8\ Al_{10.67}\ O_{32})(\text{spinel})$$

Path B

$$Al_2O_3.2SiO_2 \rightarrow (1-3y)\,\gamma\text{-}\ Al_2O_3 + y(3Al_2O_3.2SiO_2)\,(\text{mullite}) + x\ SiO_2(\text{cristo.}) + (2-2y-x)\ SiO_2\,(\text{amorph.})$$

Path C

$$Al_2O_3.2SiO_2 \rightarrow x\,\gamma\text{-}Al_2O_3 + y(3Al_2O_3.2SiO_2)\,(\text{mullite}) + [(1-x-3y)\ Al_2O_3.(2-2y)\ SiO_2]\,(\text{amorph.})$$

Path A corresponds to the hypothesis of Brindley and Nakahira (1959). Path B corresponds to the parallel formation of γ-Al_2O_3, mullite, and silica both amorphous and crystalline. Path C corresponds to the simultaneous decomposition of metakaolinite to γ-Al_2O_3, mullite, and a residue composed of a mixture of amorphous Al_2O_3 and SiO_2. They computed the theoretical density values concerning the respective amount of the various products (i.e., x and y) of varying transformation of path B and path C. On comparing three paths of transformations it is clear that path A could not explain the low-density value ($d = 2.56$) for kaolinite heated to 900 °C/oh). Even when kaolinite was mineralized by Mg^{+2} which generally promotes spinel formation shows density approaching 3.20 g/cc. This result discards the transformation suggested by Brindley and Nakahira which essentially suggests Al–Si mixed spinel. Concerning path B, it is clear that, according to the assumed proportions of various products, densities range from 2.63 to 2.99. These values are compatible with the experimental ones unless one allows for the presence of mullite. Path C allows densities lower that 2.65 even when a small amount of mullite is present. They noted decrease in density of metakaolinite between 800 and 900 °C/oh and assumed that this effect may be due to segregation of Al_2O_3 and SiO_2. Finally, they concluded that metakaolinite pseudo-lattice collapse between 800 and 900 °C and gives way to extensive segregation of alumina and silica which progressively crystallizes to mullite during prolonged heating. The remaining part of the fired metakaolinite is possible composed of Al_2O_3 and quartz.

10.2 Surface Area Measurement

Using nitrogen absorption apparatus McConnell and Fleet (1970) measured the surface area of isothermally heat treated kaolinite from a standard B.E.T. plot of the absorption data. The specific surface area data of them give an idea about the solid-state reaction of kaolinite decomposition.

A sudden decrease in surface area occurs during dehydroxylation later on a consistent and gradual decrease in surface area is noted and it is a function of increasing temperature of treatment. According to them, sudden reduction in

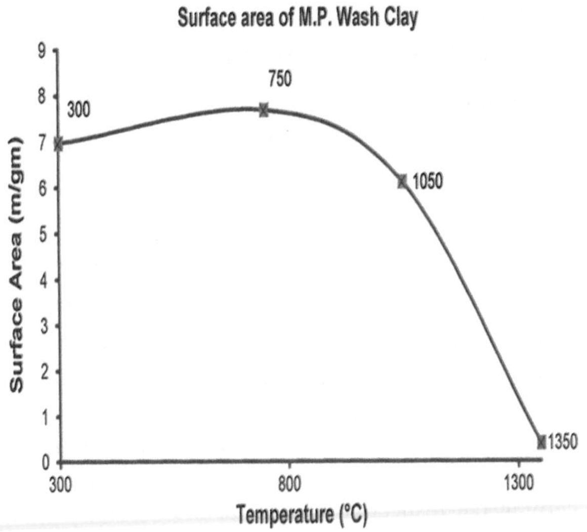

Fig. 10.3 Change in surface area of M.P. wash clay versus temperature plot

surface area above 1,200 °C might be the incidence of the clinkering mechanism. Chakraborty measured the changes in surface area of M.P. kaolinite (Fig. 10.3).

10.3 Summary

Various authors noted both changes in density and surface area values of kaolinite on heat treatment. Harmon and Parmelee (1942) noted two steps of density change. First one occurs during dehydration at ~450 °C and second step occurs during crystallization at ~950 °C. In addition to two density steps, Bleininger and Montgomery (1972) noted fall in density due to fusion at the third stage. Slaughter and Keller (1959) made systematic study on changes in density during heating clay of different mineralogical compositions. Lemaitre et al. (1975a) measure both changes in density and surface area of pure and mineralized kaolinite on heating and tried to correlate measured value with those of theoretical density change for three paths of kaolinite transformations. Comparative results indicated segregation of Al_2O_3 and SiO_2 prior to mullitization.

Nelson and Hendric (1944) indicated abrupt changes in surface area during dehydration and crystallization steps.

Lemaitre et al. showed that in three cases of kaolinite pure or mineralized by Ca or Mg cations, an abrupt decrease in surface area occur due to crystallization of either γ-Al_2O_3 or mullite or both. Generally, two or three steps of density changes are noted during heating kaolinite.

(i) During dehydration stage, sp gravity drops suddenly. Thereafter, it again increases as the temperature rises.
(ii) At about 950–1,000 °C specific gravity increases abruptly again to formation of aluminous phase/aluminosilicate phases. Consequently, surface area drops to a great extent.
(iii) On further heating to 1,200 °C and above, density decreases. Glassy phase formation and crystallization of mullite take place here with further decrease in surface area, Lemaitre et al. (1975a, b) measured density values of heat treated kaolinite mineralized with Ca^{+2} and Mg^{+2} salts and attempted to forecast three different theoretical reaction paths. Finally, they suggested that metakaolinite segregates to free Al_2O_3 and SiO_2 other than forming hypothetical Si–Al spinel phase.

References

A.V. Bleininger, E.T. Montgomery, Tech. Paper. U.S. Bur. Stand. No 22, 21 (1972)

M.J. Brown, S.J. Gregg, A study of the effect of heat on Kaoinite by adsorption Method. Clay Miner. Bull. **1**, 228 (1952)

G.W. Brindley, M. Nakahira, The Kaolinite–Mullite reaction series: I, A survey of outstanding problems. J. Am. Ceram. Soc. **42**(7), 311–314 (1959)

W.L. De Keyser, Note concerning the exotherm reaction of Kaolinite and formation of spinel phase preceding that of Mullite. in *Proceedings of International Clay Conference*, Pergamon Press, 91–96 (1963e)

C.G.Harman and C.W.Parmelee, Testing and classification of ball clays thermal history. Bull. Am. Ceram. Soc. **21**, 280–286 (1942)

J. Lemaitre, M. Bullens, B. Delmon, Influence of mineralizers on the 950 °C exothermic reaction of Metakaolinite. in *Proceedings of the International Clay Conference*, Mexico City, Mexico, July 1975a, ed. by S.W. Bailey. Applied Publishing Ltd., Wilmette, pp. 539–544

J. Lemaitre, A.J. Leonard, B. Delmon, The sequence of phases in 900–1,050 °C transformation of Metakaoinite. in *Proceedings of International Clay Conference*, ed. by S.W. Bailey, Applied Publishing Ltd., Wilmatte, p. 545 (1975b)

J.D.C. McConnell, S.G. Fleet, Electron optical study of the thermal decomposition of Kaolinite. Clay Miner. **8**, 279–290 (1970)

R.A. Nelson and S. B. Hendricks, Specific surface of some clay minerals, soils and soil colloids. Soil Sci. **56**, 285–296 (1944)

R. Rieke, L. Mauve, Zur Frage des Nachweises der mineralischen Bestandteile der Kaoline (Indications as to Mineral Constituents of Kaolin). Ber. Deut. Keram. Ges. **23**, 119–150 (1942)

M. Slaughter, W.D. Keller, 'High temperature transformation from impure Kaolin clays. Am. Ceram. Soc. Bull. **38**, 703–707 (1959)

Chapter 11
Crystallization Studies of Preheated Metakaolinite

It is well-known that metakaolinite forms mullite and β-cristobalite at high temperatures. The alumina component of metakaolinite never crystallizes into corundum during heating. But corundum forms in a kaolinite if it contains alumina minerals as associated phase. Bratton and Brindley (1962) reported that corundum forms easily on heating different kaolinites containing free alumina in the form of diaspore and boehmite.

11.1 Pretreated with Acids e.g., HF and Fuming H_2SO_4

According to Chakraborty and Ghosh (1978), γ-Al_2O_3 forms from metakaolinite when its SiO_2 layer was removed by HF treatment prior to heating at 1,200 °C (Table 11.1).

It shows that corundum crystallization is perceptible on heating sample marked MK_2. Its formation becomes more and more pronounced when HF treatment was continued from 5 to 10 min duration. The intensities of XRD peaks of corundum increase with the gradual increase of severity of hydro-fluoridation from samples MK_2 and MK_3 and lastly to MK_4. This observation leads one to believe that free alumina must have been liberated during the treatment of metakaolinite with HF. Since silica is only removed by HF with the formation of volatile SiF_4. In the case of MK_2 samples, the extent of HF attack was such that a portion of metakaolinite was attacked with the liberation of amorphous alumina and rest of metakaolinite remained unaffected. On subsequent heat treatment, this newly formed free alumina crystallizes into corundum and the unattacked portion of metakaolinite converts into mullite and cristobalite in the usual course. The reaction sequence is shown in Eq. (11.1).

$$\text{Metakaolinite} \underset{\text{Treatment}}{\overset{\text{HF}}{\longrightarrow}} \begin{matrix} \text{(i) } SiF_4 \\ \text{(ii) } AlF_3 \\ \text{(iii) Metakaolinite} \\ \text{(un-decomposed)} \end{matrix} \underset{1{,}200\,^\circ\text{C}/1\,\text{h}}{\overset{\text{Heat treatment}}{\longrightarrow}} \begin{matrix} \text{(i) } \gamma\text{-}Al_2O_3 \\ \text{(ii) Mullite} \\ \text{(iii) Cristobalite} \end{matrix} \quad (11.1)$$

A. K. Chakraborty, *Phase Transformation of Kaolinite Clay*,
DOI: 10.1007/978-81-322-1154-9_11, © Springer India 2014

Table 11.1 Crystalline phases developed after heat treatment of metakaolinite after hydrofuorization (after Chakraborty 1978)

Sample mark	Condition of hydrofluorization		Phases formed on heating hydrofluorized samples at 1,200 °C/1 h analyzed by XRD
	Amount of HF used (ml)	Time per gm of metakaoline (min)	
Metakaoline	0	0	Mullite and Cristobalite
MK_2	2	5	Mullite, Cristobalite, and Corundum (minor)
MK_3	2	10	Mullite, Cristobalite, and Corundum (major)
MK_4	2	20	Corundum only

Therefore, the above result suggest that HF attack of metakaolinite has opened up the metakaolinite structure by breaking chemical linkages that existed between silica and alumina layers of metakaolinite. Therefore, the hypothesis of Colegrave and Rigby (1952) that metakaolinite is a mechanical mixture of alumina and silica and the alumina in metakaolinite complex 'protects' the SiO_2 does not appear to be true in the light of the work presented above.

Chakraborty and Ghosh (1978) first heat treated Bhandak kaolinite to 1000/4 h and then fumed with a few drops of concentrated H_2SO_4 in a platinum crucible, and finally heat treated to 1,200 °C/1 h. X-ray studies showed the presence of corundum along with mullite and cristobalite as is found in the usual course. When fuming is prolonged, corundum and cristobalite developed, but not mullite. On continued heating, mullite forms gradually, whereas the amounts of corundum and cristobalite decrease. This result suggests that H_2SO_4 attacks the cubic phase and liberates free Al_2O_3 as $Al_2(SO_4)_3$ which decomposes first to γ-Al_2O_3 on heating. This γ-Al_2O_3 phase, although it is associated with large amounts of free SiO_2 does not transform directly to mullite on heating, rather crystallizes first into γ-Al_2O_3 and then reacts with cristobalite and SiO_2(A) to form mullite gradually. This evidence proves that 980 °C heated cubic spinel phase is Al–Si spinel rather than simple γ-Al_2O_3.

11.2 Pretreated with NaOH

Occurrence of edge share tetrahedral in alumina layer may be the reason for showing unusual high reactivity toward basic oxides.

11.2.1 Reaction Behavior of Metakaolinite with Alkali Solution

Pauling (1960) pointed out that, metakaolinite possesses two necessary requirements for developments of zeolite formation. Primarily, the electrostatic valance rule is satisfied in metakaolinite structure. For example, electrostatic bond strength of oxygen ion (=2) is satisfied by two silicon bonds and or one silicon bond plus two octahedral aluminum bonds in kaolinite. In metakaolinite, Al^{+3} of co-ordination number (4) may be linked to Si^{+4} through oxygen.

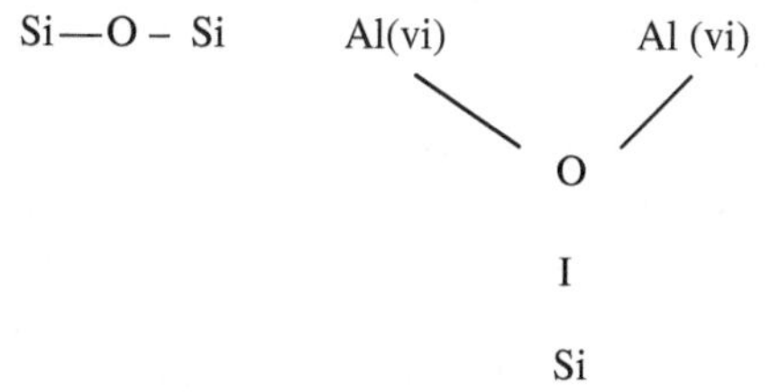

So oxygen ions-common to aluminum and silicon, tetrahedrons are then by bonds of total strength (1 + 3/4) = 7/4 and thus it required a bond of strength (=2 − 7/4) = 1/4 for saturation from outside the silica/alumina unit. Such a bond is not provided by a cation with large charge and small radius. It is necessary that large univalent or bivalent cations namely alkali or alkaline earth ion for every quadri co-ordinate aluminum ion is required.

Second, if the ratio of the numbers of oxygen atoms to total number of aluminum and silicon atoms is 2:1, there would be a possibility of formation of a complete tetrahedral framework. Further, ordering will be good only when Al and Si tetrahedral is connected alternatively and the ratio of Al/Si should not exceed 1; but should be equal to 1 instead of two adjacent aluminium tetrahedral sharing a common oxygen atoms. In metakaolinite, the ratio of oxygen to total number of aluminum and silicon atoms is = (7/(2 + 2)) = 7/4, approximately. With the addition of 1 mole of Na_2O, the ratio will be = ((7 + 1)/4) = 2, i.e., this condition is equal to the formation of complete tetrahedral framework. The ratio of Al/Si = 2/2 = 1 in metakaolinite. Accordingly, Pauling (1960) pointed out the formation of a zeolitic type material out of metakaolin which will form a network of good ordering. Now the question is, how the interaction between metakaolinite with alkali takes place in aqueous solution. Regarding such solid–liquid reaction, it is to be considered that acidic nature/or site has been developed during the formation of metakaolinite. Thus, the first reaction prior to zeolitic formation is an acid–base reaction to form adduct.

$$Al_2O_3.2SiO_2 \times H_2O + M^+OH^- \xrightarrow{\text{Heating}} \begin{array}{l}\text{Alkali}\backslash\text{Alkaline earth}\\ \text{Aluminum-silicate hydrate Zeolite A}\end{array} \quad (11.2)$$

where M^+ is alkali or alkaline earth cation.

In the second stage i.e., on heating at about boiling temperature, amorphous alkali/alkaline earth alumino-silicate hydrate crystallizes to zeolite A. For the development of acidity, one has to consider the formation of oxygen surface after the removal of H_2O from the hydroxyl groups remaining after dehydroxylation.

These oxygen sites are behaving as Lewis acid as described by Fripiat (1964) in Fig. 11.1.

Four field co-ordination of Al is in some way related to evolution of such acid site. It is conceived that in aqueous medium Na^+ and OH^- ions are preferentially absorbed on the oxygen sites and will simultaneously diffuse and form adduct which latter crystallizes to zeolite. In order to form adduct, M^+ ions are required to penetrate through the stacked layers. In original kaolinite structure, it is difficult to penetrate alkali or alkali earth cations through interlayer. On dehydroxylation, metakaolin is accessible to cations for absorption. Since hydrogen bonds existing between hydroxyl surface and oxygen atoms of the next silica layer are disrupted and there is no force connecting the successive layers. Therefore, there is every possibility of random stacking of layers as conceived by Freund (1960).

For this reason, an organic dye absorbs on oxide surface of metakaolinite in a homogenous way (Agafonoff 1924). Glycerine is also found to be absorbed in a similar way and generate heat. Chakraborty and Ghosh (1976) also showed the reaction or LiOH with metakaolinite in aqueous solution. The washed mass forms eucriptite followed by an exotherm.

11.3 Pretreated with Basic Oxides

De Keyser (1963e) heat treated kaolinite with CaO and noted the formation of gehlenite at about 900 °C; while at the same temperature, a mixture of SiO_2 and beyerite formed C_2S only. With this observation, he concluded that a chemical bond between SiO_2 and Al_2O_3 is existed on heating kaolinite at about 900 °C.

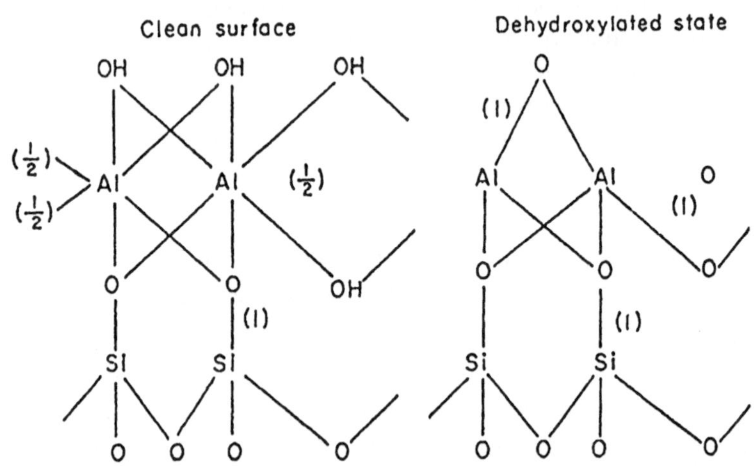

Fig. 11.1 Surface Structure of Kaoinite and Metakaolinite. *Left* clean surface of kaolinite. *Right* dehydroxylated surface of metakaolinite

Kubo and Yamada (1969) showed that kaolin reacts with alkali carbonates (R_2CO_3 where R = Li, Na or K) at about 700–800 °C and give alkali-alumino silicates. Similarly, Nakahira and Kato (1964) showed that even on heat treatment of metakaolin with salts of alkalis different crystalline phases form which have alumino-silicate frame work and are related to those of silica minerals namely quartz, tridymite, and cristobalite types. Because of the difference in their ionic radii, alkali ions stabilize alumino-silica frameworks with the formations of different type of phases. For example, with Li_2O, metakaolinite forms β-eucriptite (Quartz type structure) with Na_2O, it forms nepheline (Tridymite type structure) and/or carnegieite (cristobalite type structure); with K_2O, forms kalsilite (Tridymite type structure). In kaolin minerals the layer sequence is determined by the hydrogen bonding between the unit layers. The loss of protons upon dehydroxylation destroys the ordering of layer sequence. These authors showed that order of the layer sequence in alumino-silicate phases is restored again by the diffusion of sodium ions into metakaolinite structure. Considering the open space around the original hydroxyl layer in metakaolin and its topotactic relationship to the alumino-silicate product, they believed that the diffusion of Na^+ ions into metakaolinite is not accompanied by a counter diffusion of Si^{+4} or Al^{+3} ions but by diffusion of oxygen ion into metakaolin in the ratio of one oxygen ion per two sodium ions to maintain electrical neutrality. Likely, the diffusion of sodium and oxygen ions takes place simultaneously rather than separately to maintain local neutrality of charge along the diffusion paths.

11.4 Summary

Chakraborty (1978) studied the crystallization behavior of pure and HF treated metakaolinite to reveal the nature of bonding between SiO_2 and Al_2O_3 components of dehydrated kaolinite. Study of HF treatment conforms that metakaolinite is a compound. Further, Chakraborty and Ghosh (1978) studied the crystallization behavior of H_2SO_4 treated kaolinite heated previously to 980 °C where spinel phase was already developed to ascertain the nature of the cubic phase. The result of conforms that spinel is a silicon-bearing phase.

Pauling (1960) first conjectured that metakaolinite should form zeolite by considering its structural frame work. Fripiat (1964) pointed out the existence of active acid site in metakaolinite which are the centers for NaOH adsorption to develop zeolite materials.

Breck (1974) and then Chakraborty (1978) showed the details reaction behavior of metakaolinite with alkali solution to form zeolite phases in laboratory scale and commercial scale. Besides solid–liquid reaction, metakaolinite also found to react with alkalis/alkaline earths in solid state during heating. De Keyser (1963e) showed the formation of gelignite by reacting metakaolinite with CaO. Kubo and Yamabe (1969) showed the solid-state reaction of metakaolinite with R_2CO_3.

Likely, Nakahira and Kato (1964) showed the formation of -eucriptite, nepheline, carnegie, and kalsilite during heating metakaolinite with basic oxides.

They believed that both Na^+ and O^{-2} diffuse simultaneously into metakaolinite structure during heating and react at the strained O^{-2} site of it and form necessary adduct, and finally crystallized to an oxide phase (a crystalline silica modification).

References

V. Agafonoff, W.I. Vernadsky, Compt. Rend. **178**, 1082 (1924)

D.W. Breck, *Zeolite Molecular Sieves* (Wiley, New York, 1974)

R.J. Bratton and G. W. Brindley, Structure controlled reactions in kaolinite–diaspore-boehmite clays. J.Am.Ceram.Soc. **45**(11) 513–16 (1962)

A.K. Chakraborty, D.K. Ghosh, Kaolinite–Mullite reaction series. Cent. Glass Ceram. Res. Instt. Bull. **23**(2), 86–88 (1976)

A.K. Chakraborty, D.K. Ghosh, Comment on the interpretation of the Kaolinite–Mullite reaction sequence from infra-red absorption spectra. J. Am. Ceram. Soc. **61**(1–2), 90–91 (1978a)

A.K. Chakraborty, D.K. Ghosh, Re-examination of the kaolinite to mullite reaction series. J. Am. Ceram. Soc. **61**(3–4), 170–173 (1978b)

E.B. Colegrave and G.R. Rigby, The Decomposition of kaoliite by heat. Trans. Brit. Ceram. Soc. **51**(6), 355–367 (1952)

W.L. De Keyser, Note concerning the exotherm reaction of Kaolinite and formation of spinel phase preceding that of Mullite. in *Proceedings of International Clay Conference*, Pergamon Press, pp. 91–96 (1963e)

F. Freund, Die Deutungder Exothermen Reaktio Des Kaolinite Als Reaktio Des Aktiven Zustandes. Ber. Deut. Keram. Ges. **37**, 209–218 (1960)

J.J. Fripiat, Surface properties of aluminosilicates. in *Proceedings of 12th National Clay Conference*, pp. 327–358 (1964)

Y. Kubo, T. Yamada, Mechanism of solid state reactions between Kaolin and alkali (Lithium, Sodium and Potassium) carbonates. in *Proceedings of International Clay Conference*, vol. I, pp. 915–922 (1969)

M. Nakahira, T. Kato, Thermal transformation of Pyrophyllite and Talc as revealed by X-ray and electron diffraction studies. in *Proceedings of 12th National Conference on Clay and Clay Minerals*, pp. 21–27 (1964)

L. Pauling, in *The Nature of the Chemical Bond*, 3rd edn. (Oxford & IBH Publishing Co, India (1960) . Indian Edition 1967), p. 549

Chapter 12
Thermodynamic Approach

12.1 Introduction

Thermodynamic calculations have been done for the factors e.g., free energy (ΔF), enthalpy (ΔH), and lattice energy (U) by the various researchers namely Vaughn (1955), Schicltz and Soliman (1966), and Mazumdar and Mukherjee (1983) in predicting the nature of intermediate spinel phase, and second the course of mullite formation reaction of kaolinite.

12.2 Calculation of Free Energy and Enthalpy

Vaughn (1955) calculated the enthalpy and free energy changes for some possible theoretical transformation of kaolinite at 980 °C exotherm. If the dehydration products of kaolintie are a mixture of silica and alumina then reactions are as follows.

$$Al_2O_3\ (\text{Amorph}) = \gamma\text{-}Al_2O_3 \tag{12.1}$$

$$3\gamma\text{-}Al_2O_3 + 2SiO_2\ (\text{glass}) = Al_6Si_2O_{13} \tag{12.2}$$

If the breakdown product is a compound then possible modes of break down are as follows.

$$Al_2Si_2O_7 = Al_2O_3\ (\text{Amorph}) + 2SiO_2\ (\text{glass}) \tag{12.3}$$

$$Al_2Si_2O_7 = \gamma\text{-}Al_2O_3 + SiO_2\ (\text{glass}) \tag{12.4}$$

$$3(Al_2Si_2O_7) = Al_6Si_2O_{13} + 4SiO_2\ (\text{glass}) \tag{12.5}$$

$$4(Al_2Si_2O_7) = Al_6Si_2O_{13} + \gamma\text{-}Al_2O_3 + 6SiO_2\ (\text{glass}) \tag{12.6}$$

ΔH_{298} and ΔF_{298} are calculated based on the data given by Kroser (1953) and Avgustinik and Petrosyan (1952). They showed that at 980 °C, the greatest free

A. K. Chakraborty, *Phase Transformation of Kaolinite Clay*,
DOI: 10.1007/978-81-322-1154-9_12, © Springer India 2014

energy change accompanied reaction ~12.6. These approximate calculations implied that the most probable mode of decomposition, if compound is produced in dehydration is into mullite, γ-Al_2O_3 and SiO_2. If the dehydration product is a mixture this conclusion is invalid.

Schieltz and Soliman (1966) choose six possible transformations of metakaolin e.g., Eq. 12.4 as 1; Eq.12.7 as 2; Eq.12.5 as 3; Eq. 12.8 as 4; Eq. 12.6 as 5; Eq. 12.9 as 6. They calculated change in free energy and change in enthalpy at 1250 K instead of only $\Delta G_{298\ K}$, $\Delta H_{298\ K}$ as like *Vaughn*. The calculated values of ΔG of the different transformation have been plotted against temperature and are shown in Fig. 12.1.

Generally, the reaction which has the maximum $-\Delta G$ is the most stable. Transformation 12.8 being the one with highest $-\Delta G$ is the most stable.

$$Al_2Si_2O_7 \rightarrow \gamma\text{-}Al_2O_3 + 2SiO_2 \tag{12.4}$$

$$Al_2Si_2O_7 \rightarrow \gamma\text{-}Al_2O_3 + 2SiO_2\ (\text{crist}) \tag{12.7}$$

$$Al_2Si_2O_7 \rightarrow \frac{1}{3}Al_6Si_2O_{13} + \frac{4}{3}SiO_2\ (\text{glass}) \tag{12.5}$$

$$Al_2O_3.\ 2SiO_2 \rightarrow \frac{1}{3}(3Al_2O_3\ 2SiO_2) + \frac{4}{3}SiO_2\ (\text{crist}) \tag{12.8}$$

$$Al_2Si_2O_7 \rightarrow \frac{1}{4}(Al_6Si_2O_{13}) + \frac{1}{4}(\gamma\text{-}Al_2O_3) + \frac{3}{2}SiO_2\ (\text{glass}) \tag{12.6}$$

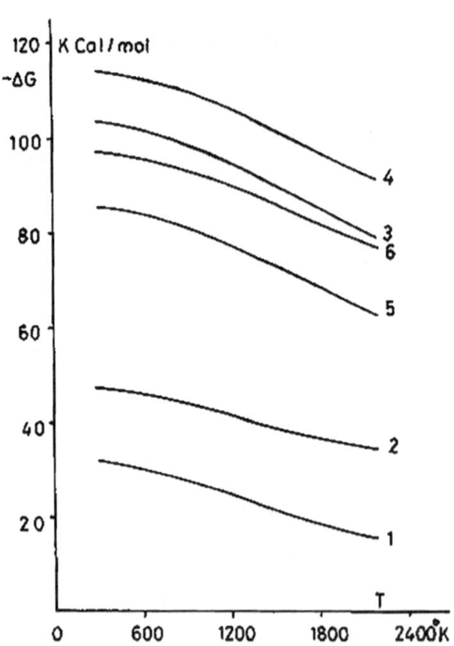

Fig. 12.1 Change of free energy with temperature for all possible transformations (After Scheltz and Soliman 1966). Free available

$$Al_2Si_2O_{7-} \rightarrow \frac{1}{4}(Al_6Si_2O_{13}) + \frac{1}{4}\gamma\text{-}Al_2O_3 + \frac{3}{2}SiO_2\ (\text{crist}) \quad (12.9)$$

According to Schieltz and Soliman by heating, metakaolin may transform to alumina and silica; then to alumina, silica, and mullite; at the end of its transformation series, it yields mullite and silica that is in excess of the silica required for mullite.

Mazumdar and Mukherjee (1983) calculated the enthalpy of reaction for the conversion of metakaolin into γ-Al_2O_3 at 980 °C according to the equation below.

$$Al_2Si_2O_7 \rightarrow \gamma\text{-}Al_2O_3 + 2SiO_2\ (\text{crist}),\quad \Delta H_{1253\,K} = -54.31\ \text{kJ/mole} \quad (12.7)$$

They assumed that this heat involvement is the net effect of (i) formation of spinel phase and (ii) crystallization of amorphous SiO_2 into cristobalite. Although cristobalite is not found to occur on heating kaolin at 1,000 °C. This value is in fairly good agreement with that of experimentally obtained value (−40.18 kJ/mole). They also calculated enthalpy of reaction for the conversion of mixture of γ-Al_2O_3 and cristobalite into mullite at about 1,200 °C as follows.

$$3(\gamma\text{-}Al_2O_3) + 2SiO_2\ (\text{crist}) \rightarrow Al_6Si_2O_{13} \quad \Delta H_{1473\,K} = -54.4\text{kJ/mole} \quad (12.10)$$

This reaction is exothermic and the heat evolved is in fairly good agreement with the experimental value (−30.17 ± 7.1 kJ/mole) as per Shvetsov and Gevorkyan (1942).

Chakraborty (1986) first recalculated ΔF and ΔH for all theoretically possible reactions path of kaolinite.

(i) Calculation of change in free energy of formation for various transformation paths of metakaolinite.

Standard free energy changes of all possible theoretical transformations vide equations below were calculated based on the published free energy data (Table 12.1a, b) and are shown in Table 12.2.

Table 12.1a Free energy of formation data (K.J/mole) of some substances

Substances 1	Free energy of formation at 298 °K KJ/mole Vaughn and Budnikov et al. 2 −ΔF	Free energy of formation at 298 °K KJ/mole Avgustinik & Mchedlov-Petrosyan & Budnikov and Mchedlov-Petrosyan 3 −ΔF
A. Metakaolin	3011.01	-
B. γ-Al_2O_3	1560.63	1544.74
C. SiO_2 (g)	799.14	799.55
D. SiO_2(crist)	-	803.74
E. SiO_2(quartz)	-	805.04
F. Mullite	7361.04	-
G. Al_2O_3 (amorp)	1535.52	-
H. α-Al_2O_3		1576.40

Table 12.1b Heat of formation data (K.J/mole) of some substances

1	4		5			6				7	
Substance	Avgustianik & Mchedlov-Petrosyan (1952)		Kroger (1953) referred by Vaughan (1955)			Budnikov & Mched-lov-Petrosyan (1960) referred by Schieltz and Soliman (1966)				Waldbaum (1965) referred by Holm and Kleppa (1966)	
	$-\Delta H_{298^\circ K}$	$-\Delta H_{1253^\circ K}$,	$-\Delta H_{298^\circ K}$	$-\Delta H_{1253^\circ K}$	$-\Delta H_{1523^\circ K}$,	$-\Delta H_{298^\circ K}$	$-\Delta H_{1253^\circ K}$	$-\Delta H_{1473^\circ K}$	$-\Delta H_{1523^\circ K}$, Cp	$-\Delta H_{298^\circ K}$ $-\Delta H_{1253^\circ K}$ $-\Delta H_{1523^\circ K}$	
1	4		5			6				7	
A.	3211.22	-	3359.75	3116.95	3041.46	3212.75	2969.97	-	-	$0.2296 + 0.0366 \times 10^{-3}T$ $-0.01456 \times 10^{5}T^{-2}$	-
B.	1637.94	1557.94	1662.68	-	-	1653.47	1553.69	1524.70	-	$0.0685 + 0.46 \times 10^{-3}$ T	-
C.	-	-	-	-	-	845.57	784.29	767.60	-	$0.56 + 0.154 \times 10^{-3}$ T -0.01444×10^{5} T^{-2}	-
D.	-	-	-	-	-	877.17	813.38	798.16	-	$0.602 + 0.085 \times 10^{-2}T$	-
E	-	-	-	-	-	878.64	815.05	-	-	$0.603 + 0.0081 \times 10^{-3}$ T	-
F.	7547.93	-	-	-	-	7551.54	-	7078.36	7156.2	$0.3524 + 0.0837 \times 10^{-3}$ -0.1045×10^{5} T^{-2}	6643.87 6178.0 6047.74
G.	-	-	-	-	-	-	-	-	-	-	-
H.	-	-	-	-	-	1670.59	-	-	-	$0.1149 + 0128 \times 10^{-3}$ T -0.0354×10^{5} T^{-2}	-

Table 12.2 Change in free energy (ΔF) calculated for different possible transformations of metakaolinnite

Reaction	−ΔF 298 (kJ/mole)
12.3	123.11
12.4	148.22
12.5	441.52
12.6	368.11
12.7	157.66
12.8	447.65
12.9	375.01
12.10	–
12.11	–
12.12	–
12.13	156.66
12.14	163.67
12.15	172.87
12.16	293.37

1 Avgustinik and Mechedlov-Petrosyan (1952)
2 Kroser (1953)
3 Scheiltz and Soliman (1966)
4 Waldbam (1965)

$$\begin{array}{l} Al_2O_3\,2SiO_2 \rightarrow 2Al_2O_3 3SiO_2 \rightarrow 3Al_2O_3 2SiO_2 \\ \quad \text{Al-Si Spinel} \qquad\qquad \text{Mullite} \\ \quad + \\ \quad SiO_2\ (\text{amorph}) \end{array} \tag{12.11}$$

$$Al_2O_3\,2SiO_2 \rightarrow Al_2O_3\ (\text{amorph.}) + 2SiO_2\ (\text{amorph.}) \tag{12.3}$$

$$Al_2O_3\ 2SiO_2 \rightarrow \gamma\text{-}Al_2O_3 + 2\ SiO_2\ (\text{glass}) \tag{12.4}$$

$$Al_2O_3\,2SiO_2 \rightarrow \gamma\text{-}Al_2O_3 + 2\ SiO_2\ (\text{crist.}) \tag{12.7}$$

$$\gamma\text{-}Al_2O_3 + 2\ SiO_2\ (\text{crist.}) \rightarrow \frac{1}{3}(3Al_2O_3 2SiO_2) + \frac{4}{3}\ SiO_2\ (\text{crist.}) \tag{12.12}$$

$$Al_2O_3 + 2\ SiO_2 \rightarrow \frac{1}{3}(3\ Al_2O_3 2SiO_2) + \frac{4}{3}\ SiO_2\ (\text{glass}) \tag{12.5}$$

$$Al_2O_3 + 2SiO_2 \rightarrow \frac{1}{3}(Al_2O_3\ 2SiO_2) + \frac{4}{3}\ SiO_2\ (\text{crist.}) \tag{12.8}$$

$$Al_2O_3 + 2\ SiO_2 \rightarrow \frac{1}{4}\ (3\ Al_2O_3\ 2SiO_2) + \frac{1}{4}\ \sim\ Al_2O_3 + 3/2\ SiO_2\ (\text{glass}) \tag{12.6}$$

$$Al_2O_3 + 2\ SiO_2 \rightarrow \frac{1}{4}\ (3\ Al_2O_3\ 2SiO_2) + \frac{1}{4}\ \sim\ Al_2O_3 + \frac{3}{2}\ SiO_2\ (\text{crist.}) \tag{12.9}$$

Table 12.3 Calculated enthalpy change (kJ/mole) for five hypothetical reactions at two exothermic peak temperatures by using different $-\Delta H_{298}$ values from literature

Equation	First exothermic peak temperature				Second exothermic peak temperature	
	$-\Delta H$value[1]	$-\Delta H$ value[2]	$-\Delta H$ value[3]	Table 8[3]	$-\Delta H$ value[3]	$-\Delta H$ value[4]
Equation 12.4	+10.43	−5.32	−151.52	−152.80	–	–
Equation 12.7	−47.89	−63.50	−210.50	−213.20 SiO_2 (Tri) is considered in place of SiO_2 (Crist.)	–	–
Equation 12.16	−51.23	−66.84	−213.84	–	–	–
Equation 12.2	–	–	–	–	−323.01	−4.75
Equation 12.10	–	–	–		−302.64	−22.50

$$Al_2O_3 + 2SiO_2 \rightarrow \alpha\text{-}Al_2O_3 + 2SiO_2\ (\text{glass}) \quad (12.13)$$

$$Al_2O_3 + 2\ SiO_2 \rightarrow \alpha\text{-}Al_2O_3 + 2\ SiO_2\ (\text{crist.}) \quad (12.14)$$

$$\gamma\text{-}Al_2O_3 + 2SiO_2\ (\text{glass}) \longrightarrow \frac{1}{3}(Al_2O_3 2SiO_2) + \frac{4}{3}\ SiO_2\ (\text{glass}) \quad (12.15)$$

$$Al_2O_3\, 2\ SiO_2 \xrightarrow{980^\circ C} \gamma\text{-}Al_2O_3 + SiO_2\ (\text{quartz}) \quad (12.16)$$

$$\gamma\text{-}Al_2O_3 + 2SiO_2\ (\text{glass}) \xrightarrow{1250^\circ C} 3\ Al_2O_3\, 2SiO_2 \quad (12.2)$$

$$\gamma\text{-}Al_2O_3 + 2\ SiO_2\ (\text{crist}) \xrightarrow{1250^\circ C} 3\ Al_2O_3\, 2\ SiO_2 \quad (12.10)$$

ii **Enthalpy calculations at the two exotherms.**

With the help of available heats of formation data of metakaolinite γ-Al_2O_3 and mullite from above, the enthalpy changes for the hypothetical reactions vide Eqs. 12.4, 12.7 and 12.16 occurring at 980 °C and the corresponding enthalpy change followed by recombination of the products at 1250 °C formed from metakaolinite at 980° vide two hypothetical reactions 12.2 and 12.11 were, respectively, calculated and are shown in Table 12.3.

12.3 Calculation of Lattice Energy

An attempt was made to evaluate lattice energies separately for γ-Al_2O_3 and hypothetical Al–Si spinel ($2Al_2O_3{\cdot}3SiO_2$). The phase that will correspond to minimum energy between the two will be the desired phase.

Mazumdar and Mukherjee (1983) calculated lattice energy by using Born and Mayer (1932) equation as follows.

$$U = \left(-\mathrm{Na}\ e^2 Z^2/R\right)\left[1 - (1/n)\right] \quad (12.17)$$

where Ze = Charge of each ion, n = Born exponent
N = Avogadro's number R = Lattice constant
A = Madelung constant

Madelung factor AZ^2 was determined by Ewald method for both the structures for one formula unit. The lattice energy value obtained for Al–Si spinel (~10339 kJ/mole) is definitely more than that obtained for γ-Al_2O_3 (=12602 kJ/mole). Accordingly, they believed that the intermediate spinel phase is γ-Al_2O_3 as the lattice energy of this phase is lower.

Chakraborty et al. (1986) recalculated lattice energy value for spinel of mullite like composition by Kapustinsky's method (1943) since heat of formation data of spinel is not known.

12.3.1 Kapustinsky's Method

Vaughn (1955) first calculated the lattice energy values of Al_2O_3 and SiO_2 by adopting Kapustinsky's method and by B.H. cycle as well. According to Kapustinsky the lattice energy (U) of a compound could be approximately calculated as follows.

$$U = 287.2 \frac{n\, n_1\, n_2}{r_1 + r_2} \times \left[1 - \frac{0.345}{r_1 + r_2}\right] \quad (12.18)$$

where r_1 = ionic radius of the cation, r_2 = ionic radius of the anion, n = number of ions in the molecule, n_1 = valence of the cation, n_2 = valence of the anion.

This method was applied to a complex lattice e.g., Al–Si spinel by splitting the structure into simpler units.

Lattice energy values of 2 mol of silica and 3 mol of alumina were first of all found out and then these were added to obtain the lattice energy of one mole of Al–Si spinel vide Table 12.4.

The calculated value of Al–Si spinel of mullite composition by Kapustinsky's method (−70,827.1 kJ/mole) is less than the value for orthorhombic mullite (−59, 464.63 kJ/mole) and seems unrealistic. This may be due to nonavailability of heats of formation data in Kapustinsky's equation. Thus, the value obtained by this method is of theoretical importance since the equation is an empirical one.

Table 12.4 Lattice energy of different compounds. data taken from Vaughn 1955. Reproduced with kind permission of The Clay Minerals Society, publisher of Clays, and Clay Minerals

Compounds	Lattice energy value (−kJ/mole)
1. γ-Al_2O_3	12,403
2. SiO_2 (glass)	11,051
3. Mullite	59,464.63
4. Al–Si spinel ($3Al_2O_32SiO_2$, calculated by Kapustinsky's equation)	70,827.1
5. Al–Si spinel ($2Al_2O_33SiO_2$, calculated by Kapustinsky's equation)	68,534.0
6. Metakaolinite	34,468
7. Al–Si spinel ($3Al_2O_32SiO_2$, shown by Chakraborty et al. 1986; After correction of data given by Mazumdar and Mukherjee 1983)	41,315

12.4 Summary

Thermodynamical calculations namely changes in free energy, and changes in enthalpy for various ways of transformations have been calculated to predict the nature of intermediate spinel phase. Vaughn (1955) calculated ΔH and ΔF at 298°K only for some transformation and showed that greatest free energy change occur when mullite, Al_2O_3 and amorphous silica are the product of decomposition. Schieltz and Soliman (1966) calculated free energy change for some decomposition product of kaolinite at 1250°K. Instead of formation of both mullite and spinel they noted the highest free energy change for reaction leading to mullite and β-cristobalite. In a newer approach, Mazumder and Mukherjee evaluated lattice energy both for γ-Al_2O_3 and Al–Si spinel by Born and Mayer method and then compared. *Chakraborty's* finally recalculated free energy and enthalpy changes of all possible theoretical transformation of metakaolinite and pointed out the practical discrepancies in predicting the nature of spinel phase and second ruled out the theoretical possibility of formation of γ-Al_2O_3/SiO_2 segregation and recombination to form mullite thereof.

References

A.I. Avgustinik, M. Petrosyan, O.P. J. Appl. Chem. U.S.S.R. **25**, 216 (1952)

M. Born and J.E. Mayer, Lattice Theory of Ionic Crystals. Z. Phys. **75**,1–18 (1932)

A.K. Chakraborty, D.K. Ghosh, P. Kundu, Comment on Structural characterization of the spinel phase in the Kaolin–Mullite research series through lattice energy. J. Am. Ceram. Soc. **68**(8), C-200–C-201 (1986)

A.F. Kapustinsky, J. Gen, Kemp. Moscow **13**, 497 (1943)
C. Kroser, Theoretischer warme bedraf der glaschme. Z.Proze Se (Glastech. Ber.) **26**, 202 (1953)
S. Mazumdar and B. Mukherjee, Structural characterization of the spinel phase in the kaolinmullite reaction series through lattice energy. J. Am. Ceram. Soc. **66**(9), 610–12 (1983)
N.C. Schieltz, M.R. Soleman (1966) *Thermodynamics of various high temperature transformation of Kaolinite.* in ed. by Earl I Proceedings of the 13th Natational Conference on Clays, Madison, Wisconsin, Pergamon press, Monograph, No. 25 pp. 419–425 (1966)
B.S. Shvetsov and C. O. Gevorkyan, Thermochemical investigation of the kaolin heating process. J.Appl.Chem.(U.S.S.R), **15**, 302–18 (1942)
F. Vaughan, Energy changes when kaolin minerals are heated. Clay Miner. Bull. **2**(13), 265–274 (1955).
D.R. Waldbaum, Thermodynamic properties of mullite, sillimanite, andalusite, Kyanite. Am. Miner **50**, 186–195 (1965)

Chapter 13
MAS NMR

13.1 Characterization of Phases

With recent development of high-resolution MAS NMR, it is now possible to study the atomic environment in a very disordered state. Accordingly, it is applied to study the local structure of Si and Al environments in intermediate products of K–M reaction series where neither XRD nor TEM is able to identify those phases. In order to characterize the high-temperature reaction products of kaolinite, Meanhold et al. (1985) and later Brown et al. (1985) applied solid-state NMR spectroscopy along with usual X-ray. They took BDH kaolinite of 'light' variety and heated it to various temperatures, cooled, and analyzed. The ^{29}Si and ^{27}Al NMR spectra of kaolinite heated to various temperatures are shown in Figs. 13.1 and 13.2.

Unheated kaolinite shows a single sharp peak at −91.5 ppm relative to tetramethyl silane. When heated to 650–800 °C i.e., after dehydroxylation, silica resonance broadens and shifts to −100 ppm consistent with the presence of a number of silicon sites of differing bond lengths, bond angles, or cation oxygen bond strengths. Heating to 970 °C, the Si resonance progressively shifts up field to −110 ppm, reflecting a decrease in the number of aluminum groups bonded to SiO_4 groups.

They conceived a separation of SiO_2 from alumino-silicate phase at this stage. This process is completed essentially at about 1050 °C. Second, a similar peak appeared at −90 ppm in Fig. 13.2. Such peak was predominantly found in samples heated at 1,200 °C. This peak could be resolved into at least two resonances at −99 and −105 ppm corresponding to silicon associated with mullite phase which was shown by XRD to be present even at 970 °C. Thus, ^{29}Si peak was characteristic of pure silica resonance in the ^{29}Si spectra; this feature could be seen even at 970 °C and corresponded to the main silicate resonance in mullite. Smaller Si peaks occurring under this probably results from the other less-populated Si sites in mullite. Heating to 1,450 °C cristobalite virtually disappeared due to verification which was reflected in the broadening of the free silica peak at −109 ppm. The principal resonance in unheated kaolinite at −13 ppm as shown in the Fig. 13.1, is the expected position for the octahedral aluminum co-ordinated to hydroxyl groups.

A. K. Chakraborty, *Phase Transformation of Kaolinite Clay*,
DOI: 10.1007/978-81-322-1154-9_13, © Springer India 2014

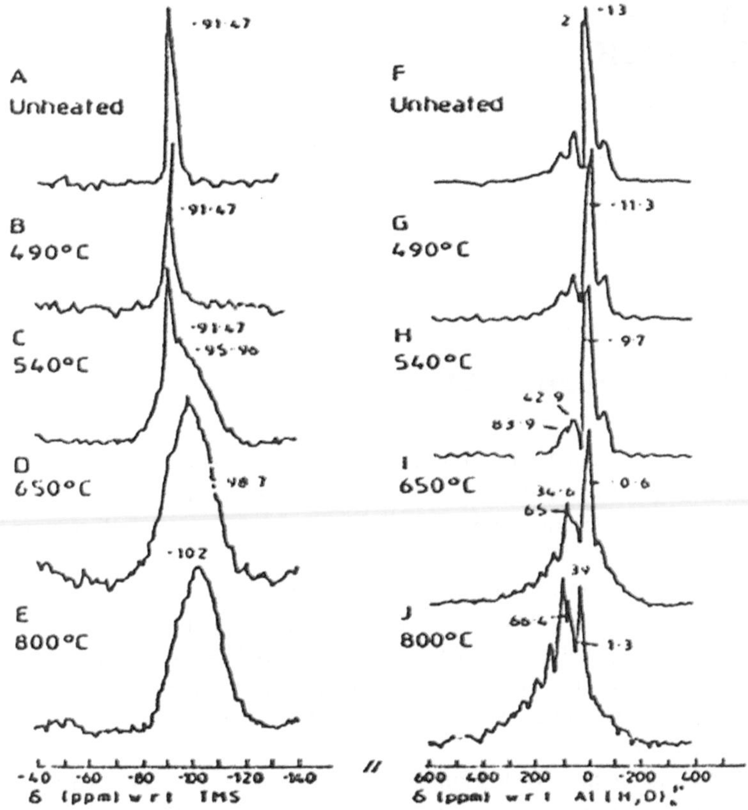

Fig. 13.1 High-resolution solid state NMR spectra of kaolinite heated to temperatures indicated; spinning speed ~2600 rps : (**A–E**) ^{29}Si spectra; (**F–J**) ^{27}Al spectra (after Mac Kenzie et al. 1985a, b). Reprinted by permission of the American Ceramic Society

Heating the kaolinite to 490 °C (just below dehydroxylation) produces no change in the ^{27}Al spectrum. On further heating to 540 °C, spectrum signal was reduced to 30 % of its original intensity. After dehydroxylation at 650 °C, the intensity of the aluminum signal is reduced to about 8 % of its unheated intensity tetrahedral aluminum resonance at 0–1.5 ppm. On heating to, 970 °C and onwards, only the relative proportions of tetrahedral and octahedral aluminum changed slightly. At about 1,050 °C, the two tetrahedral Al resonances merged and the chemical shift of the octahedral resonance assumed a value approaching that of well-formed mullite. A slight change in the ratio of octahedral to tetrahedral Al occurred from the previously heated sample. Final ratio of approximately 0.6 is comparable to the theoretical value of 0.71 for 2:1 mullite. Brown et al. (1985) finally clarified that the broad ^{29}Si resonance peak at −99 to −102 ppm of metakaolinite to a range of Si–O–Si(Al) bond angles. ^{27}Al spectra of Fig. 13.1 also show in addition to the octahedral Al peak at −0.6 ppm, two new non-octahedral peaks. The peak at

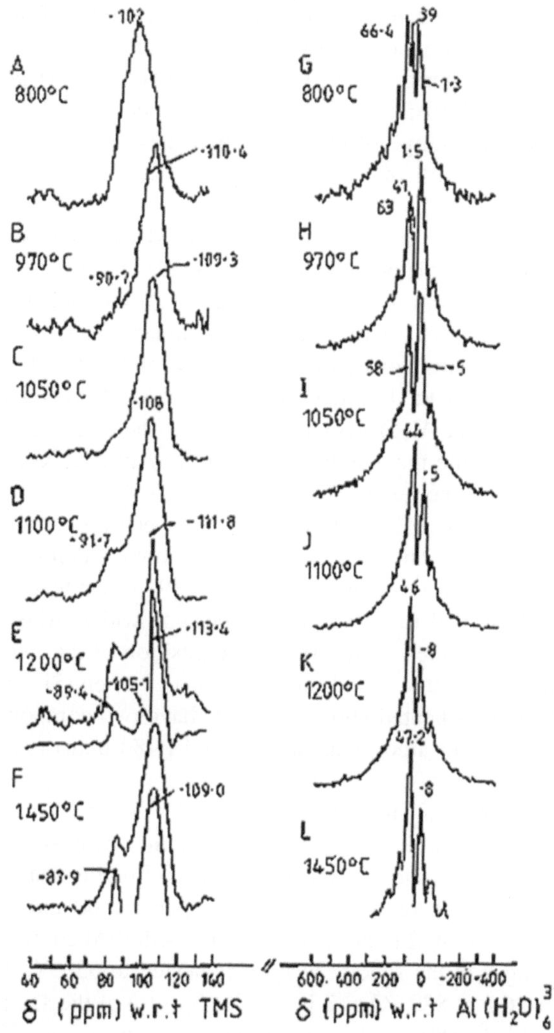

Fig. 13.2 High-resolution solid–state NMR spectra of kaolinite heated to temperature indicated for 1.0 h. Spinning speed ~2600 rps : (**A–F**) ^{29}Si NMR spectra: (**G–I**) ^{27}Al NMR spectra (after Brown et al. 1985). Reprinted by permission of the American Ceramic Society

65–66 ppm is characteristically tetrahedral and the other peak occurs at 34–39 ppm. They explained the NMR observations to be due to tetrahedral Al resonances in terms of two regular tetrahedral sites, one co-ordinated with four oxygen's and the other with three oxygens and hydroxyl and then proposed metakaolinite structure.

Watanabe et al. (1987) used Georgia kaolinite whose crystallinity index was 1.15 (as per Hinckley 1963) and heat treated it to 637, 703, 841, 915 and 1,034 °C, respectively for 1 h only. Heated samples were examined for ^{29}Si and ^{27}Al magic angle spinning by high-resolution NMR method. In the original kaolinite, a sharp absorption of ^{29}Si NMR was observed at −91.5 ppm which corresponded to the signal of Si nuclei in the Q^3 polymerization state i.e., three Si atoms bonded to

SiO_4 tetrahedron. Kaolinite heated between 637 and 841 °C showed a broad absorption around −104 ppm. When heated at 915 °C (it is the temperature of the faint endotherm and just before the exotherm as per the available kinetic data given by Tsuzuki (1961) the Si resonance peak shifted to −108 ppm. This value falls in the range of $Q^4(0)$ and indicates that Si is dominantly in a three-dimensional network of SiO_4 tetrahedral i.e. amorphous silica. At 1,034 °C, Si NMR further shifted to the higher field side and a small peak appeared at −91 ppm and is ascribed to mullite formation. Al MAS NMR spectrum of original kaolinite showed a single peak which corresponded to the six co-ordinated Al atom. On heat treatment, both 4 and 6 co-ordination Al atoms developed and at 1,034 °C and 6 coordinated Al again became predominant. The ^{27}Al MAS NMR spectra resolved into two parts corresponding to Al(4) and Al(6). Proportions of Al(4) and Al(6) nuclei are plotted. In the original kaolinite, Al was mainly in six co-ordination. On dehydration Al(4) nuclei increased at the expenses of Al(6). This increase continued in the metakaolin state from 637 to 841 °C and at 841 °C the Al(4):Al(6) ratio was maximum.

Sanz et al. (1988) showed the ^{29}Si spectrum of kaolinite and its heated products. Kaolinite contains one peak at −91.5 ppm which is associated with Si surrounded by three Si (Q^3 in the Lippmaa et al. 1983). On heat treatment above 450 °C the Si NMR shifted to −98 ppm and became wider. Above 600 °C, it shifted to more negative values. Near 900 °C the position of the line is −110 ppm. This value is the same as that found for Si in silicates when (Si surrounded by four Si). Spectra of samples heated above 1,050 °C have two components; the first (−110 ppm) is attributed to amorphous and second (−88 ppm) is assigned to poorly crystalline mullite.

Figure 13.3 shows the variation of ^{29}Si signal with rise of temperature. The ^{27}Al NMR spectrum of kaolinite also contains one peak at 0 ppm characteristic of Al in octahedral co-ordination. When the kaolinite is heated above 400 °C, the intensity of the line decreases and two lines appear at 60 and 30 ppm. At 725 °C the spectra shows minimum intensity of the octahedral Al component. There are no significant changes in the spectra when it is heated to 850 °C. Spectra of the sample heated for 36 h at 880 °C or 3 h 980 °C. Spectra of the sample heated for 36 h at 880 °C or 3 h to 980 °C show considerable changes e.g., 30 ppm resonance disappears, the peak at 0 ppm becomes pronounced and at the same time a new line at 42 ppm appears in the spectra and the band at 60 ppm decreases. The 60 and 42 ppm lines increase in intensity when kaolinite is heated to higher temperatures. Sample heated at 1,150 °C have an ^{27}Al NMR spectrum similar to that of 3.2 mullite. Based on the above observation, they characterized metakaolinite and predicted its structure based on penta co-ordinated Al. In contrast to the believe of Brown et al. (1985), Sanz et al. (1988) are of the opinion that the position and relative intensity of the components detected in the ^{27}Al NMR signal (60, 43 and 1 ppm) are different and thus exclude the presence of γ-Al_2O_3 phase. At temperatures 980–1,150 °C, the band components located at 42 and 60 ppm in the ^{27}Al NMR signal increase progressively at the expense of the band associated with hexa co-ordinated Al (about 0 ppm). Finally, the ^{27}Al NMR spectra of samples in which

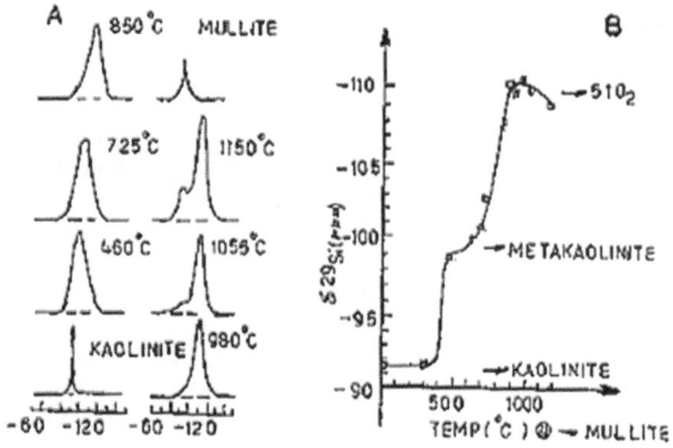

Fig. 13.3 Variation of the chemical shift of the signal versus temperature (after Sanz et al. 1988). Reprinted by permission of the American Ceramic Society

mullite is the main crystalline component were compared with a sample of well-crystallized mullite. The relative intensities of the peaks at about 40 and 6.0 ppm of the sample heated at 1,150 °C are similar to those of mullite and ^{29}Si NMR show the presence of two peaks, one associated with the mullite and the other with segregated silica (A) phase.

13.2 Summary

In recent advancement of high-resolution MAS NMR, various researchers thought that it is very adequate to study disordered intermediate phases like metakaolinite and spinel in K–M reaction series to reveal the local Al and Si environment. Since both XRD and TEM are unable to identify and evaluate their structural compositions. Meinhold et al. (1985), Brown et al. (1985), Watanabe et al. (1987) and Sanz et al. (1988) noted changes in ^{29}Si and ^{27}Al NMR spectra of clays heated to various temperatures. According to Watnabe (1987), Si in metakaolinite is dominantly present in Q^3 state. Based on shifting of both spectra, Brown et al. proposed a new structural model for metakaolinite with fulfillment of different criteria and suggested segregation of free silica on heating at 970 °C. Regarding identification of intermediate spinel phase they showed that ^{27}Al spectra of heated kaolinite and γ-Al_2O_3 are not identical. Predicted chemical shift for the presence of tetrahedral silicon sites in spinel structure is −80 ppm which is, however, absent in heated ^{29}Si spectrum. Sanz et al. noted 30 ppm peak due to AlO_5 group in ^{27}Al peak in the spectra of metakaolinite sample in addition to 60 ppm peak. ^{27}Al NMR spectra of clay heated to 980 °C disagrees with that of γ-Al_2O_3. On further heating to 1,150 °C, the spectra resemble that of 3:2 mullite. ^{29}Si NMR spectrum show two peak—one is associated with mullite and the other is that for segregated silica phase.

References

I.W.M. Brown, K.J.D. MacKenzie, M.E. Bowden, R.H. Meinhold, Outstanding problems in the kaolinite–mullite reaction sequence investigated by 29Si and 27Al solid-state nuclear magnetic resonance: II, high-temperature transformations of metakaolinite. J. Am. Ceram. Soc. **68**(6), 298–301 (1985)

D.N. Hinckley, Variability in crystallinity values among the kaolin deposits of the coastal plain of Georgia and South Carolina. Clays Clay Miner. **11**, 229–235 (1963)

K.J.D. MacKenzie, I.W.M. Brown, R.H. Meinhold, M.E. Bowden, Outstanding problems in the kaolinite–mullite reaction sequence investigated by 29Si and 27Al solid-state nuclear magnetic resonance: I, metakaolinite. J. Am. Ceram. Soc. **68**(6), 293–297 (1985a)

K.J.D. MacKenzie, I.W.M. Brown, R.H. Meinhold, M.E. Bowden, The thermal reactions of pyrophyllite studied by high-resolution solid-state 27A1 and 29Si nuclear magnetic resonance spectroscopy. J. Am. Ceram. Soc. **68**(5), 266–272 (1985b)

R.H. Meinhold, K.L.D. Mackenzie, I.W.M. Brown, Thermal reactions of kaolinite studied by solid state 27Al and 29Si NMR. J. Mater. Sci. Lett. **4**, 163–166 (1985)

J. Sanz, A. Madani, J.M. Serratosa, J.S. Moya, S. Aza, 27Aluminum and 29Silicon-magic-angle spinning nuclear magnetic resonance study of the kaolinite–mullite transformation. J. Am. Ceram. Soc. **71**(10), C-418–C-421 (1988)

Y. Tsuzuki, Mechanism of the 980 °C exothern of kaolin minerals. J. Earth Sci. Nagoya Univ. **9**(2), 305–344 (1961)

T. Watanabe, H. Shimizu, K. Nagasawa, A. Masuda, H. Saito, 29Si and 27Al-MAS/NMR study of the thermal transformations of kaolinite. ClayMiner. **22**, 37–48 (1987)

Chapter 14
QXRD Studies of Phases Formed

14.1 Introduction

As shown earlier 980 °C reaction products of kaolinite constitute poorly crystallized spinel phase, weakly crystallized mullite, and amorphous SiO_2. The SiO_2 phase was estimated by alkali-leaching technique as described in Chap. 17. The estimation of the other two weakly crystalline phases was attempted by different researchers using QXRD technique. Both crystallinity and quantity of mullite develop largely near 1250 °C exotherm. β-cristobalite usually developed at 1200 °C. The growth rates of both mullite and cristobalite were subsequently determined by the same QXRD technique.

14.2 Estimations of 980 °C Spinel Phase and Mullite Phase

Tsuzuki et al. (1969) used QXRD technique to estimate spinel and mullite phases only. They heated one halloysite and two other kaolinites isothermally and noted the diffraction peaks of the high-temperature phases namely spinel and mullite repeatedly by oscillatory scanning with a Geigerflex X-ray diffractometer provided with a high-temperature attachment. The amounts of high-temperature phases were estimated by comparing with standards. The change in the amount of high-temperature phases with time is shown in Fig. 14.1.

The amounts of high-temperature phases are of semi-quantitative accuracy because of weak intensity and broad shape of the diffraction peaks. It was noted that spinel crystallized from halloysite and disordered kaolinites, whereas both spinel and mullite crystallized simultaneously from kaolinite. In the high-temperature runs, the rate of crystallization was the largest at the beginning and then it slowed down with time. Whereas, in the low temperature runs, the crystallization began after an induction period and then the rate became larger following a sigmoid shape of the curve. Quantitative assessment of 970 °C heated product out of Mese Alta kaolinite is as follows. Amount of spinel = 35 %

A. K. Chakraborty, *Phase Transformation of Kaolinite Clay*,
DOI: 10.1007/978-81-322-1154-9_14, © Springer India 2014

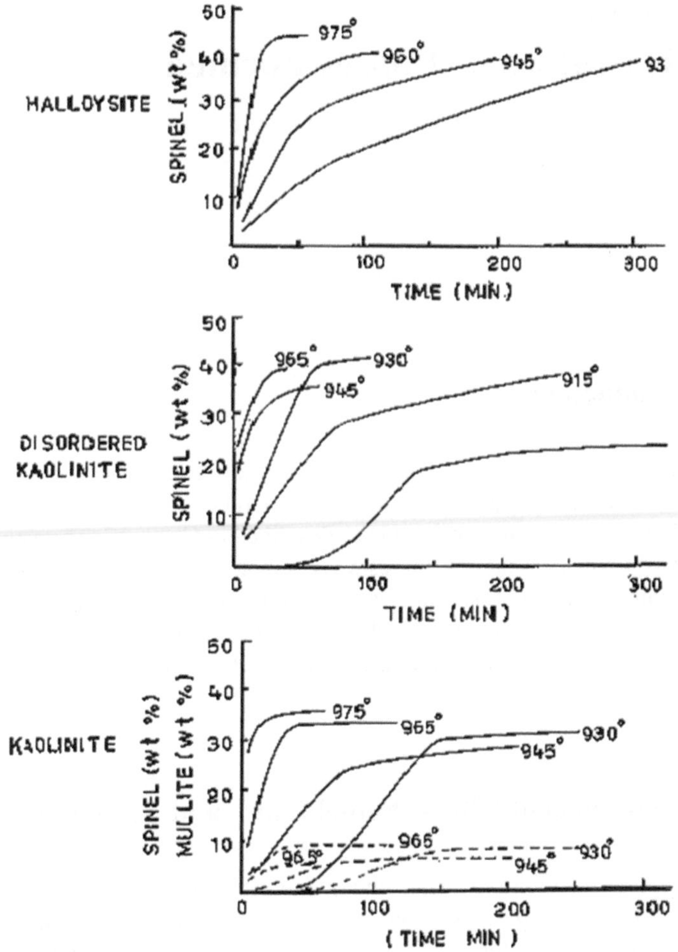

Fig. 14.1 The result of isothermal X-ray diffraction experiments at the temperatures indicated. The changes in the amounts of high-temperature phases are shown as a function of time (After Tsuzukiand Nagasawa 1969). Reproduced by permission of The Clay Science Society of Japan

approximately. Amount of mullite = 5 % approximately. Total amount of crystalline phase = 40 % approximately.

Okada et al. (1986) estimated spinel, mullite, and cristobalite by heating to various temperature from several kaolin groups of minerals by X-ray quantitative analysis. X-ray measurements were made using a graphite monochromatized (Cuk$_\alpha$ radiation).

For X-ray quantitative analysis CaF_2 was used as an internal standard. A step scanning technique of 0.05° in 2θ intervals and fixed time of 40 s was applied for measurement of areas of spinel phase because of the faint and diffused reflections. Reflections used for quantitative analysis were (440) of the spinel phase, the (121) of mullite, the (101) of cristobalite, and the (220) of CaF_2. Substances used for calibration of each phase were synthetic ones: spinel phase (fired Al_2O_3 gel at 900 °C for 24 h) mullite (fired $3Al_2O_3$ $2SiO_2$ xerogel at 1600 °C for 24 h) and cristobalite (fired SiO_2 gel at 1500 °C for 24 h). Figure. 14.2 shows the results of X-ray quantitative analysis of the specimens.

Chakraborty and Ghosh (1991a, b) measured semi-quantitative values of spinel and mullite phases by QXRD technique. The spinel phase was estimated by comparing the 0.137 nm. Bragg diffraction peak height of 980 °C heat-treated kaolinite with that of a standard γ-Al_2O_3 and by using pure.

MgO as an internal standard according to the formulae used by Verduch and Estrade (1962). The calibration curve for spinel was first obtained by measuring mixtures of 10–40 wt%. γ-Al_2O_3 with SiO_2 (A) as diluents. Thereafter, R values of different heat-treated samples were determined separately from their individual XRD runs and then R values versus % spinel was plotted. Where,

$$\mathbf{R} = \frac{\mathbf{H}_{\mathrm{sp}} \times \mathbf{W}_{\mathrm{sp}}}{\mathbf{H}_{\mathrm{MgO}} \times \mathbf{W}_{\mathrm{MgO}}}$$

where H_{sp} = Height of spinel peak, W_{sp} = Width at half height; and H_{MgO} and W_{MgO} are the corresponding values for standard MgO.

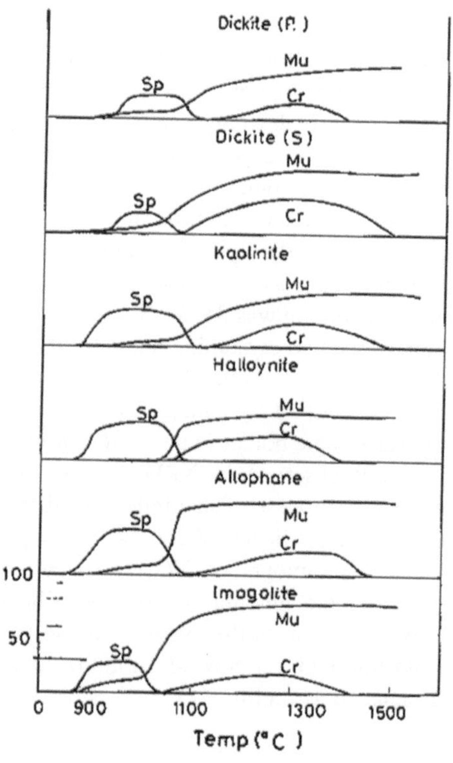

Fig. 14.2 Amount of spinel phase, mullite, and cristobalite in each specimen as a function of firing temperature. Calculated amounts for spinel phase based on the compositions of (1) $3SiO_2.2Al_2O_3$, (2) $2SiO_2.3Al_2O_3$, and (3) Al_2O_3 are represented by arrows with their numbers, and those for mullite and cristobalite are represented by Mu_c and Cr_c, respectively, Sp: spinel phase; Mu: Mullite; Cr: cristobalite (after Okada et al. 1986). Reprinted by permission of the American Ceramic Society

Table 14.1 Semi-quantitative estimation of Ai–Si spinel phase DERIVED from kaolinites heated at 980 °C

Type of kaolinite	Calculated value of *R*	Al–Si spinel phase (%)
Bhandak	1.20	30.0
English China clay	1.16	29.0
Zettlitz	1.33	34.0
Rajmohol	1.40	35.0
Georgia (highly crystalline)	1.45	36.0

From the calibration curve, approximately, values of spinel content formed on heating different clays have been estimated. Table 14.1 shows that the content of spinel in different kaolinite varies. The average value lies within 30–40 wt%.

Mackenzie and Brown (1987) calculated the spinel content formed out of kaolinite marked as BDH "light'' by quantitative X-ray diffraction measurements of the 440 reflection, using γ -Al_2O_3 as celebrant phase and elemental Si as internal standard. Their estimated value is ~19 wt%, which is considerably less than that found by Okada et al. (1986).

14.3 Calculated Value of Aluminosilicate (A) Phase

Weakly crystalline mullite formed during decomposition of kaolinite at 980 °C is different from that of well-crystalline mullite which is usually taken as a standard. Even in poorly developed stage the intensity of the 0.537 nm Bragg reflection peak varies and it is dependent on the origin of the clay. Keler and Leonov (1955) observed more spinel and little mullite, where as Mikheev and Stulov (1955) noted comparatively more mullite and less spinel. Under these circumstances, the percentage of poorly crystalline mullite which forms in different kaolinites at 980 °C may be taken as 4–6 wt%, approximately. The amount of SiO_2 (A) liberated during kaolinite was 35–37 % as estimated by Chakraborty and Ghosh (1978a, b). Sum of the two crystalline phases and SiO_2(A) is still short of 100, which predicts the existence of another amorphous phase in 980 °C heated products of kaolinite. It was identified by taking powder diffraction photograph using Lindamen glass tube as sample holder in an 11.46 cm Philips circular camera. An amorphous band was still observed in XRD patterns of the alkali leached residue during the extraction process of kaolinite heated to 980 °C. The approximate percentage of this aluminosilicate phase (A) is = [100−(amount of spinel) + (amount of mullite) + (amount of SiO_2 (A))]. By substituting the estimated values in the equation, the amount of alumino silicate (A) is calculated as 30–40 %. The varying percent of this phase is due to the variations of spinel and mullite formed from different varieties of kaolinite found in nature. This aluminosilicate (A) is designated as mullite (A).

14.4 Estimation of Mullite and Cristobalite Phases

Konopicky (1959, 1962) identified mineral phases formed during heating wide range of German and other continental commercial fire bricks and then estimated the amounts of mullite and cristobalite by X-ray methods and the glass content by difference. Grimshaw (1971) reviewed the works of the above authors. The amount of mullite content is found to be related to the alumina present in the clay and the relative ratio of cristobalite to glass may be dependent upon the content of fluxing oxides.

Johnson and Andrews (1962) studied the mineralogical compositions of a range of British alumino-silicate fired products made from high- and low-grade fire clays. Later on they compared the composition of fired clay products with fired products made out of calcined Kyanite or Sillimanite and calcined Bauxite as raw materials.

Mcgee (1966) heat-treated three clays, a high duty plastic fire clay, a flint fire clay, and a kaolin at 1350 °C for 5 h after which these were cooled in variety of ways. The mineralogical composition of the fired products namely mullites and cristobalite were quantitatively calculated by X-ray diffraction method using CaF_2 and BaF_2 as internal standards. It is shown those mullites content in respective clays are not affected by difference in cooling schedule. Further it also noted that the rate of crystallization from glassy silica phase to β-cristobalite is very much influenced by the cooling rate. Generally, rapid cooling resulted in a high-glass content in contrast to the more crystalline phase developing under slow conditions. Four classes of fire clay refractories, viz., (1) high heat duty, (2) intermediate heat duty, (3) moderate heat duty, and (4) low heat duty is tentatively defined by some physical tests under the A.S.T.M. Performance of these bricks might be predicted out of the knowledge of the crystal development of mullite out of fire clay. Harvey and Birch (1936) fired super heat duty bricks at 1280, 1320, and 1430 °C and examined for petrographic microscope for microstructure developments. Percentages of mullite formed were ascertained by eye estimating the relative intensities of X-ray lines of the known and unknown specimens. Qualitative X-ray tests conclusively show mullite formation at all three successive heat-treated specimens. The percentage of mullite was, approximately, 50 % which is independent of firing temperatures. Petrographic examination showed no indication of mullite development at 1280 °C but indication of incipient crystallization in the form of minute skeletal outlines was noted at 1320 °C. On high temperature calcination at ~1430 °C, they observed an interlocking mass of very fine mullite crystals of needle like habit in the matrix. The sizes of them are ~0.01 mm in length and 0.0007 mm in width. Thus, they came to the conclusion that the size of the mullite crystals increases with increase of calcination temperature but probably not the total amount of mullite content.

Chakraborty (1993) estimated the semi-quantitative amount of cristobalite formed in Rajmohol kaolinite heated to different temperatures with 2 h soaking by XRD technique. The standard cristobalite used was prepared by heating pure quartz at 1650 °C for 8 h followed by leaching with 5 % HF in an ice both for 5 min and lastly by X-ray analysis of the washed residue. This process is repeated till the height of the 0.404 nm XRD peak of *β*-cristobalite became nearly constant. Figure. 14.3 shows the growth curve of *β*–cristobalite.

14.5 Assessment of Vitreous Phase

Kaolinite generally contains some associated impurity oxides namely Fe_2O_3, FeO, TiO_2, CaO, MgO, K_2O, and Na_2O to the extent about 2.5 % by weight. These oxides may undergo solid state reaction during heating with free SiO_2 liberated out of kaolinite and forms a silica-rich vitreous phase. The vitreous phase may have played a role in mullite. Major fluxing oxides reduce the refractoriness of kaolinite body during heating process. These fluxing oxides control the viscosity of the resultant vitreous phase. Finally, the impurity oxides also influence the quantity of the glassy phase developing. Thus, the nature and quantity of impurities exert influence on the formation process of mullite and its crystal growth. Generally higher the kaolinite content and lower the flux concentration, higher will be the content of mullite. However, the amount and nature of vitreous phase developed and the way it reinforced the mullite needless is the determining factor in obtaining a definite microstructure. The vitreous phase improves homogeneity of the matrix containing various sizes of mullite. Thus, mullite/glass ratio is an important index in determining the strength characteristics, since the matrix contains a well-growth nest like interwoven network of mullite crystals in glassy mass (Fig. 14.4).

Mcgee (1966) calculated the percentage of liquid-phase by subtracting total crystalline phases from 100. The result shows a constant value of about 60 % of liquid-phase in three types of clay Fig. 14.5. The chemical composition of the liquid-phase was calculated based on the assumptions that Al_2O_3 to SiO_2 ratio of

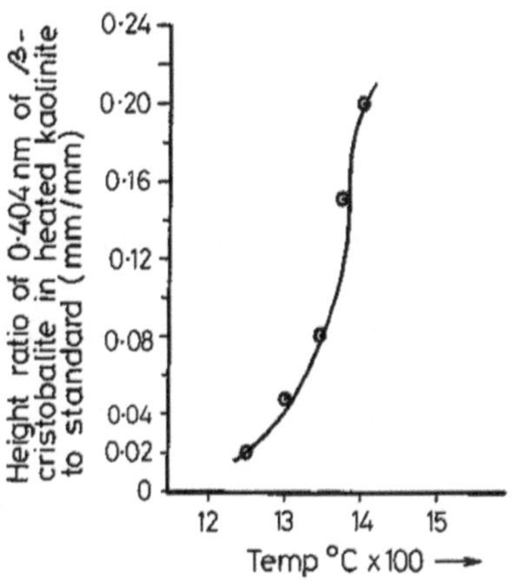

Fig. 14.3 Amount of *β*-cristobalite formed versus temperature of heating of Rajmohol kaolinite under static heating conditions

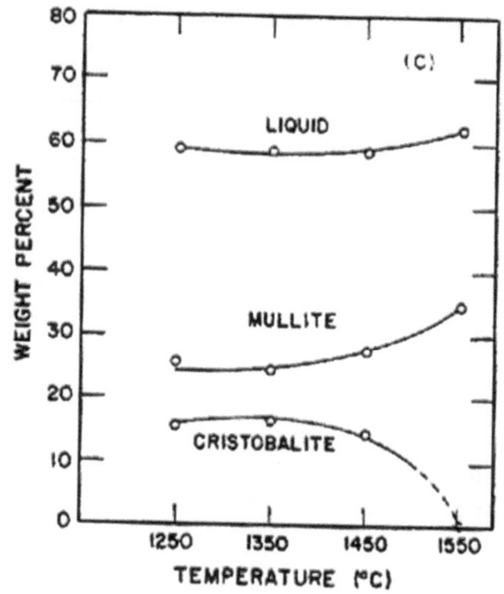

Fig. 14.4 Mineralogical composition versus firing temperature of pioneer kaolin (after McGee1966). Reprinted by permission of the American Ceramic Society

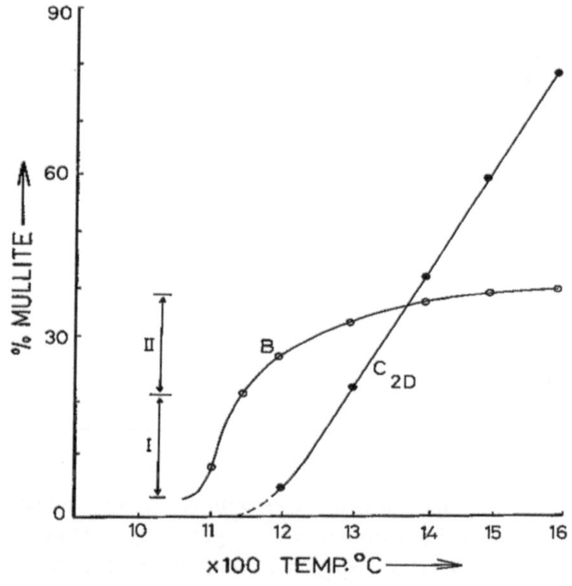

Fig. 14.5 Mullite development versus temperature from different sources : (C2D) mixture of Al_2O_3(A) and SiO_2 (A), (B) Bhandak kaolinite (after Chakraborty and Ghosh 1991). Reprinted by permission of the American Ceramic Society

mullite is 3:2, Fe_2O_3 and TiO_2 replaced alumina only and is shown in Table 14.2. A few problems are noted in estimation process. There are variations in both mullite and cristobalite contents of fired kaolinites. Mcgee (1966) showed % mullite is in the range 27–33 which is lower than the mullite developed in a commercially fired aluminosilicate products reported by Konopicky (1962) *and* Johnson and Andrew (1962). The main source of error in estimation of mullite is the variation of degree of crystallinity of mullite formed from kaolinite during

Table 14.2 Composition of liquid phase of fireclays and kaolin (After McGee, 1966) Reprinted with kind permission of American Ceramic Society

Firing temp. (°C)				
	1250	1350	1450	1550
Missouri plastic fireclay				
SiO_2	61.6	69.4	71.9	
Al_2O_3	26.8	20.2	18.4	
Fe_2O_3	2.7	2.2	2.2	
TiO_2	1.7	1.5	1.6	
CaO	1.9	1.8	1.5	
MgO	1.2	1.2	1	
K_2O	1.5	1.4	1.2	
Na_2O	2.6	2.5	2.1	
	100	100.2	99.9	
Missouri flint fireclay				
SiO_2	55	55.8	64.8	61.2
Al_2O_3	36.2	35.7	25.9	30.2
Fe_2O_3	1.9	1.9	2.1	1.9
TiO_2	1.3	1.2	1.2	1.5
CaO	1.2	1.2	1.4	1.2
MgO	2.2	2.1	2.3	2
K_2O	0.8	0.8	0.9	0.7
Na_2O	1.4	1.3	1.5	1.3
	100	100	100.1	100
Pioneer kaolin				
SiO_2	50.7	49.2	51.1	65.2
Al_2O_3	44	45.6	43.4	29.6
Fe_2O_3	0.1	0.2	0.2	0.2
TiO_2	2.2	2.2	2.4	2.2
CaO	1.1	1.1	1.1	2
MgO	1.3	1.2	1.3	1.2
K_2O	0.3	0.3	0.3	0.4
Na_2O	0.4	0.4	0.4	0.4
	100.2	100.2	100.2	101

different stages of heating as pointed out by McGee (1966). In contrast, standard mullite used in estimation work generally of sintered/or fused variety. Konopicky and Kohler (1958) suggested that for mullite determination in heated clays, the 100 % mullite should be obtained from alumino silicate in question by separation process. McGee (1966) synthesized cristobalite to be used as a standard by calcining reagent grade silicic acid at 1600 °C for 4 h. Chakraborty and Ghosh heat-treated pure quartz at 1650 °C/1 h. to develop major cristobalite.

Chakraborty and Ghosh (1991a, b) compared the mullite formation behavior of kaolinite with that of synthetic mixture. An artificial mixture (C_{2D}) of Al_2O_3 and SiO_2 (ratio 3:2) was first prepared by mixing amorphous alumina and amorphous SiO_2 in an agate mortar for 2 h. This mixture and Bhandak kaolinite were heat

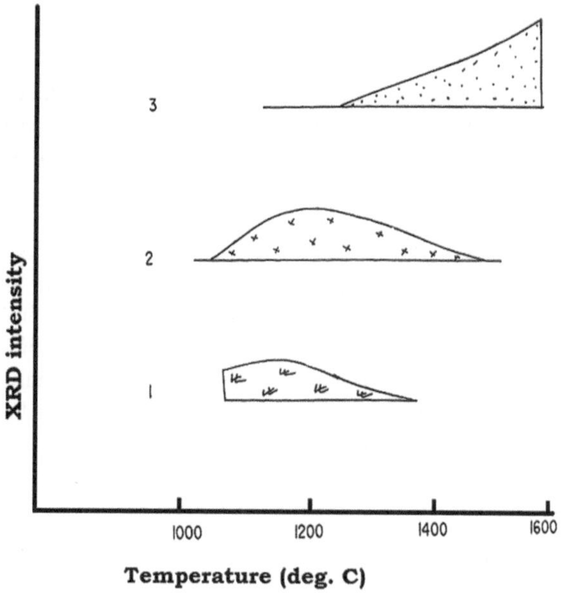

Fig. 14.6 Sequential phase developments during heating of artificial mixture of amorphous Al2O3 and amorphous SiO_2: (1) cristobalite, (2) corundum, (3) mullite (after Chakraborty and Ghosh 1991). Reprinted by permission of the American Ceramic Society

treated at successively higher temperatures, cooled, ground to −325 mesh, and analyzed by XRD for quantitative identification of the phases present as well as far quantitative estimation of mullite. In the synthetic mixture C_{2D}, γ-Al_2O_3 is the first crystalline phase and it exists between 700 and 900 °C. On further heating, γ-Al_2O_3 gradually starts to disappear and transforms to corundum (major) at 1100 °C, leaving behind a minor amount of unconverted α-Al_2O_3 phase. More and more corundum crystallizes, as is evident from the increase in intensity of the 0.208 nm Bragg reflection of the corundum pattern with an increase of temperature up to 1200 °C and then it decreases. The development of corundum takes place independently whether amorphous Al_2O_3 is present in a mixture with silica (A) or alone. Cristibalite appears at about 1075 °C and goes on increasing in between 1200 and 1300 °C after that it decreases. Mullite can be identified at 1200 °C from its faint X-ray pattern, but with a progressive increase of temperature both the intensity and the crystallinity of mullite increase (Fig. 14.5).

Thus, the results show that component oxides first crystallites into their respective high-temperature crystalline forms, e.g., alumina (A) to γ-Al_2O_3 and then to α-Al_2O_3. Side by side silica (A) to β–cristobalite. Afterward, as the temperature rises, the intensities of the two newly formed crystals, namely, corundum and cristobalite decrease gradually with the formation of an aluminosilicate phase (A). Thereafter, mullite nucleation takes place and then steady crystallization as recorded in the sequential phase developments versus temperature curves (Fig. 14.6). In comparison, quantitative mullite formation from Bhandak clay shows different growth rate.

14.6 Summary

Amount of spinel phase formed in K–M reaction series was measured by QXRD technique. Tszuki et al. (1969) noted difference in formation behavior of spinel phase and its quantity in ordered/disordered kaolinite and halloysite. Instead of dynamic heat-treated kaolinite, Chakraborty and Ghosh (1991a, b) made quantitative study of spinel phase on soaked kaolinite and noted a variation in the spinel content from 30–40 %.They indicated the existence of 35–40 % mullite (A) phase in the 980 °C decomposition product of kaolinite. Okada et al. (1986) also estimated spinel phase formed in different kaolin minerals by QXRD technique. They compared the estimated value of spinel formed with theoretical amount expected for three types of spinel formation models, and finally showed that observed value of spinel content best fitted with calculated ones based on the assumption of γ-Al_2O_3 spinel formation as intermediary.

References

A.G. Verduch, D.A. Estrade, *in Science of Ceramics*, ed by G.H. Stewart. In *the Formation of Mullite from Sericite and Its Mixtures with Alumina and Kaoin*, vol.1 (Academic Press for the British Ceramic Society, New York, 1962), pp. 285–94

A.K. Chakraborty, Application of TMA and DTA studies on the crystallization behavior of SiO2 in thermal transformation of kaolinite. J. Therm. Anal. **39**, 280–299 (1993)

A.K. Chakraborty, D.K. Ghosh, Comment on the "interpretation of the kaolinite—mullite reaction sequence from infra—red absorption spectra". J. Am. Ceram. Soc. **61**(1–2), 90–91 (1978a)

A.K. Chakraborty, D.K. Ghosh, Re-examination of the kaolinite to mullite reaction series. J. Am. Ceram. Soc. **61**(3–4), 170–173 (1978b)

A.K. Chakraborty, D.K. Ghosh, Interpretation on the changes of co-ordination number of al in the thermal changes of kaolinite. Clay. Sci. **8**, 45–57 (1991a)

A.K. Chakraborty, D.K. Ghosh, Kaolinite—mullite reaction series. The development and significance of a binary aluminosilicate phase. J. Am. Ceram. Soc. **74**(6), 1401–1406 (1991b)

K. Konopicky, Allgemeies zum aufbau der schamottesfiene. Ber. Deut. Keram.Ges. **35**, 367–71 (1959)

F.A. Harvey, R.E. Brich, Mullite development in super duty fireclay brick. J. Am. Ceram. Soc. **19**, 322–327 (1936)

K. Konopicky, Constant silica-alumina preps. & factors on which it depends. Silic. Ind. **27**(1), (1962).

K. Konopicky, E. Kohler, Determination of mullite and glass content of ceramic materials. Ber. Deut. Keram. Ges. **35**(6), 187–93 (1958)

K. Okada, N. Ostuka, J. Ossaka, Characterization of spinel phase formed in the kaolin–mullite thermal sequence. J. Am. Ceram. Soc. **69** (10), C-251–C-253 (1986)

K.J.D. Mackenzie, I.W.M. Brown, Comment on characterization of spinel phase formed in the kaolin–mullite thermal sequence. J. Am. Ceram. Soc. **70**(9), C-222–C-223 (1987)

R.W. Grimshaw, *The Chemistry and Physics of Clays*, 4th edn. (Wiley-Interscience, New York, 1971)

T.D. McGee, Contn. of fireclays at high temperatures :I, methods of analysis; II, mineralogical composition, III, deformation characteristics. J. Am. Ceram.Soc. **49**(2), 83–94 (1966)

W. Johnson, K.W. Andrews, Quantitative X-ray examination of alumino-silicates. Trans. Brit. Ceram. Soc. **61**(11), 724–52 (1962)

Y. Tsuzuki, K. Nagasawa, A transitional stage to 980°C exotherm of kaolin minerals. Clay. Sci. **3**(5), 87–102 (1969)

E.K. Keler and A.I. Leonov, Effects of contaminationin kaolinon the reactions in its calcination. Doklady Akad.Nauk S.S.S.R. **1**, 137–39 (1955)

V.I. Mikheev and N.N. Stulov, Products of high temperature calcination of layer structure silicates, Zapiski Vsesoyuz. Mieralog. Obshchestva, **84**, 3–29 (1955)

Chapter 15
Scanning Electron Microscopic (SEM) Study

Segnet and Anderson (1971) indicated the nature of changes occurring in the physical structures of a well and a poorly crystallized kaolinite when heated to different temperatures by scanning electron microscopic studies.

Unfired Mt. Egerton kaolinite shows hexagonal crystallites and most of the crystals are of the order of 1 μm in width.

At 900–1,200 °C
Apparently no change in morphology is seen, partial collapsing of the booklets occurred. However, in this temperature range a great change took place in the atomic structure leading to the formation of silicon spinel phase from breakdown of metakaolin and thereafter the subsequent formation of poorly crystalline mullite at ~1,200 °C as revealed by X-ray study and apparent increase in hardness of the test material occurred.

At 1,300 °C
Pseudomorphic crystallite shapes disappear, coalescence, and rounding of particles occurred due to sintering phenomenon. Recrystallization eventually took place, mullite crystallites in the form of small needless with diameters of the order of 30–50 nm is observed in higher magnification. At this stage, X-ray study confirmed the presence of well crystallized mullite. In addition, some amount of cristobalite although recognized in X-ray but it remains undistinguishable in the micrograph.

At 1,400 °C
The micrograph showing more or less uniform and dense structure of crystalline phases are not seen.

In comparison, Anna Kaolinite showed poorly formed crystallites. At 800–1,100 °C, coalescence of crystallites occurred. At this stage, XRD result indicated weak broad peaks of poorly crystallized mullite. At 1,200 °C, sintering occurred which led to a dense microstructure. Thus, scanning electron microscopic study in complementary to XRD study concludes that no significant change in morphology occurs during metakaolinite formation and its subsequent changes in crystal structure to Al–Si spinel formation. However, a radical change occurs

A. K. Chakraborty, *Phase Transformation of Kaolinite Clay*,
DOI: 10.1007/978-81-322-1154-9_15, © Springer India 2014

during latter sintering stage when mullite and crystabalite develops abundantly. In well crystallized kaolinite this recrystallization occurs at 1,200–1,300 °C, whereas the same for poorly crystallized kaolinite occurs at lower temperature between 1,150 and 1,200 °C.

Comer (1960) investigated mullite development in heat treated poorly crystallized as well crystallized kaolinites by electron microscopic studies to note down the conditions of random to oriented growth of it relative to the parent kaolinite.

At 950–1,000 °C/20 h
The particles of the poorly crystallized clay altered in appearance, adjacent particles have merged, boundaries of the individual particles are vanished and subsequently flow occurred with development of rod shaped mullite particles in random orientation. Many protrusions are noted, these particles are appeared to be cubic and they vary from 10–60 nm in sizes.

At 1,100 °C
Mullite needless are shown to be randomly oriented. At 1,200 °C the size of the mullite increased with no orientation effect. In comparison, well crystallized kaolinites at 1,000 °C showed mullite needless are in preferred orientation. Outlines of hexagonal forms are still observed in certain regions. At 1,100 °C oriented mullite needless are noted in those areas where the sheets of the original kaolinite plates are still intact.

At 1,200 °C
Orientation effect is well noticed. Finally, they indicated that in two well crystallized kaolinite showed stacks of plates ranging in size from 5 to 40 μ (specimen C), and from stacks of plates 0.5–1 mm in size (specimen D) showed mullite formed with an orientation related to that of the original kaolinite crystals. But two poorly crystallized kaolinites showed thin plates ranging in size from 200 to 4200 a.u. (specimen A) and thin plates ranging in size from 0.3 to 30 μ (specimen B) loosed their hexagonal outline at ~1,000 °C, however mullite developed in a random manner only on further heat treatment to 1,200 °C (Fig. 15.1).

Schuller and Kromer (1976) fired some china clay and ball clays and showed that the shape of original kaolinite is mostly unaffected during firing to high temperature when mullite developed. The type of kaolinites they studied were originally scaly in appearance in the electron micrographs. XRD pattern of the scaly mullite obtained after firing of raw kaolinite at 1,400 °C resembles with mullite formed in porcelain.

Chen and Tuan (2002) carried out morphology evolution during heating of a highly textured kaolin prepared by tape casting technique which is superior to die casting technique to words texture control of the green compact. The microstructures studied using SEM or Field Emission Scanning Electron Microscopy (FE-SEM) are shown in (Fig. 15.2).

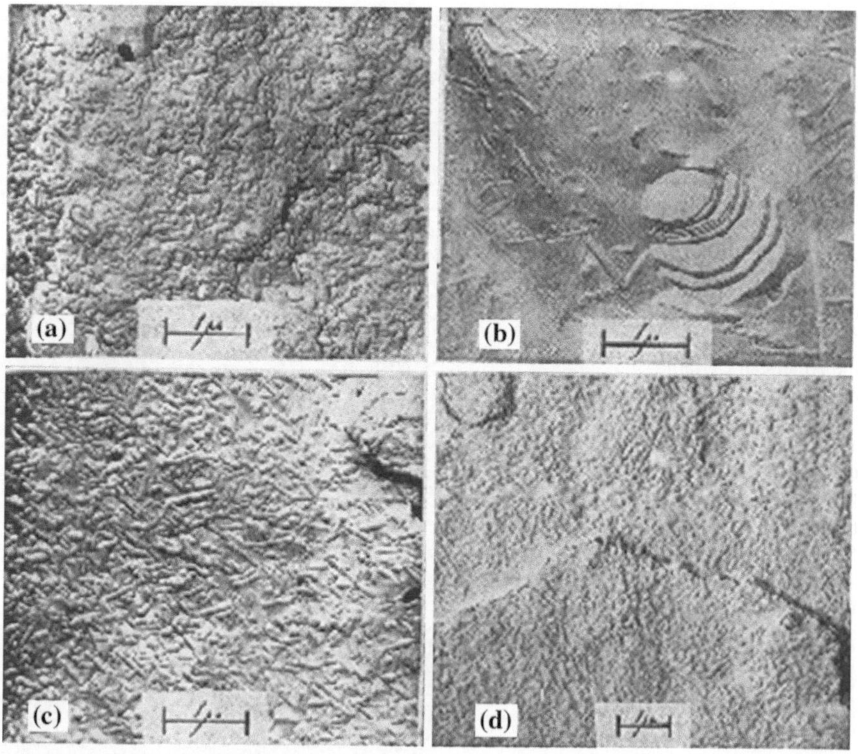

Fig. 15.1 Replicas of kaolinites fired to 1,200 °C. **a** Specimen A, random orientation of mullite; **b** specimen B, random orientation of mullite; **c** specimen C, preferred orientation of mullite; and **d** specimen D, preferred orientation of mullite (After Comer 1960). Reprinted by permission of the American Ceramic Society

At 1,100 °C
Flaky nature of particles are seen. On the perpendicular plane, particles occur in the range 0.5–1 μm.

Within these particles, ~50 nm equiaxed mullite nuclei are embedded.

At 1,200–1,400 °C
The morphology of mullite grains change from equiaxed to aciculars.

At 1,500 and 1,600 °C
The mullite aciculars grow considerably. Many aciculars aligned parallel. This is confirmed by the XRD analysis of calcined kaolinite at 1,600 °C. The intensity of mullite at 1,600 °C (hkl) l $\neq$ 0 peaks in kaolin tape is lower than that of peaks in the loose packed kaolinite.

The c-axis of the mullite is parallel to the long axis of the aciculars. The three-dimensional micrograph demonstrates a strongly textured microstructure.

Fig. 15.2 **a** Morphology of the particles on the perpendicular plane of the of the specimen fired at 1100/1 h, washed with HF (After Chen and Tuan 2002). Reprinted by permission of the American Ceramic Society. **b** Microstructure of the specimen fired at 1200/1 h, washed with HF (After Chen and Tuan 2002). Reprinted by permission of the American Ceramic Society. **c** Microstructure of the specimen fired at 1400/1 h, washed with HF (After Chen and Tuan 2002). Reprinted by permission of the American Ceramic Society. **d** Microstructure specimen fired at 1600/1 h, washed with HF (After Chen and Tuan 2002)

15.1 Morphology of Mullite

There are two varieties of mullite namely "primary" and "secondary" in nature. These are also termed as scaly as per Schuller and Kromer (1976) and ordinary needle-like mullite. Transition from primary to secondary form take place on firing in presence of flux\glass. Based on two step of mullite formation curves, Okada et al. (1991) termed the weakly crystalline mullite phase which is formed during heating of kaolinite at about 1,000–1,100 °C (Fig. 15.3) as primary mullite.

Secondary mullite is the recrystallized mullite developed by transition from primary mullite in presence of viscous phase. Thus primary mullite loses its shape and its habit changes during further heating and is characterized as needle-like.

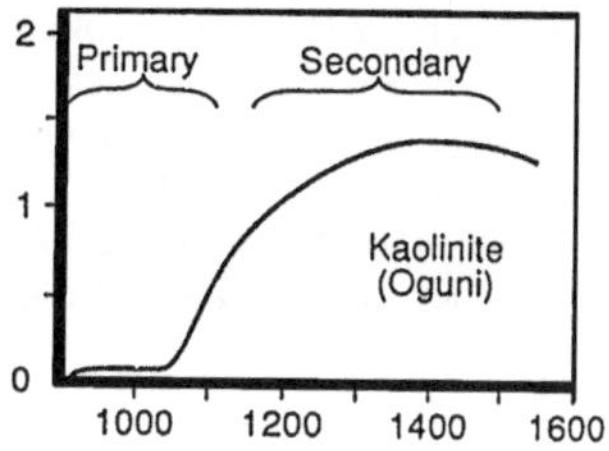

Fig. 15.3 Formation curve of mullite from kaolinite as function of firing temperatures (After Okada et al. 1991). "Used with permission. *ACerS Bulletin*, The American Ceramic Society"

15.2 Summary

Kaolinite undergoes a series of transformation of phases on heating. The morphological evolution during transition of each phase was investigated by SEM, TEM, HRTM, and finally by EF-TEM using kaolinite either in the powder form or prefferly oriented powder compact made by extruding and or tape casting technique.

SEM study by Segnet and Anderson (1971) concludes that no significant change in morphology occurs during metakaolinite formation and during subsequent Al–Si spinel formation. A radical change occurs during latter sintering stage, hardness changes, recrystallization takes place and mullite I needles form crystallites.

Comer (1960) used both poorly crystallized as well crystallized kaolinites for electron microscopic studies. Poorly crystallized kaolinite develops needle shaped mullite crystals at ~1,000–1,100 °C/20 h in a random manner. On further calcinations to 1,200 °C/20 h growth of mullite occurs with no orientation effect. In case of well crystallized kaolinites, mullite needless showed preferred orientation ~1,000 °C. Outlines of hexagonal forms are still observed in certain regions at 1,200 °C. Oriented mullite needless are noted.

In HRTEM, the morphology of spiel crystallites in the phase separated amorphous matrix is observed by Sonuparlak et al. (1987).

A special processing of course shows unique microstructure Chen et al. (2002). Die cast or tape cast kaolinite powder on calcination showed mullite needles in preferred orientation.

References

E.R. Segnit, C.A. Anderson, Electron microscopy of fired kaolinite. Ceram. Bull. **50**(5), 480–483 (1971)

J.J. Comer, Electron microscope studies of mullite development in fired kaolinite. J. Am. Ceram. Soc. **43**(7), 378–84 (1960)

K.H. Schuller, H. Kromer, in *Primary Mullite as a Pseudomorph After Kaolinite*, ed. by S.W. Bailey. In Proceedings of the International Clay Conference (Mexico City, 1975) (Applied Publishing, Wilmette, 1976), pp. 533–538

C.Y. Chen, W.H. Tuan, Evolution of mullite texture on firing tape-cast kaolin bodies. J. Am. Ceram. Soc. **85**(5), 1121–1126 (2002)

K. Okada, N. Otsuka, S. Somiya, Review of mullite synthesis routes in Japan. Ceram. Bul. **70**(10), 1633–1640 (1991)

B. Sonuparlak, M. Sarikaya, I.A. Aksay, Spinel phase formation during the 980 °C exothermic reaction in the kaolinite-to-mullite reaction series. J. Am. Ceram. Soc. **70**(11), 837–842 (1987)

Chapter 16
Hot Pressing (RHP) Study

16.1 Introduction

Hot pressing characteristics of clays have been done for the following aspects. (i) In achieving high density of the products; (ii) To reveal the mechanism of densification; (iii) To study the mineralogical changes in comparison to normal phase change taking place during usual heating. RHP was first noted by Hedvall (1938) and is commonly called "Hedvall effect". Pressure is applied during a phase change where there is enhanced reactivity and consequently it assists densification.

16.2 Application of Pressure on Transformation of Kaolinite

Carruthers and Wheat (1965) reported the application of Hadvall effect in transformation behavior of kaolinite. They pressed kaolinite clay in metal dies in the temperature range covering the main endothermic dehydroxylation of clay. At a pressure exceeding 5 tonf/in^2 at 750 °C when mullite phase formed, which at atmospheric pressure, did not generally form below 1,000 °C. When it is formed, the material no longer shows the strong exothermic reaction in at 980 °C.

Effect of pressure on the endothermic and exothermic reaction of a well crystallized and a-b axis disordered clay have been studied by Carruthers and Scott (1968). They applied pressure ranging from 2.5 to 15 tonf/in^2 on clay specimens in a specially designed pressing apparatus containing die heated to 750 °C and 1,000 °C, respectively. Pressed pellets were measured for density and analyzed by XRD. DTA and dilatomatric measurements were also made under pressed conditions during the successive runs. Near the dehydroxylation temperature (e.g., 750 °C), both types of clays densified to a maximum extent at a pressure of about 7.5 tonf/in^2. With increase of temperature from 750 to 1,000 °C (i.e., at a temp. of 1st exotherm) the maximum densities of 2.63 gm/cc and 2.70 gm/cc were attained, respectively for well and poorly crystallized clays at a reduced pressure of only

A. K. Chakraborty, *Phase Transformation of Kaolinite Clay*,
DOI: 10.1007/978-81-322-1154-9_16, © Springer India 2014

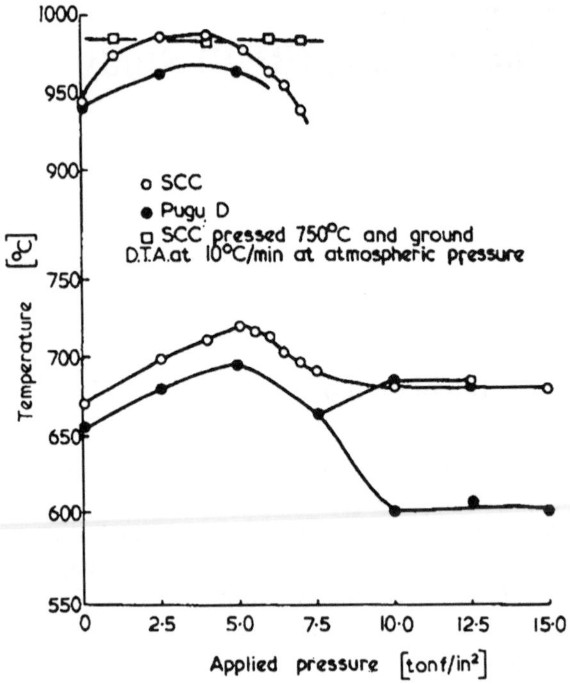

Fig. 16.1 Variation of endothermic and exothermic peak temperature with applied pressure (After Carruthers and Scott 1968)

5 tonf/in^2. The temperature of the maxima of the endothermic and exothermic DTA peaks changes with pressure (Fig. 16.1). Both endothermic and exothermic peaks for two clays increase with increase of pressure at the commencement of the peaks and thereafter decreased. The dilatometric curves obtained during pressing of both clays with super imposition of endo and exothermic DTA peaks are shown in Fig. 16.2. With increase of pressure more contractions are noted in endo and exothermic peaks. Phase developed during hot pressing of the two clays with increase of pressure are summarized.

They suggested that "Hedval Effect" is playing a significant role when pressure was applied on clays during its transformation at exothermic peak temperature. For both calcined and raw clays density was close to the theoretical value which suggests that the exotherm is associated with a major reconstructive change. With further increase of pressure from 0 to 6 to 8 tonf/in^2, the exotherm at 980 °C gradually shortens and finally disappears. Size of contraction decreases at 980 °C with its concomitant increase during the earlier endotherm. It also increases with increase of pressure. With pressure greater than 8 tonf/in^2 a sudden expansion occurs during endothermic peak. Thus, application of pressure results in lowering down of formation of temperature of phases which normally formed at high temperature at normal atmosphere.

Weiss et al. (1970) noted the densification behavior of pressurized metakaolinite. Behavior of metakaolinite transformation was unaltered even with increase of

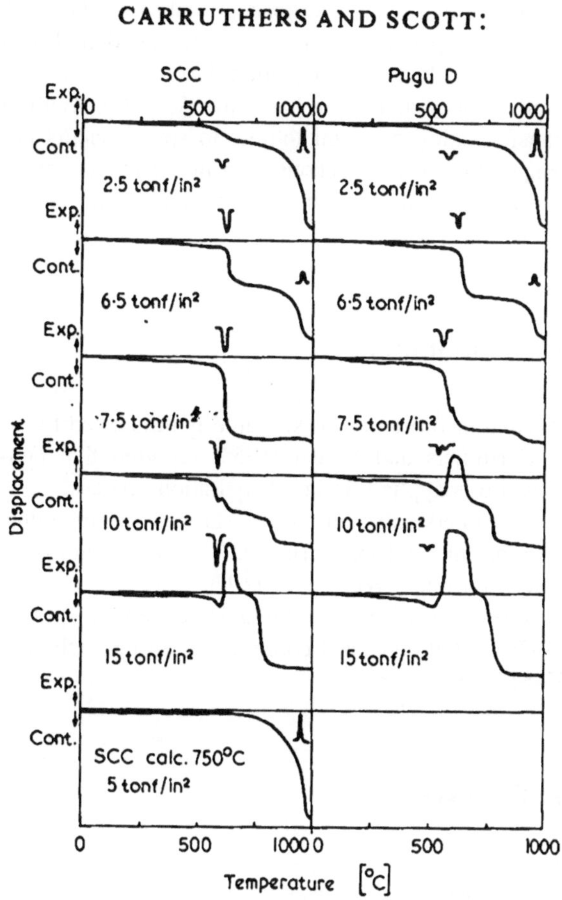

Fig. 16.2 Dilatometric traces taken during hot pressing of kaolinites. Corresponding DTA peaks are also marked (After Carruthers and Scott 1968)

pressure up to 90 kbar. When metakaolinite pre pressurized at 150 kbar, yields some quartz at 800 °C in association to normally formed spinel phase. On further increment of pressure at 250 kbar, mullite forms directly as low as 600 °C. Fluoro kaolinite obtained by replacement of –OH by F^- behaves similar to previous experiment.

Chaklader (1976) investigated the compaction behavior of kaolinite under isothermal condition at constant heating rate. He noted enhanced compaction during each stage of phase transformation in K–M reaction series as follows.

(i) First compaction occurs during the dehydroxylation reaction at 550–650 °C.
(ii) Second enhanced compaction starts at about 835 °C and finished at about 980 °C is prior to the first exothermic reaction in DTA.
(iii) Enhanced compaction (which might be a two stage process) occurs above 1,160 °C and continued up to 1,400 °C. In between the temperature limits there is an exotherm at 1,300 °C in DTA.

On correlating the compaction data with DTA plots, Chaklader believed that reasonable plasticity existed during the dehydroxylation reaction of kaolinite. Where as enhanced compaction, i.e., more transformation plasticity precedes the exothermic reaction of kaolinite (990 °C). With this result he proposed that transformation of metakaolinite to spinel occurs before the exothermic reaction temperature usually encountered in the DTA plot of a kaolinite. Thus, enhanced plasticity encountered in kaolinite during pressing is directly associated with the above transformation.

16.3 Summary

Phase transformation of kaolinite is influenced by pressure.

Carruthers and Wheat (1965) reported that densification of it occurs when pressure is applied during dehydration process.

Carruthers and Scott (1968) showed that the temperature of the maxima of the endothermic and exothermic DTA peaks changes with pressure which is due to "Hedval Effect" Weiss et al. (1970) noted that the densification behavior and phase development of metakaolinite is related to application of pressure. Chaklader (1976) noted enhanced compaction during each stage of phase transformation in K–M reaction series.

References

J.A. Hedvall, *Reactivity of Solids* (Verlag Johann Ambrosious Barth, Leipzig, 1938)

T.G. Carruthers, T.A. Wheat, Hot pressing of kaolin and of mixtures of alumina and silica. Proc. Br. Ceram. Soc. **3**, 259 (1965)

T.G. Carruthers, B. Scott, Reactive hot pressing on kaolinite. Trans. Br. Ceram. Soc. **67**, 185 (1968)

A. Weiss, K.J. Range, J. Russow, The Al, Si-spinel phase from kaoinite (Isolation, chemical analysis, orientation and reactions to its low-temperature precursors), in *Proceedings of the International Clay Conference, Tokyo,* 1969, vol. 2 (Israel Universities Press, 1970), pp. 34–37

A.C.D. Chaklader, Fow properties of clay minerals during phase transformatios. Cent. Glass Ceram. Res. Bull. **23**, 5–14 (1976)

Chapter 17
NAOH Leaching Study

17.1 Introduction

It was thought earlier that kaolinite decomposes during formation of metakaolinite into free silica and alumina. Some also believed that metakaolinite transforms to spinel phase with the liberation of SiO_2. Further, the amount of SiO_2 exolvation indirectly predicts the nature of spinel phase formed i.e., whether it is γ-Al_2O_3 or Al–Si spinel. Since the composition of γ-Al_2O_3 and proposed Al–Si spinel is different. So it is obvious that the amount of amorphous SiO_2 would be different. Thus, it is worthwhile to detect and estimate the amount of silica liberated during the formation and subsequent decomposition of metakaolinite. Various researchers have taken recourse to SiO_2 estimation by either of the two techniques.

17.2 Dissolution of Free SiO_2 by Use of Pyrocatechol

Weiss et al. (1970) used pyrocatechol to remove amorphous SiO_2 liberated during decomposition of metakaolinite as per the scheme below:

Kaolinite
↓ 550°C
Metakaolinite
↓ 900°C
Spinel like phase
\+ — Leached with — Al-Si Spinel
Amorphous SiO_2 -----------→ Chemical analysis
(Amorphous products) Pyrocatechol — $Al_4\ 0.07Si_{2.95}\ O_{12}$
at pH – 8

Pyrocatechol reacts with silica (A) giving stable anionic Si complexes with octahedral coordination. The exhaustive leaching of heat-treated kaolinite finally

A. K. Chakraborty, *Phase Transformation of Kaolinite Clay*,
DOI: 10.1007/978-81-322-1154-9_17, © Springer India 2014

yields the Al–Si spinel, the formula of which is in excellent agreement with the theoretical formula derived by Brindley and Nakahira ($Al_4Si_3O_{12}$) with experimental $Al_40.07Si_{2.95}O_{12}$).

17.3 Dissolution of Free SiO_2 Alkali Leaching Technique

Solubilization of SiO_2 (A) by the leaching process is dependent on the concentration of alkali used and temperature of leaching bath at normal atmospheric pressure. Generally, free SiO_2 and aluminosilicate would show reactivity toward alkali in different degrees i.e., the rate of dissolution of SiO_2 from free state and from aluminosilicate materials, like aluminosilicate (A) compound, proposed Al–Si spinel, and mullite would be different. It is obvious that NaOH very rapidly reacts with free SiO_2 (A) (as it is an acidic oxide) to form soluble sodium silicate and comparatively less readily with an alumina (A) (amphoteric oxide) to form sodium aluminate and react least with a chemical compound of SiO_2 and Al_2O_3.

Eremin (1969) first heat treated kaolin to 925–1,000 °C and then employed alkali solution to extract SiO_2 (A) liberated during the 980 °C exothermic reaction. He further heated the leached residue to observe the development of mullite. Following these studies, he apparently confirmed the hypothesis that kaolinite decomposed into free oxides which reacted to give mullite as per the equation below:

$$\text{Metakaolinite} \xrightarrow[\text{Exotherm}]{980\,^\circ\text{C}} \gamma\text{-}Al_2O_3 + SiO_2\ (A) \tag{12.4}$$

$$3\,\gamma\text{-}Al_2O_3 + 2SiO_2\ (A) \xrightarrow[\text{Mullite}]{\text{Heating}} 3Al_2O_3.2SiO_2 \tag{12.2}$$

By conducting alkali leaching process methodically Chakraborty and Ghosh (1978b) estimated semiquantitatively the amount of SiO_2 liberated at each thermal transformation steps of kaolinite. They choose centrifuged Bhandak kaolinite and heat treated at 100 °C/h to various temperatures ranging from 500 to 1,300 °C in an electric furnace with a soaking of 2 h in each case. Ground samples were leached with 10 wt % NaOH solutions at boiling water bath condition at different duration of leaching time. The leached solutions were analyzed for SiO_2 and Al_2O_3. Plots of amounts of SiO_2 and Al_2O_3 extracted from heated kaolinites are shown in Fig. 17.1a–c, respectively.

17.3.1 Leaching Behavior of Metakaolinite with Alkali Solution

Figure 17.1a shows that negligible amounts of both SiO_2 and Al_2O_3 are dissolved in hot alkali solution. Instead meakaolinite is found to react with alkali solutions

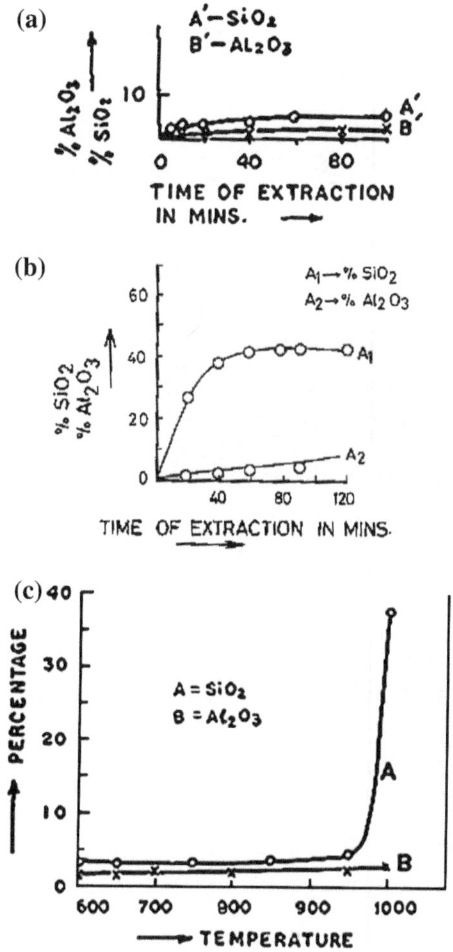

Fig. 17.1 **a** Amounts of SiO_2 (A1) and Al_2O_3(A2) extracted from Bhandak kaolinite which had been heated to 750 °C/2 h by 10 wt% NaOH solution in a boiling water bath. **b** Amounts of SiO_2 (A1) and Al_2O_3(A2) extracted from Bhandak kaolinite which had been heated to 980 °C by 10 wt% NaOH solution in a boiling water bath. **c** Amounts of SiO_2 (A) and Al_2O_3(B) extracted from Bhandak kaolinite which had been heated to 600–980 °C by 10 wt% NaOH solution in a boiling water bath

resulting in the formation of respective alumino silicate hydrate which are usually called as zeolitic type phases. For example, with 10 wt% NaOH solution at boiling condition metakaolin forms zeolite-A (Table 17.1) as per the following reactions:

$$6(Al_2O_3.2SiO_2) + 12NaOH\ (aq) \xrightarrow{\sim 100\,^{\circ}C} Na_2(AlO_2)12(SiO_2)_2.7H_2O + \text{Zeolite–A}\ 6H_2O \quad (17.1)$$

The question is: whether SiO_2 and Al_2O_3 components of metakaolinite first went into alkali solution as silicates and aluminates and then reacted to forms crystalline zeolite phase? It was also shown that metakaolinite shows high reactivity with NaOH even in cold condition. A gain in weight ~18 % took place when metakaolinite was soaked in NaOH solution for ~1.5 h. Crystallization did not take place when this soaked sample was analyzed for X-ray. But zeolite phase

Table 17.1 X-ray diffraction data of zeolite phases obtained from heated Bhandak Kaolinite at different sets of conditions (after Chakraborty)

hkl	d Å Q Series	I/I	d Å a	I/I	d Å b	I/I	d Å c	I/I
100	12.36	vs	12.30	10	12.20	10	–	
							9.75	4
110	8.77	s	8.71	7	8.30	8	–	
111	7.15	s	7.11	4	7.00	9	–	
200	6.35	w	–		6.25	4	–	
210	5.52	s	5.51	3	5.40	3	5.00	5
211	5.03	w	–		5.00	3b	–	
220	4.36	m	–		–		4.30	6
300	4.10	s	4.10	4	4.00	6	4.19	10
211								
310	3.90	w	–		–		3.85	9
311	3.71	s	3.71	5	3.65	8	–	
320	3.41	m	–		3.48	2	3.25	8
321	3.29	ms	3.29	5	3.20	6	3.00	1
400	3.07	vw	–		–		–	
410	2.98	vs	2.99	6	–		2.98	9
322								
411	2.90	m	–		2.92	6	–	
330								
420	2.75	m	–		–		–	
421	2.68	mw	–		–		–	
332	2.62	s			–		–	
422	2.51	mw	–		2.55	6b	2.56	2
431	2.41	w	–		2.48	2	–	
510								
511	2.37	mw	–		2.37	2	–	
333					2.33	8	–	

(continued)

Table 17.1 (continued)

hkl	d Å Q Series	I/I	d Å a	I/I	d Å b	I/I	d Å c	I/I
	2.75	w	–		2.28	5	–	
440	2.17	m	–		2.15	2	–	
	2.14	mw	–		2.10	2	–	
433	2.11	m	–		–		–	
530								
531	2.08	mw	–					
600	2.05	m	–		2.06	8b	2.02	1
442								
660	2.05	m	–		2.06	5b	2.02	1
442								
					1.92	2b		
					1.87	2		
					1.82	w		
					1.68	2vb		
					1.64	1		
					1.60	2		
					1.58	7b		
					1.54	2b		
					1.42	2b	1.46	3vb
					1.37	3b	1.42	2vb
					1.30	2b	1.26	2

Table 17.1 X-ray diffraction data of zeolite phases obtained from heated Bhandak Kaolinite at different sets of conditions (after Chakraborty)

hkl	d Å P Series	I/I	d Å d	I/I	d Å e	I/I	d Å f	I/I	d Å g	I/I	d Å h	I/I
									8.75		5	
110	7.10	s	–		7.25	8	–		–		7.08	4
	–		6.48 wb		–				6.40	1	–	
200	5.00	m	–		–		–		–		–	
					4.43	3					4.89	3
112	4.1	s	4.20	wb	4.38	4	–		4.40	6vb	4.07	6
							3.98	3	–		–	
							3.45	8	–		–	
			3.50	5	3.80	2	3.20	3	3.60	6b	–	
220	3.45	vw	3.30	5	3.56	5	3.16	3	–		–	
310	3.18	vvs	3.25	w	–				3.25	2	3.14	10
			3.00	5	3.00	4	3.00	2	–		3.04	
	–		–		–		–		2.95	3	–	
222	2.90	vw	–									
321	2.68	s									2.66	6
400	2.52	vw	2.56	3b	2.55	8	2.53	7	2.57	7	2.50	1
410	2.46	vw	2.47	2b	2.47	3	–		2.55	3b		
332												
411	2.36	w			2.32	4	2.36	7	2.36	1	2.36	1
204	2.23	vw	2.25	3b	2.27	2	2.26	wb	–		–	
			2.15	2	–		2.15	1	–			
			2.10	3	–		2.10	w				
			2.07	10	2.02	1	–					
	–		–		2.07	10	2.02	1	–			
			1.98	5b	1.98	4	–		1.97	1b	1.96	1

(continued)

Table 17.1 (continued)

hkl	d Å P Series	I/I	d Å d	I/I	d Å e	I/I	d Å f	I/I	d Å g	I/I	d Å h	I/I
			1.88	2vb	1.88	3	–		1.88	2		
			–		1.82	w	–					
			–		1.72	7		1.71	1			
			1.68	3b	1.65	w	–		1.67	1b	1.67	1
			1.57	3b	–		1.58	5b	–			
			1.53	w	1.53	w	–					
			1.51	wb	1.51	w	1.51	w	–			
			–			1.47	3b	–				
			1.37	10vb	1.38	10vb	1.39	9	1.39	3vb	–	
–		–		1.36	10	–						
			–		–		1.22	5b	–			

vs very strong, *s* strong, *ms* medium strong, *m* medium, *vw* very weak, *w* weak
a Sodalite
b Metakaolinite heated to 750 °C/2 h and leached with 10 wt% NaOH for 2 h at boiling water bath showing *Q*-series zeolite
c Metakaolinite treated with 10 wt% NaOH for 7 days and then analyzed for DTA up to 1,000 °C
d Bhandak Kaolinite heated to 1,000 °C/4h and alkali leached for 1 h as per Hisimato and Jackson showing *P*-series zeolite
e Bhandak Kaolinnite heated to 1,000 °C/4 h and alkali leached for 2 h as per Hisimato and Jackson showing χ- and κ-alumina in addition to *P*-series zeolite
f Alkali leached residue left after leaching as per the procedure "e" was heat treated to 1,200 °C/1 h showing α-alumina and sodium aluminosilicate phase
g bhandak Kaolinite heated to 1,000 °C/4 h and treated with 10 wt% NaOH at boiling water bath for >40 min showing *Q*-series zeolite
h Georgia kaolinite heated to 1,000 °C/3 h treated with 5 wt% NaOH at boiling water bath for 60 min showing *P*-series zeolite

formation is evident when washed alkali soaked mass was heated (Table 17.1). These two experiments further demonstrated that SiO_2 component of metakaolinite could not dissolve in cold or in hot and concentrated NaOH solutions. It was taken for granted that SiO_2 and Al_2O_3 are not present in free state in metakaolinite as conjectured by previous authors.

17.3.2 Leaching Behavior of Kaolinite Heated to 1,000 °C in Alkali Solution

It is seen from the figures above that the amount of silica (A) and alumina liberated from different heat-treated kaolinite samples is very small in amounts and change very slightly with increase of leaching time, i.e., insignificant amounts of SiO_2 and Al_2O_3 can be extracted with 10 wt% NaOH solution from metakaolinite structure. These results are quite good in agreement with that are shown earlier by Hyslop (1944). Thus during formation of metakaoline, silica (A) is not liberated. The Si–O–Al linkage remains intact and it releases silica only at the exothermic peak temperature as observed.

Figure 17.1b appears to consist of two portions. The initial step portion of the curve, a portion of silica is extracted rapidly as in the case of free silica (A). About 30 wt % silica was extracted within 30–40 min, whereas very little alumina was detected. By considering the large amount of silica which was readily extracted from the 980–1,000 °C heated kaolinite by alkali in comparison to the very insignificant amount of alumina solubilized, it is concluded that free silica (A) is present in 980–1,000 °C heat treated kaolinite (Fig. 17.1c). Otherwise, the amount of alumina like silica extracted might be expected to increase with time if silica part were extracted from aluminosilicate compound. Flat/horizontal portion of the curve, for leaching time more than 40 min, the amounts of both silica and alumina extracted increase very slowly. The slow extraction of both phases might be attributed to the attack of aluminosilicate compound and or removal of last traces of silica by alkali. The residual compounds may be Al–Si spinel, mullite, and aluminosilicate (A). Now the question is: What should be the actual quantity of free silica present in 980 °C heat treated kaolinite? It can be said that when alkali is added to 980 °C heated kaolinite, both silica as well as alumina are dissolved as detected. But it is further shown that within the leaching period, e.g., 30–40 min duration of leaching condition, the portion of silica preferentially dissolved is about 35–37 % without effecting the cubic spinel phase and mullite phase as revealed by XRD analysis of leaching residues. Therefore, it is predicted that the amount of free silica present will be 35–37 wt% and remaining silica, i.e., 18 wt% is designated as bonded to alumina. Lastly, it is also revealed that metakaolinite does not liberate any silica and alumina as amorphous oxides. It however discards silica only during the occurrence of 980 °C exotherm. Thus, by performing the alkali leaching experiments, it is considered that kaolinite at 980 °C exotherm releases approximately 35–37 wt% silica (A) along with the formation of spinel and weakly crystalline mullite phase. They carried out chemical analysis of the leached residue and studied the phase transformation behavior of it on further heating followed by XRD analysis.

The chemical composition of the leached residue corresponds to the composition of 3:2 mullite. On X-ray analysis of it shows (pronounced) XRD peaks due to spinel phase and some weakly crystalline mullite phase. Therefore, the portion of the bonded SiO_2 is present in tetrahedral part of the spinel structure. Accordingly, it is concluded that the intermediate spinel phase is a Si bearing spinel rather than simple γ-Al_2O_3 and it is of the composition analogous to 3:2 mullite. The remaining portion of the bonded silica is connected with alumina and this amorphous material also corresponds to 3:2 mullite in composition. Thus, it is designated as mullite (A).

17.3.3 Leaching Behavior of Kaolinite Heated Beyond 1,000 °C with Alkali Solution

It was noted that no further silica is extracted even when kaolinite is heated beyond first exothermic peak temperature.

17.3.4 Comparison of Alkali Leaching Data of Various Authors

17.3.4.1 IR Study

Percival et al. (1974) reported that ~67 % of SiO_2 (A) was removed from kaolinite heated to 950 °C/24 h by the NaOH extraction for 2.5 min using the procedure of Hashimoto and Jackson (1939). On analyzed by IR, they reported that the characteristic band due to free silica at 472 cm^{-1} in heated kaolinite decreased on alkali leaching. By leaching twice for 10 min, the band at 472 cm^{-1} disappeared and they conjectured that virtually all SiO_2 (A) was removed and the remaining bands corresponded to γ-Al_2O_3 formation other than Al–Si spinel suggested by Brindley and Nakahira (1959) (Fig. 17.2).

17.3.4.2 EDS Study

Okada et al. (1986) first measured the composition of spinel phase by EDS study of alkali leached minerals namely Imogolite, Halloysite, Kaolinite, and Dickite fired for 24 h at 950 °C. Alkali leaching was done by using 7 wt% NaOH solution

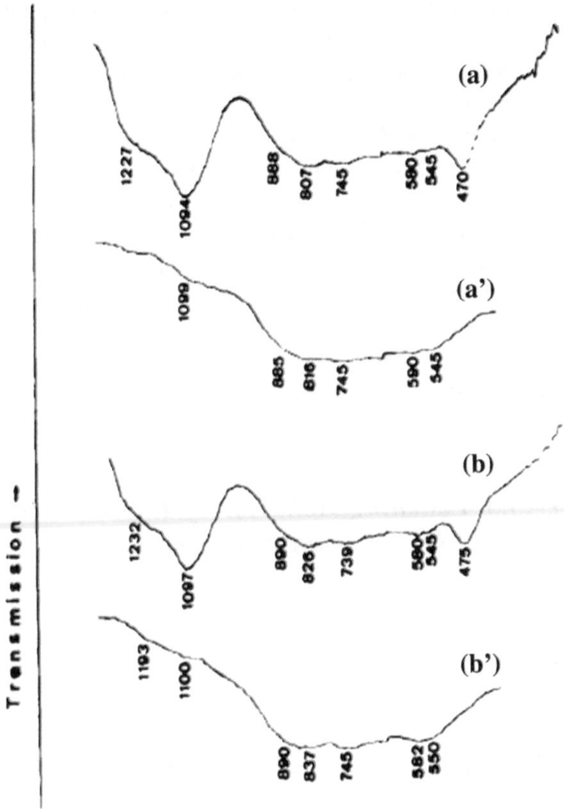

Fig. 17.2 Infrared spectra of leached fired kaolinite. **a** 950 °C for 24 h; **a′** as in **a**, leached twice; **b** 1,050 °C for 24 h; **b′** as in **b**, Leached twice (after Percival et al. 1974). Reprinted by permission of the American Ceramic Society

by boiling them for 40 min. X-ray diffractograms of fired halloysite and kaolinite before and after NaOH treatment show that the halo around 20–22° 2θ due to an amorphous material was almost absent in fired halloysite, but was still noticeable in fired kaolinite. This disappeared completely by repeating NaOH treatment twice. TEM photographs of fired and NaOH-treated kaolinite after leaching were taken, the particles were revealed to be spinel phase by electron diffraction and their sizes were as large as 10 nm. Chemical composition of spinel phase in the fired and NaOH-treated specimens obtained by analytical TEM is around 8 wt% of silica which corresponds to approximately the $SiO_2.6Al_2O_3$ ($Si_{1.6}$ $Al_{19.2}$ 0_{32}) composition. According to Okada et al. extraction of amorphous material was, however, insufficient in fired kaolinite treated by NaOH solution only once. The samples show data were observed around $2SiO_2.3Al_2O_3$ composition, a result compatible with Chakraborty and Ghosh's data. However, specimens treated twice by NaOH solution had compositions approaching to 8 wt% SiO_2.

17.3.4.3 High Resolution Electron Microscopy

Sonuparlak et al. (1987) physically isolated spinel phases by dissolving the surrounded amorphous phase by NaOH leaching and then directly made the composition analysis of spinel phase by EDS in analytical TEM. They chose Georgia kaolinite heat treated for 7 days at 850 °C treated with 10 wt% boiling NaOH at different duration of time. Figure 17.3 illustrates the results of the analysis where the amounts of silica determined directly by EDS are plotted with respect to leaching times.

The plot indicates that at the extended leaching time (>25 min), the rate of leaching of silica approaches zero below 10 wt% silica. They conjectured that the spinel crystallites remaining intact. On prolonged leaching, the amount of silica continuously decreases and the residue would be nearly γ-Al_2O_3. In evidence, they took electron micrographs of spinel crystallites in a high resolution electron microscope obtained out of the suspension of the edge of the powder. In support, they showed data of these EDS crystallites which displayed only Al peaks and very negligible Si peak.

Srikrishna et al. (1990) took electron diffraction pattern of alkali leached kaolinite after heat treatment at 1,000 °C and showed the micrograph and diffraction pattern. Energy dispersive X-ray spectroscopy revealed the spinel phase to have a composition very close to that of mullite (Fig. 17.4).

Jantzen (1990) used high dissolution temperature ($\sim$103 °C) and showed the partial crystallization when c-glass filters treated with NaOH for 2 days to Linde B, a P-type zeolite (Table 17.1). She compared this leaching behavior of c-glass with that of heated kaolinite as shown by Chakraborty and Ghosh. It was pointed out that both P- and Q-type zeolites were formed during leaching heated kaolinite at water bath as well as in boiling condition (Table 17.1). The SiO_2 content in the two zeolites are different and second stage of leaching indicated a slow dissolution of silica. With these two points, she questioned the applicability of alkali leaching technique for measuring SiO_2 quantitatively.

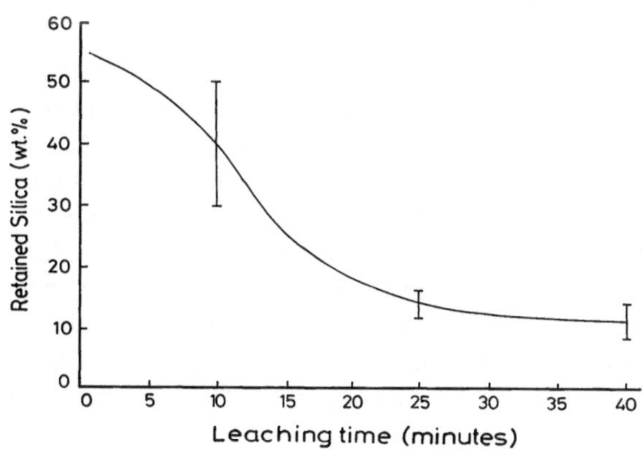

Fig. 17.3 The amount of silica left in the spinel containing specimen as a function of leaching time. The wt % silica was calculated from the amount of Si measured by using EDS analysis in the TEM (after Sonuparlak et al. 1987). Reprinted by permission of the American Ceramic Society

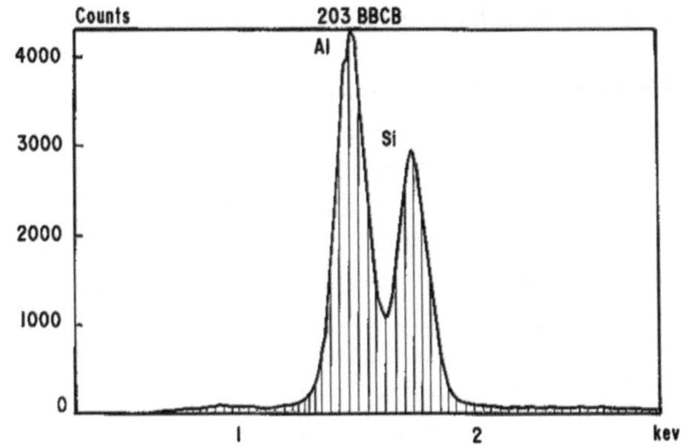

Fig. 17.4 EDX spectrum showing the presence of alumina and silica in the spinel phase (after Srikrishna et al. 1990)

17.3.5 Different Conditions of Alkali Leaching Processes used in Estimating Exolved Silica and Their Comparisons

Blasco et al. (1990) set up a reactor for studying the reaction kinetics of NaOH with heated kaolinite with a view to answer two questions namely: (i) Whether the nature or characteristics of kaolinite has any influence on its reactivity with NaOH after its calcinations. (ii) Whether the nature of phases formed on heating kaolinite at ~980 °C by way of changing maximum temperature of heating cycle or heating rate have any relation with dissolution of Al and Si components by NaOH. They noted a quantitative dependence between the reaction rates with initial surface area of kaolin for Si dissolution is concerned. Secondly, one of the important process variables, e.g., the heating cycles also affect the reactivity of the components with NaOH. It is observed that initial reaction rate increases with the specific surface area of the starting fired material for Al and Si, respectively.

Alkali leaching process was first standardized by Chakraborty and Ghosh (1978a, b). This technique was repeated subsequently by various researchers and they obtained different results. Chakraborty (1992) carried out detailed leaching studies and tried to clarify others results.

The major objectives of alkali leaching technique are to demarcate between free silica and bonded silica bearing phases present in 980 °C heat-treated kaolinite. Reactivities of those two forms of silica with NaOH generally vary to a different degree and those are dependent mainly upon the following parameters: (i) Concentration of NaOH solution, (ii) temperature of leaching bath, (iii) time of leaching period, and (iv) speed of stirring the solid with leachent.

By varying both temperature/time of leaching conditions, dissolution behaviors of both silica and alumina are shown to be different. Table 17.2 and Fig. 17.5 briefly show that there are two stages of leaching in each case of leaching condition. The free silica (A) part of 980 °C heated kaolinite dissolves very rapidly in

Table 17.2 Bhandak Kaolinnite heated to 1,000 °C/4 h and alkali leached at four different conditions (After Chakraborty)

Leaching condition	First stage of leaching/ predominant reaction	Second stage of leaching/predominant reaction	Third stage of leaching at ~2 h period
1. Drastic leaching condition			
(a) As per Hisimato and Jackson, 100 mg sample boiled in 100 ml of 0.5 N NaOH	(a) Less than 20 min	(a) Forms *P*-series Zeolite	(a) Zeolite phase, χ- and κ-alumina. Leached residue forms α-alumina on heating
(b) 10 wt% NaOH/ in direct boiling condition ~102 °C	(b) Less than 20 min	(b) Zeolite phase as above	(b) – do –
2. Less drastic leaching condition 10 wt% NaOH/at boiling water bath/occasional swirling	30–40 min/ Solubilization of siliceous phase occurs	beyond 50 min/ (i) Intensity of spinel increases (ii) *Q*-series zeolite starts forming (iii) Dissolution of alumina takes place	No new alumina phases form
3. Mild leaching condition 5 wt% NaOH solution/at boiling water bath/ intermittent swirling.	50–60 min	(i) Intensity of spinel first increases then attains constant at ~120 min of leaching (ii) Zeolite formation won't occur on leaching for 100 min (iii) Alumina dissolves very slowly	Zeolite phase starts slowly
4. Less mild leaching condition	10–15 min	(i) beyond 20 min. spinel enriches. Leached residue attains composition of 3:2 mullite	Not require
5. 5 % NaOH/at boiling water bath/continuous stirring		(ii) No zeolite formation even on 1 h of leaching (iii) Alumina dissolves very slowly	

hot NaOH solution in the first stage. In the second stage, bonded silica dissolves slowly on continued leaching along with slow dissolution of alumina either from amorphous or crystalline alumina silicate compound present. Heated kaolinite is reacted with NaOH solutions at four leaching conditions.

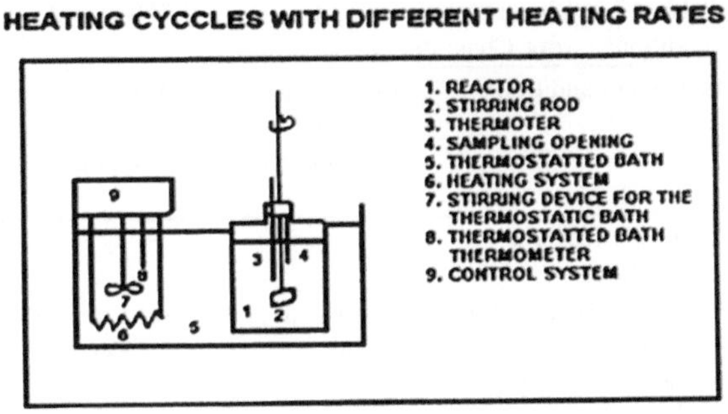

Fig. 17.5 Alkali leaching set up of heat-treated kaolinite (after Blasco et al. 1990)

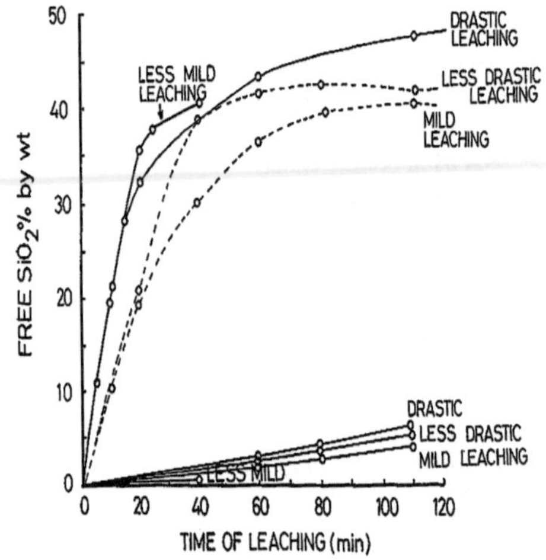

Fig. 17.6 Amounts of SiO_2 and Al_2O_3 extracted from Bhandak kaolinite which had been heated to 980 °C by four different leaching conditions

(i) **Drastic Leaching Condition.** In this case, dissolution was done at direct boiling temperature as per Hashimoto and Jackson (1960). During leaching process, free silica (A) dissolves readily at or within 20 min when spinel phase is concentrated as evidenced by the corroborative X-ray study (Fig. 17.6).

At 1 h of leaching time, aluminosilicate materials, e.g., weakly crystalline mullite is prone to attack with the formation of Q-type zeolite (Table 17.1d) followed by slow dissolution of silica. With further leaching period of 2 h weakly crystalline spinel phase is attacked with the formation of x- or κ-Al_2O_3 (Table 17.1e), P-type zeolite, and more dissolution of bonded silica. Thus on continued leaching, Q-type zeolite was converted to P-type as per the following equation stated by Break (1974).

$$\underset{\text{Zeolite–A}}{5[Na_{12}(AlO_2)12(SiO_2)_{12}].27H_2O} \xrightarrow{\text{NaOH (aq.)}} \underset{\text{Zeolite–P}}{Na_6[(AlO_2)6(SiO_2)_{12}].14H_2O} + 24NaAlO_2 \quad (17.2)$$

No marked increased in the solubility of alumina is noted during continued leaching. On the other hand, the formations of newly formed alumina polymorphs indicate the following dissolution reaction:

Kaolinite —980°C→ Al – Si spinel + Mullite (weakly crystalline) + Mullite (A) phase + SiO_2 (A) —NaOH at boiling condition→ Al –Si spinel + Mullite (A phase Q -- Type zeolite

↓ Continued leaching at boiling condition

Al – Si spinel + Mullite (A) phase + P – Type zeolite + x or κ – Al_2O_3

-----17.3

The formations of newer forms of alumina along with greater amounts of silica dissolution indicate that $[SiO_4]^{-4}$ groups at the Al–Si spinel lattice are being replaced by OH groups during drastic alkali treatment. Formation of these alumina phases are subsequently proved by XRD studies of heat-treated (1,200 °C/1 h) alkali leached residues of different time periods. Residue obtained after 5 min of leaching shows the formation of mullite and some cristobalite. At 15 min of leaching mullite only forms. At 20 min of leaching, mullite and minor amount of corundum phases form. At extended leaching time, XRD intensities of mullite reflections decrease and the intensities of reflections of corundum increase gradually. At 2 h leaching, the heated alkali leached residue does not form any mullite but shows corundum and sodium aluminosilicate phases due to the transformation of alumina phase and zeolitic phase formed earlier during drastic alkali leaching process (Table 17.1f).

(ii) **Less Drastic Leaching Condition.** In this condition, silica dissolution is done at boiling water bath instead of direct heating to boiling temperature as already discussed. At the first stage of leaching lasted for 30–40 min. Thereafter, both silica and alumina are dissolving into alkali solution as silicates and aluminates by the etching of aluminosilicates. On further leaching, both heat treated Bhandak and Georgia kaolinites show the formation of additional Q-type zeolite phase (Table 17.1g).

(iii) **Mild Leaching Condition.** Table 17.2 shows that the time of dissolution of free silica (A) increases from ~15 min in direct leaching condition (i) to ~40 min in less drastic leaching process, (ii) to 50–60 min in mild leaching process. But the other reaction i.e., dissolution of alumina is found to have reduced considerably which is most desirable. XRD analysis of alkali-leached residues for different duration of leaching time shows the following observations:

1. The amorphous band in the region 22° 2θ Cuk$_\alpha$ decreases with increase of leaching time. These indirectly suggest that the associated free silica (A) is going into alkali solution. At ~50–60 min of leaching time, the amorphous band almost disappears. This time is taken to be the first indication of the end-point of the SiO_2 (A)–NaOH interaction.
2. Al–Si spinel gives two characteristic intense bands in XRD recordings, one at ~47° 2θ and the other at 67° 2θ with Cuk$_\alpha$ radiation. The intensities of both these bands increase with increase of leaching time from 0–10 to 20–40 to 50 min and then attains more or less a constant value. The leaching time of ~50 min time is also be taken as a second indication of the end-point of SiO_2 (A)—NaOH reaction. These results suggest that there is no necessity for continued leaching beyond this specific leaching time say 50 min.

Another question arises regarding the solubilization of SiO_2 in alkali followed by the enrichment of Al–Si spinel concentration: Whether the hot alkali solution reacts with Al–Si spinel and mullite (A) phases present in 980 °C heated kaolinite?

To reveal it, the 60-min alkali leached mass was further alkali treated with 10 wt % NaOH solution in a hot water bath for additional duration of time. XRD analysis of this alkali leached residue shows the following observations.

(i) The areas of the peaks corresponding to Al–Si spinel phase at 67.2 and 47° 2θ Cuk$_\alpha$ now decreased with increase of leaching time. This indicates that the spinel phase was attacked by NaOH and further indicates the futility of the leaching process.

(ii) Peaks corresponding to zeolitic phase gradually develops as noted by a gradual increase of it intensities from 20 to 40-min time. At 60-min leaching time zeolite phase was fully developed. Chemical analysis of the leached residue showed an increase in the concentration of Na_2O (SiO_2 = 14.59 %, Al_2O_3 = 62.95 %, TiO_2 = 6.37 %, Fe_2O_3 = 0.54 %, MgO = 1.53 %, CaO = 2.88 %, K_2O = 0.08 %, Na_2O = 4.37 %, and LOI = 6.37 %) in contrast to very little concentration of Na_2O present originally in heated Georgia kaolinite itself (SiO_2 = 50.75, Al_2O_3 = 42.93 %, TiO_2 = 4.02, Fe_2O_3 = 0.46 %, MgO = trace, CaO = 0.51 %, K_2O = 0.04 %, Na_2O = 0.08 %, and LOI = 0.85 %) (Fig. 17.7).

The gradual diminution of amorphous band due to solubilization of silica (A) in NaOH as noted by Okada et al. (1986) is verified by the XRD analysis of leached residues. The remaining amorphous band which dissolves rather slowly accounts

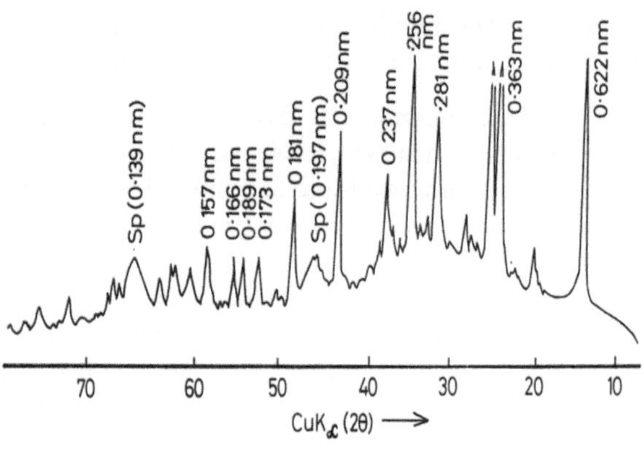

Fig. 17.7 Shows the XRD reflection of the zeolitic phase formed during NaOH leaching of georgia Kaolinite heat treated to 1,000 °C (after Chakraborty and Ghosh 1989, Chakraborty 1992). Reprinted by permission of the American Ceramic Society

Table 17.3 d(nm) Values of zeolitic phase formed during NaOH leaching of Georgia Kaolinite heat treated to 1,000 °C (after Chakraborty 1992). Reprinted by permission of Springer

d(nm)	I/I_0
0.622	80
0.444	30
0.363	100
0.313	5
0.281	60
0.256	80
0.237	30
0.209	80
0.198	5
0.188	5
0.181	30
0.173	40
0.163	5
0.157	30
0.148	30
0.144	30
0.140	5
0.137	30
0.131	30

for the presence of an aluminosilicate phase (A). The major advantage noticed in this mild leaching process is that the remaining mullite (A) phase is found to attack to a lesser extent than in two previous cases. Since, intensity of spinel increases gradually with leaching time and then attains constant. With change of leaching condition from mild to drastic in nature, intensities of Al–Si spinel phase increase due to liberation of new x- and κ-Al_2O_3 phases (these phases have common XRD peaks). On continued leaching period, intensities decrease again with increase in the formation of zeolitic phase (Table 17.3).

XRD analysis of heat treated alkali leached residue shows the development of mullite, corundum, and cristobalite depending upon their previous leaching conditions. The relative development of phases on heat treatment at 1,200 and 1,300 °C of different leached residues shows the following observations:

1. *Formation of corundum*: Traces of α-Al_2O_3 are detected during heating sample leached for 1 h at both 1,200 and 1,300 °C by the appearance of 0.208 nm peak in XRD recordings. More and more development of corundum is observed on heating the alkali-leached residues collected with increase of leaching time. From these observations, the end-point of alkali–silica reaction is predicted to be less than 60 min. Beyond this time, alkali leaching leads to the attack of both Al–Si spinel and mullite (A) phases with liberation of an aluminous phase from which corundum develops on further heating.
2. *Cristobalite formation*: The amount of its formation decreases with increase of leaching time. This shows that free SiO_2 (A) is going into alkali solution during leaching process. The residue left after leaching ~40 min time on further heating just shows the absence of 0.404 nm peak of cristobalite. These results suggest that complete removal of SiO_2 (A) takes place within 40–50 min of leaching time.
3. *Mullite formation*: It is observed that the amount of mullite developed on heating the leached residue increased with increase of leaching time up to 40-min duration, after which it decreased even on further leaching. Therefore, roughly 40–50 min time is enough to leach out the associated free SiO_2 (A) present in Georgia kaolinite heat treated to 980 °C. Furthermore, on comparing the mullite formation of alkali leached residues at two different temperatures, namely 1,200 and 1,300 °C, it was observed that the amount of mulllite so developed on firing at 1,200 °C decreased on further heating to 1,300 °C. It is noted previously that with an increase of leaching time, gradually more and more Na_2O is introduced into the residue and it hinders mullite development or it facilitates conversion of mullite into glass.

(iv) **Less Mild Leaching Condition.** Instead of occasional swirling as done in previous cases, in this case continuous stirring was provided by use of electric stirrer installed into the leaching bath. As a result, extraction of SiO_2(A) takes place within 15–20 min (Fig. 17.8).

The question of attack of NaOH on mullite (A) phase is largely eliminated even on 1 h of leaching. Neither zeolite forms nor much of Al_2O_3 dissolves into alkali solution. Since no P-type or even Q-type zeolite forms, it is believed that free silica is more selectively leached out at the first stage of leaching and thus serves the objective of leaching process.

The nature of the curve (wt % SiO_2 vs. leaching time) for different residues was found to be analogous to the direct compositional analysis by EDS in an analytical TEM used by Sonuparlak et al. (1987). The dissolution SiO_2 (A) under these continuous stirring conditions was found to be comparatively more rapid than in the leaching procedure followed previously. From the extraction curve Fig. 17.3,

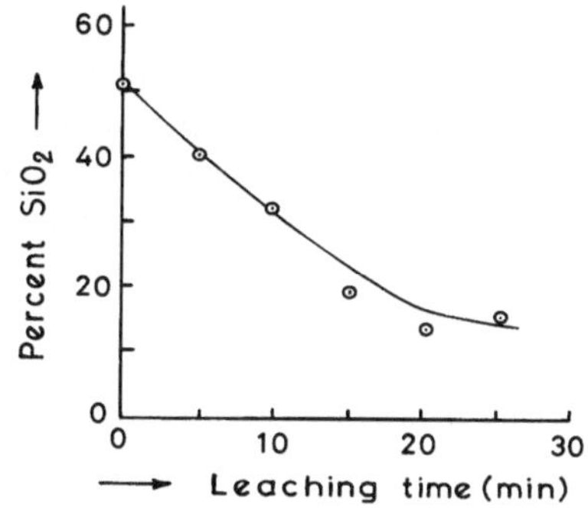

Fig. 17.8 Extraction of silica in mild leaching condition

they suggested that at extended leaching times ~25 min the rate of leaching of SiO_2 approached zero, below 10 wt % SiO_2. Accordingly, they concluded that the spinel contained not more than 10 wt % SiO_2. Question is: What was the justification or reason behind the choice of 25 min leaching time as the end-point of SiO_2 (A) –5 % NaOH solubilization reaction? To answer the above question, the crystallization behavior of alkali-leached residue was studied by heat treatment followed by XRD analysis.

(i) *Cristobalite formation* : The development of cristobalite during heating of the leached residue is a direct index of the presence of free SiO_2 (A). Accordingly, alkali-leached samples collected at different durations show a gradual diminution in their cristobalite content on heating. On heating the residue to ~1,200 °C/1 h, the 0.404 nm peak height decreases from its maximum peak height and disappears completely at ~15 min leaching. Therefore, the amount of SiO_2 leached out in the time interval of 10–15 min is considered as free SiO_2 present in 980 °C heated Georgia kaolinite. In other words, the amount of SiO_2 present in the leached residue after the above time interval is regarded as chemically bonded SiO_2. The composition of the residues in between the leaching period encompasses the composition of mullite. In this way, the amounts of free SiO_2 and bonded SiO_2 present in 980 °C heated clay have been differentiated. The percentage of SiO_2 leached out beyond 15-min time could be accounted for by the breakdown of both Al–Si spinel and mullite (A) phases. To corroborate the above findings, the second procedure was followed.

(ii) *Mullite formation* : Mullite forms from Al–Si spinel and mullite (A) phases respectively as suggested by Chakraborty and Ghosh (1991). The accumulation of mullite during heating of leached residue is an indirect index of the presence or absence of free SiO_2. For example, the concentration of mullite increases on heating the residues collected after 5–10 min and lastly up to 15-min leaching time; after that it decreases. This result definitely indicates that with removal of SiO_2, the concentration of mullite has almost attained a maximum within 15-min leaching time. Afterwords, with increase of time, SiO_2 further exolves out by break up of both the aluminosilicates and ultimately the formation of mullite decreases. Therefore, a leaching time not exceeding 15 min may be regarded as the end-point of alkali–SiO2 (A) reaction in this complex situation. The advantage of this leaching technique is that zeolitic type phase formation is completely eliminated. A wt% of Na_2O in the leached residues as found out by chemical analysis is as follows: Zero min –0.08 wt%, 5 min –0.37 wt%, 10 min –0.57 wt%, 15 min –0.84 wt%, etc. The increment in the Na_2O concentrations in comparison to the other leaching process is much less. Secondly, as per expectation, the alkali SiO_2 reaction is quite predominant but there remains the possibility of the attack of the Al–Si spinel phase. Nevertheless, the percentage of SiO_2 in the Al–Si spinel phase has been confirmed to be ~28 wt% unlike the value of 10 wt% obtained by both Sonuparlak et al. (1987) and Okada et al. (1986).

17.4 Summary

It is believed that kaolinite liberates silica on heating. It is worth while (i) to detect and estimate the amount of silica liberated during decomposition of metakaolinite and (ii) to study how 980 °C phase freed from exolved silica transforms on further heating. Two leaching processes namely use of pyrocatechol, and use of Na0H solution have been employed. Weiss et al. claimed the quantity of free silica would be ~14 %.

Chakraborty and Ghosh (1978) carried out alkali leaching study in a methodical way and showed two observations:

1. Silica and alumina components of metakaolinite do not dissolve in hot NaOH solution. Instead metakaolinite gains in weight. During heating in water bath a zeolitic phase is formed.
2. Approximately, 35–37 % silica and ~5 % Al_2O_3 have been dissolved in alkali solution during leaching kaolinite heated to 980 °C. It is predicted that metakaolin decomposes to silica rich alumina phase. Secondly, it is noted that the composition of alkali-leached residue is analogous to composition of 3:2 mullite. Later on *Chakraborty* compared leaching behavior of siliceous phase at different conditions and noted different side reaction of heated kaolinite with hot alkali e.g., formation of insoluble aluminous phase out of silica rich aluminous phase and formation of zeolitic phase out of other coexisting aluminosilicate phases (A).

Chakraborty (1992) pointed out different ways of determining end-point of alkali-free silica dissolution of heated kaolinite and finally standardized a less mild leaching condition which is found to be most effective to extract siliceous phase without reacting too much with associated mullite (A) phases.

References

A. Blasco, A. Barba, F. Negre, A. Escardino, Obtaining materials with a high specific surface area and variable Al/Si atomic ratio from Kaolin, I–Influence of the nature of the kaolin and the heating cycle to which it is subjected. Ceram. Trans. J. **89**, 28–31 (1990)

D.W. Breck, *Zeolite Molecular Sieves*, (John Wiley & Sons, New York, 1974)

G.W. Brindley and M. Nakahira, The Kaolinite-Mullite reaction series: I, II, and III. J. Am. Ceram. Soc. **42**(7), 311–324 (1959)

A.K. Chakraborty, D.K. Ghosh, Comment on the interpretation of the Kaolinite–Mullite reaction sequence from infra-red absorption spectra. J. Am. Ceram. Soc. **61**(1–2), 90–91 (1978a)

A.K. Chakraborty, D.K. Ghosh, Re-examination of the kaolinite to mullite reaction series. J. Am.Ceram. Soc. **61**(3–4), 170–173 (1978b)

A.K. Chakraborty, D.K. Ghosh, Comment on spinel formation during 980 °C exothermic reaction in the Kaolinite–Mullite reaction series. J. Am. Ceram. Soc. **72**(8), 1569–1570 (1989)

A.K. Chakraborty, D.K. Ghosh, Kaolinite–Mullite reaction series. The development and significance of a binary aluminosilicate phase. J. Am. Ceram. Soc. **74**(6), 1401–1406 (1991)

A.K. Chakraborty and D.K.Ghosh, Supplementary alkali extraction studies of 980°C heated-kaolinite by X – ray diffraction analysis. J. Mater. Sci. **27**(8), 2075–2082 (1992)

N.J. Eremin and M. Erusdimskii, Thermal transformation of kaoline studied by chem. Method. Zh. Prikikhim, 42(3) Kmchem. Abstr. 71(4) 211, Abstr. 15592j , July, 497–501 (1969)

I. Hashimoto and M.L. Jackson, in *Clays and Clay Minerals* vol. 10. ed. bv A. Swineford. (Pere-a mon Press. Inc., New York, 1960) pp. 102–73

J.F. Hyslop, Decomposition of clay by heat. Trans. Brit. Ceranz. Soc. **43**(3), 49–51 (1944)

C.M. Jantzen, Formation of zeolite during caustic dissolution of fibre glass : Implications for studies of kaolinite to mullite transformation. J. Am. Ceram. Soc. **73**(12) 3708–11 (1990)

K. Okada, N. Ostuka, J. Ossaka, Characterization of spinel phase formed in the Kaolin–Mullite thermal sequence. J. Am. Ceram. Soc. **69**(10), C-251–C-253 (1986)

H.J. Percival, J.F. Duncan, P.K. Foster, Interpretation of the Kaolinite–Mullite reaction sequence from infrared absorption spectra. J. Am. Ceram. Soc. **57**(2), 57–61 (1974)

K. Srikrishna, G. Thomas, R. Martinez, M.P. Corral, S. Aza, and J.S. Moya, Kaolinite–mullite reaction series: A TEM study. J. Mater. Sci. **25**, 607–12 (1990)

B. Sonuparlak, M. Sarikaya, I.A. Aksay, Spinel phase formation during the 980 °C exothermic reaction in the Kaolinite–Mullite reaction series. J. Am. Ceram. Soc. **70**(11), 837–42 (1987)

F. Vaughan. Energy changes when kaolin minerals are heated. Clay Mineral Bull.**2**(13), 265–274 (1955)

A. Weiss, K.J. Range, J. Russow, The Al, Si-spinel phase from Kaoinite (isolation, chemical analysis, orientation and reactions to Its low-temperature precursors), in *Proceedings of the International Clay Conference*, Tokyo, 1969, vol. 2 (Israel Universities Press, 1970), pp. 34–37

References

Chapter 18
Various Ways of Characterization of Alkali Leached Residue

18.1 Heat Treatment of the Leached Residue Followed by XRD Study

Chakraborty and Ghosh (1978a, b) showed that on heat treatment of alkali leached residue at 1,250 °C exotherm, the XRD reflections of spinel disappear with concomitant increase in the intensities of mullite reflections as shown in equation below:

$$\begin{aligned} \text{Kaolinite} \xrightarrow[\text{therm}]{\text{Endo}} \text{Metakaolinite} \xrightarrow[\text{exotherm}]{980\,^\circ\text{C}} & \ \text{Spinel phase} \\ & + \text{Mullite(A)} \\ & \quad \text{phase} \\ & + \text{SiO}_2\text{(A)} \\ \xrightarrow[\text{leaching}]{\text{NaOH}} \begin{array}{c}\text{Spinel phase} \\ + \text{Mullite(A)} \\ \text{phase}\end{array} \xrightarrow[\text{exotherm}]{1250\,^\circ\text{C}} & \ \text{Mullite phase} \end{aligned} \tag{18.1}$$

This result shows that alkali leached residue is a spinel comprised of silica and alumina whose structural transformation is responsible for the formation of mullite at 1,250 °C exotherm. In this mullitization process, neither silica liberates out nor any silica reacts with spinel phase. It may be a case of polymorphic transformation other than solid state reaction between γ-Al_2O_3/SiO_2(A) to form mullite. It is concluded that essential continuity, i.e., Si–O–Al linkages from metakaolinite to spinel and finally to mullite prevails during transformation of kaolinite.

18.2 MAS NMR

Mackenzie et al. (1996a, 1999) carried out the MAS NMR studies of kaolinite heated to 1,000 °C and leached to different duration of times. XRD trace of unleached Zettlitz kaolinite heated to ~1,000 °C contains a significant proportion of amorphous material, evidenced by the broad base line hump centered at about

A. K. Chakraborty, *Phase Transformation of Kaolinite Clay*,
DOI: 10.1007/978-81-322-1154-9_18, © Springer India 2014

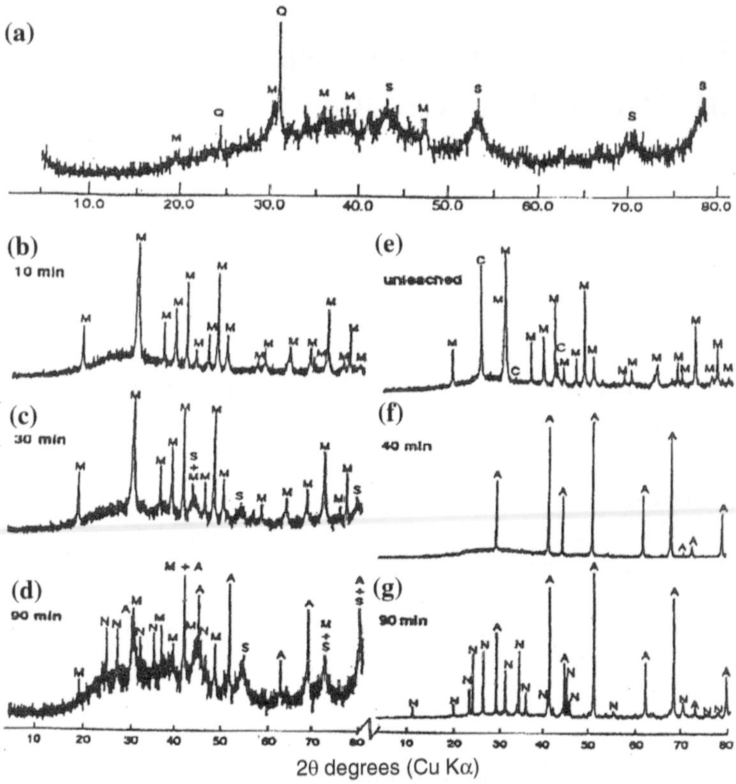

Fig. 18.1 A XRD traces of Zettlitz Kaolinite heated to 980 °C/4 h, then leached for indicated times (After MacKenzie et al. 1996a). Reproduced by permission of The Royal Society of Chemistry (RSC) on behalf of the Centre National de la Recherche Scientifique (CNRS) and the RSC

25° 2θ, broad patterns corresponding to a cubic spinel phase and small amount of quartz as impurity (Fig. 18.1).

Leaching for 10–30 min, progressively reduces the amount of amorphous material, but never entirely removes it. This indicates that some amount of alkali resistant siliceous phase may be the constituent of 980 °C decomposition product of kaolinite as observed in case of taking X-ray pattern of alkali leached kaolinite by using lindemann glass tube. Considerable changes in ^{29}Si NMR spectra of leached samples at different duration are shown in Fig. 18.2. Unleached sample show resonance at about −110 ppm and a small intensity peak at ~−88 ppm. On leaching for 10 min., the total Si intensity decreases and the −88 ppm peak becomes proportionately more intense. This trend continues with further leaching. The integrated intensities of the ^{29}Si spectra fall off with leaching time up to ~40 min leaching time (Fig. 18.3). Therefore, XRD date and ^{29}Si NMR data both indicate that there are two amorphous phases namely a silica rich phase, the silica content of which is readily soluble, and other is alumina rich silica phase.

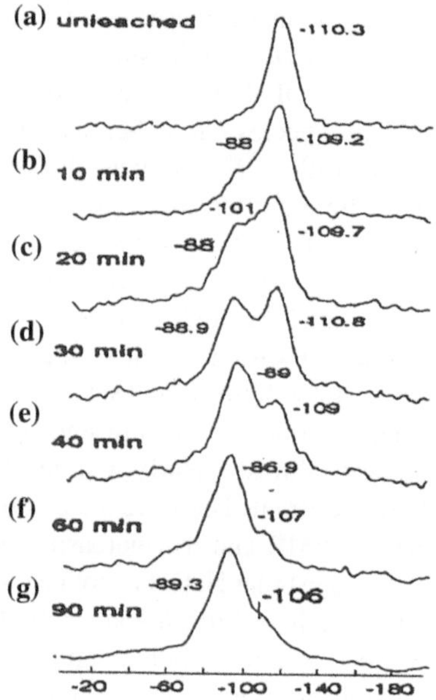

Fig. 18.2 A part of ^{29}Si MAS NMR spectra of Zettlitz Kaolinite heated at 1000°/ 4 h, then leached for different duration of times (after MacKenzie et al. 1996a). Reproduced by permission of The Royal Society of Chemistry (RSC) on behalf of the Centre National de la Recherche Scientifique (CNRS) and the RSC

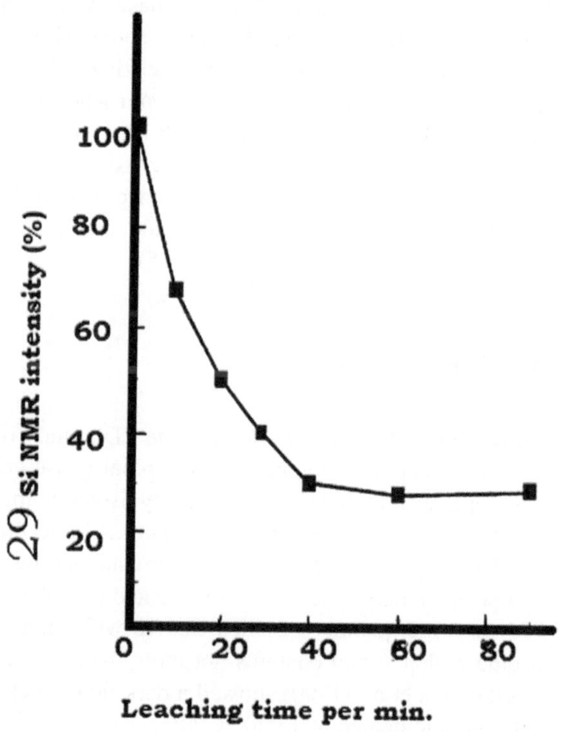

Fig. 18.3 Integrated intensity of Total ^{29}Si MR signal in the leached Zettlitz Kaolinite as a function of leaching time (after MacKenzie et al. 1996a). Reproduced by permission of The Royal Society of Chemistry (RSC) on behalf of the Centre National de la Recherche Scientifique (CNRS) and the RSC

The gradual decreases in −110 ppm peak leading to zero indicate two things. As described portion of the silicon resonance spectra is due to silica rich alumina phase. After removal of this phase by NaOH leaching, the some portion of the spectra is still left. This species has a chemical shift suggesting a degree of Al substitution and is similar to the ^{29}Si chemical shift of metakaolinite. This species is disappeared at longer time and on the contrary −88 ppm peak becomes progressively more significant with leaching, represents sodium alumino silicate (Nephelene type) structure which crystallizes from the amorphous phases in the leached residue for 60–90 min is heated. ^{23}Na spectra of heat treated leached sample confirms the presence of strong signal resulting from the incorporation of significant concentrations of sodium (Fig. 18.4) apparent even after leaching for 10–20 min.

The amount of Na present, estimated from NMR peak area measurements, increases almost linearly with leaching time, up to 40 min after which it becomes almost constant. There is a good correlation between the decreasing total Si intensity seen by NMR and the increasing Na NMR peak intensity as leaching progresses. This suggest that NaOH is not just dissolving silica and minor amount of alumina to form soluble sodium silicate, soluble aluminate from silica rich alumina phase, but develop insoluble sodium aluminosilicate out of alumina rich silica phase and insoluble aluminous phase out of silica rich aluminous phase concurrently.

Therefore, XRD and ^{29}Si NMR data both indicate that there are two amorphous phases namely a silica rich (A) phase, the silica content of which is readily soluble, and other is alumina rich silica (A) phase. MAS NMR study more convincingly showed that later phase really exists and it showed its occurrence at a broad resonance peak at ∼−101 ppm. Mackenzie et al. (1996b) showed a well-resolved ^{29}Si MAS NMR resonance at −77.5 ppm when Georgia kaolinite heat treated at 950 °C for 24 h and then leached with KOH at 90 °C for 1 h. This was explained by them as owing to Si incorporated γ-Al_2O_3 spinel. Considering the results of Mackenzie et al. (1996a, b) the resonance at −77.5 ppm is more probable to the formation of potassium aluminosilicate (A) phase instead of Si incorporated γ-alumina spinel as assumed (Fig. 18.5).

18.3 Microscopy

Okada et al. (1986) first carried out the EDS study of alkali leached kaolinite and showed the apparent composition of spinel phase corresponds to $SiO_2.6Al_2O_3$.

Sonuparlak et al. (1987) physically isolated the spinel phase from the surrounding phase by treating the Georgia kaolinite preheated to 850 °C for 7 days with 10 wt% boiling NaOH. On microanalysis by using EDS they showed that amount of silica decreases with leaching time. At extended leaching time (>25 min) the quantity of silica was 10 wt% approximately. With this data they concluded that spinel contains not more than 10 wt% silica (Fig. 17.3).

Srikrishna et al. (1990) showed a dark field electron micrograph corresponding to the sample heated at 1,020 °C after NaOH leaching (Fig. 6.5). The analysis of

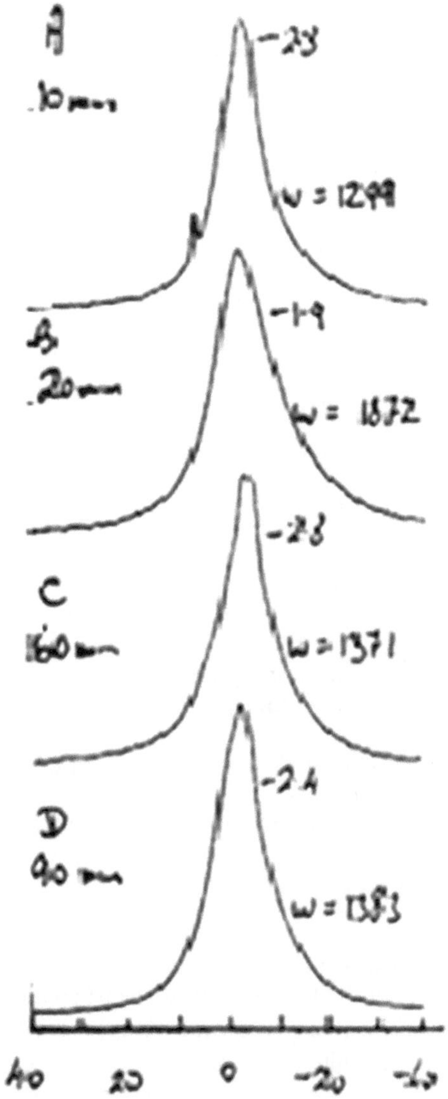

Fig. 18.4 ^{23}Na shift (ppm) with respect to NaCl solution (after MacKenzie et al. 1996a). Reproduced by permission of The Royal Society of Chemistry (RSC) on behalf of the Centre National de la Recherche Scientifique (CNRS) and the RSC

diffraction pattern reveals the crystallites to be a spinel phase very close to lattice parameter (a = 7.88 nm). At the same temperature, some mullite single crystal was also observed. This was confirmed by diffraction and on heating to temperature beyond the second exothermic peak the transformation to mullite goes to completion and acicular crystals of mullite from. The mullite crystals thus formed tend to have texture in that the c-axis lying out of the plane. It is due to the fact that metakaolinite consists of sheets of alumina and silica tetrahedral and mullite forms on these sheets. The c-axis is the smallest dimension of the mullite unit cell. Therefore, energetically it would be favorable for it to lie perpendicular to the plane of the metakaolinite sheets.

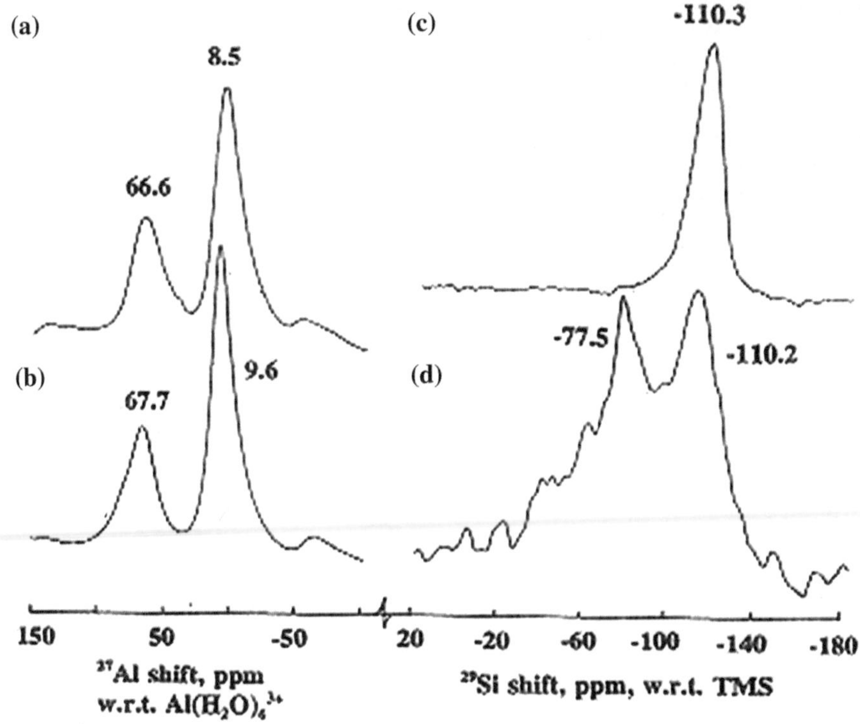

Fig. 18.5 11.7T MAS NMR Spectra of (A.C) Georgia Kaolinite dehydroxylated for 24 h and (B.D) sample A, leached with KOH at 90 °C/1 h, after MacKenzie et al. (1996a). Reprinted by permission of the American Ceramic Society

18.4 DTA

Srikrishna et al. (1990) also leached out exsolved silica out of kaolinite heat treated to 1,020 °C by using 10 % NaOH solution at 96 °C for 15 min. DTA analysis of the etched sample showed the existence of second exotherm as happened in case of raw kaolinite (Fig. 18.6).

18.5 IR

Percival et al. (1974) in their IR study showed the presence of two sets of bands due to poorly crystallized spinel and mullite phase, and amorphous silica on kaolinite heat treated to 950 °C. The distinctive band at 472 cm^{-1} in the spectrum due to amorphous silica disappears on alkali leaching. This indicates a confirmatory proof of the release of free silica during decomposition of metakaolinite (Fig. 17.2).

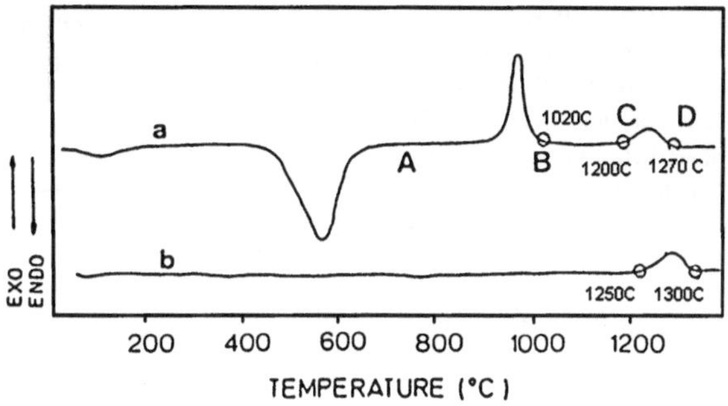

Fig. 18.6 DTA Studies of unleached (**a**) and leached Kaolinite (**b**) after Srikrishna et al. (1990)

18.6 DTMA

Chakraborty (1993) carried out DTMA studies of raw kaolinite and alkali leached kaolinite and showed the removal of differential contraction peak at ~1,400 °C and also noted the enhancememt of differential expansion peak at ~1,250 °C due to mullitization (Fig. 18.7).

18.7 IR Study of Leached Residue

Moya et al. (1985) showed that during Al–Si spinel (premullite) to mullite transformation, the IR spectrum changes. A portion of Si–O stretching band shifts from 1100 cm^{-1} to higher wave number 1160 cm^{-1}, and band at 575 cm^{-1} assigned to AlO_6 vibration increases in intensity (Fig. 18.8).

They conjectured that some tetrahedral aluminum changes to octahedral coordination by a temperature controlled diffusive process in going from premullite to mullite transformation.

18.8 Transmission Electron Microscopic Study of Leached Residue

Comer (1960) and later Tsuzuki and Nagasawa (1969) presented electron micrographs of kaolinite heated to ~1,000 °C which showed numerous small dots representing the crystallites of spinel phase after first exothermic reaction. McConnell and Fleet (1970) showed the reaction product of kaolinite heated to 900 °C was a defect spinel with strongly preferred orientation and a microporous structure. Comer (1960) noted many protrusions and these particles are appeared to be cubic and they vary from 10 to 60 nm in sizes (Fig. 18.9). Okada et al. (1986) carried out the TEM study of alkali leached kaolinite. They first showed the

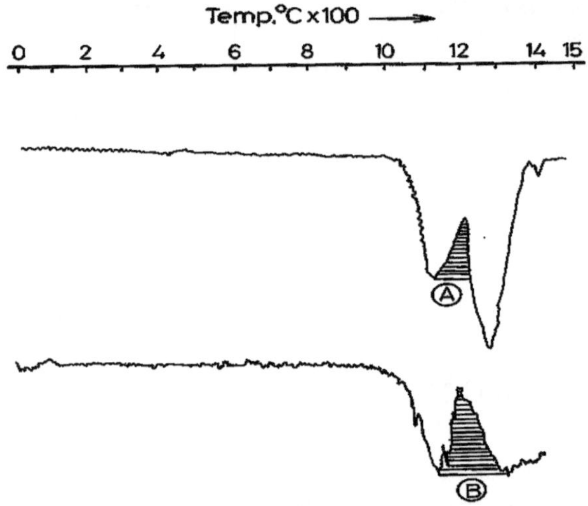

Fig. 18.7 Differential dilatometric curve (*A*) of 980 °C heated Rajmohol Kaolinite and (*B*) of 980 °C heated Rajmohol Kaolinite leached with alkali. Sample size 2 mm/4 mm dia, rate of heating 10 °C/min, sensitivity ±50 μm and chart drive 1.25 mm/min in both cases

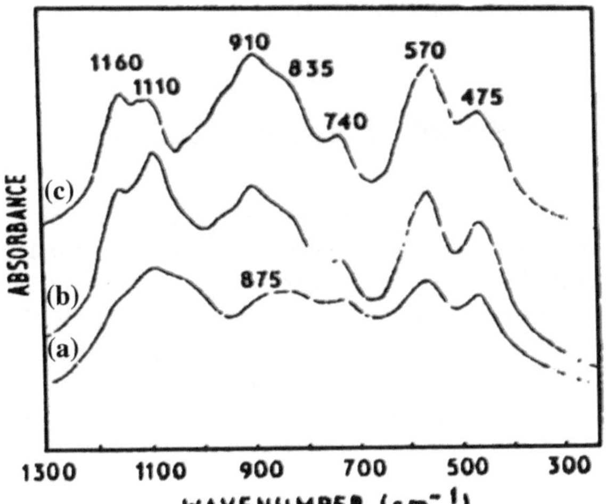

Fig. 18.8 Infrared spectra of premullite held at different temperatures: **a** 1,000 °C, 4 h, **b** 1,100 °C, 46 h, **c** 1,200 °C, 2 h (After Moya et al. 1985). Reprinted by permission of Springer

particles of spinel phase with its diffraction pattern and further showed the development of high degree of porosity surrounding spinel particles and aluminous phase after extraction of silica (A) material. Srikrishna et al. (1990) showed a dark field electron micrograph corresponding to the sample heated at 1,020 °C after NaOH leaching. The analysis of diffraction pattern revealed the crystallites to be a spinel phase very close to lattice parameter ($a = 7.88$ nm). At the same temperature, some mullite single crystal was also observed. This was confirmed by diffraction and on heating to temperature beyond the second exothermic peak the transformation to mullite goes to completion and acicular crystals of mullite form. The mullite crystals thus formed tend to have texture in that the c-axis lying out of

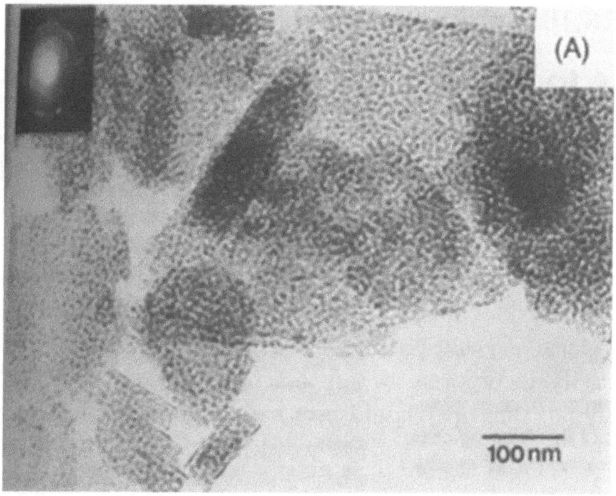

Fig. 18.9 a TEM photograph of Kaolinite fired at 950 °C and treated twice by NaOH solution (after Okada et al. (1986)). Reprinted by permission of the American Ceramic Society

the plane. It was assumed that metakaolinite consists of sheets of alumina and silica tetrahedral and mullite forms on these sheets. The c-axis is the smallest dimension of the mullite unit cell. Therefore, energetically it would be favorable for it to lie perpendicular to the plane of the metakaolinite sheets.

18.9 Microstructure of Leached Residue

The homogeneous pore structure as noted in electron micrograph (Fig. 18.9) of the alkali leached residue due to leaching out of siliceous phase indirectly predicts the transformation of metakaolinite in the following steps.

(i) With the observation of a characteristic pore structure by TEM study, it is revealed that a phase separation phenomenon had occurred previously. It suggests that metakaolinite first decomposes into two forms (i) alumina rich silica (A) phase, and (ii) siliceous (A) phase by phase separation during final dehydroxylation at the endothermic point just prior to first exotherm. During alkali treatment, the siliceous (A) phase is leached out and developed a regular and homogeneous porous structure (Fig. 18.9).
(ii) A portion of alumina rich silica (A) phase crystallizes to Al–Si spinel and traces of mullite during the exhibition of 980 °C exothermic peak.
(iii) Remaining portion of alumina rich silica (A) phase with siliceous (A) phase as solid solution remains as residue. Formation of it is conformed by MAS NMR study which shows a broad resonance at ~101 ppm (MacKemzie et al. 1996).

Both XRD and MASNMR studies showed that a considerable quantity of amorphous phases is present when kaolinite heated to ~1,000 °C. This explains why electron micrograph is very dark when kaolinite heated in the range 900–1,100 °C until much mullite crystallizes.

18.10 Summary

Alkali leached preheated kaolinite has been characterized by subsequent techniques, e.g., XRD, MAS NMR, TEM. EDS. DTA. IR DTMA, etc.

Chakraborty and Ghosh argued that when free silica part is leached out then mullite formation from alkali extracted residue which is enriched with spinel occurs rapidly due to polymorphic transformation of it. Srikrishna et al. also noted that leaching does not affect the exhibition of second exotherm. Therefore, there is no role of exolved silica in mullitization. Contrary, Sonuparlak et al. leached extensively and put forward different composition of spinel phase. Sodium could easily incorporate during leaching process as noted in MAS NMR study. This signals that one should be very cautious in performing leaching studies.

References

A.K. Chakraborty, Application of TMA and DTA studies on the crystallization behavior of SiO_2 in thermal transformation of Kaolinite. J. Therm. Anal. **39**, 280–299 (1993)

A.K. Chakraborty, D.K. Ghosh, Comment on the interpretation of the Kaolinite–Mullite reaction sequence from infra-red absorption spectra. J. Am. Ceram. Soc. **61**(1–2), 90–91 (1978a)

A.K. Chakraborty, D.K. Ghosh, Re-examination of the Kaolinite to Mullite reaction series. J. Am. Ceram. Soc. **61**(3–4), 170–173 (1978b)

J.J. Comer, Electron microscope studies of mullite development in fired Kaolinite. J. Am. Ceram. Soc. **43**(7), 378–384 (1960)

K.J.D. MacKenzie, J.S. Hartman, K. Okada, MAS NMR evidence for the presence of silicon in the alumina spinel from thermally transformed Kaolinite. J. Am. Ceram. Soc. **79**(11), 2980–2982 (1996b)

K.J.D. MacKemzie, R.W. Meinhold, A.K. Chakraborty, M.H. Dafader, Thermal reactions of alkali–leached aluminosilicates studied by XRD and solid state ^{27}Al, ^{29}Si and ^{23}Na MAS NMR. J. Mater. Chem. **6**(5), 833–841 (1996a)

K.J.D. Mackenzie, R.H. Meinhold, A.K. Chakraborty, M.H. Dafadar, Solid state MAS NMR study of the thermal reactions in alkali–leached aluminosilicates. J. Aust. Ceram. Soc. **35**, 34–38 (1999)

J.D.C. McConnell, S.G. Fleet, Electron optical study of the thermal decomposition of Kaolinite. Clay Miner. **8**, 279–290 (1970)

J.S. Moya, C.J. Serna, J.E. Iglesias, On the formation of mullite from kandaties. J. Mater. Sci. **20**, 32–36 (1985)

K. Okada, N. Ostuka, J. Ossaka, Characterization of spinel phase formed in the Kaolin–Mullite thermal sequence. J. Am. Ceram. Soc. **69**(10), C-251–C-253 (1986)

H.J. Percival, J.F. Duncan, P.K. Foster, Interpretation of the Kaolinite–Mullite reaction sequence from infrared absorption spectra. J. Am. Ceram. Soc. **57**(2), 57–61 (1974)

B. Sonuparlak, M. Sarikaya, I.A. Aksay, Spinel phase formation during the 980 °C exothermic reaction in the Kaolinite-to-Mullite reaction series. J. Am. Ceram. Soc. **70**(11) 837–842 (1987)

K. Srikrishna, G. Thomas, R. Martinez, M.P. Corral, S. Aza, J.S. Moya, Kaolinite–Mullite reaction series: a TEM study. J. Mater. Sci. **25**, 607–612 (1990)

Y. Tsuzuki, K. Nagasawa, A. Transitional stage to 980 °C exotherm of kaolin minerals. Clay Sci. **3**(5), 87–102 (1969)

Part II

Preface

Large volume of publications of kaolinite to mullite and cristobalite reaction process as shown in Chaps. 3 – 18 by the authors of different disciplines are most worthful in crystallo-chemical aspects of view. Application of many experimental techniques have been elaborated and these studies focus three major information.

A. Steps of Decomposition of Kaolinite

Overall study of progressive heat-treated kaolinite shows the following steps of transformations.

1st Step—Endothermic region at (500–600 °C). Formation of metakaolinite by major dehydroxylation.

2nd Step—Endo- to exo-region of 980 °C reaction. Final minor dehydroxylation. Decomposition of metakaolinite structure with the evolution of following phases.

(i). Evolution of siliceous (A) phase.

(ii). Crystallization of Al–Si Spinel phase (major amount) and weakly crystalline mullite (traces).

(iii) .No crystallized mullite (A) phase.

3rd Step—Two exothermic regions of mullite formation.

(i). Disappearance of Al-Si Spinel and formation of mullite.

(ii). Nucleation and crystallization of mullite (A) phase. 4th Step—Crystallization of silica (A) to β-Cristobalite.

B. Correlation of Thermal Events with Other Data

Different studies indicate that thermal events of kaolinite e.g., DTA, TGA, TMA or Shrinkage data,DTMA, etc are related with other physico–chemical data generated by various authors, e.g., intensities of Al–Si spinel, mullitization processes, wt % of silica extracted from heated kaolinite at different temperatures and finally intensity of silica crystallization processes, etc.

Various studies of progressive heat treated kaolinite are noted by some miscellaneous studies, e.g., TMA, DTMA data Density, Surface area, Change in pore size, XRD, IR. EM, MAS NMR, RED, etc.

Changes in the peak position and intensity of amorphous band related to formation of metakaolinite (Lee et al, 1999) in Fig. 6.6, release of siliceous phase by ^{29}SiMAS MNR studies by (Sanz et al .1988), as shown in (Fig. 13.3) , formation of mullite (A) phase, and its subsequent transformation as noted by XRD and also by electron microscopy.

Auxiliary studies are also performed with some objectives, e.g., XRF. RED and MAS NMR to note changes in coordination number of Al during heating process of kaolinite; EDS to note compositional analysis of enriched Al–Si Spinel phase; solubility study of Al_2O_3 in mineral acid and dissolution study of SiO_2 in alkali solution were made to test the nature of metakaolinite and estimation of liberated free silica.

Some of these data are shown in correlated diagram (Fig. 1).

This diagram shows that some correlations are noted in the following thermal transformation steps.

1. Dehydroxylation of kaolinite to formation of metakaolinite—This phenomenon is revealed by DTA/ DDTA, TGA/ DTGA, TMA/ DTMA, IR, etc.

2. Decomposition of metakaolinite structure. It is understood by XRD, I R, Electron microscopy, and Electron diffraction experiments.

3. Characterization of 980 °C exotherm.

(i). Evaluation of siliceous (A) phase by alkali extraction.

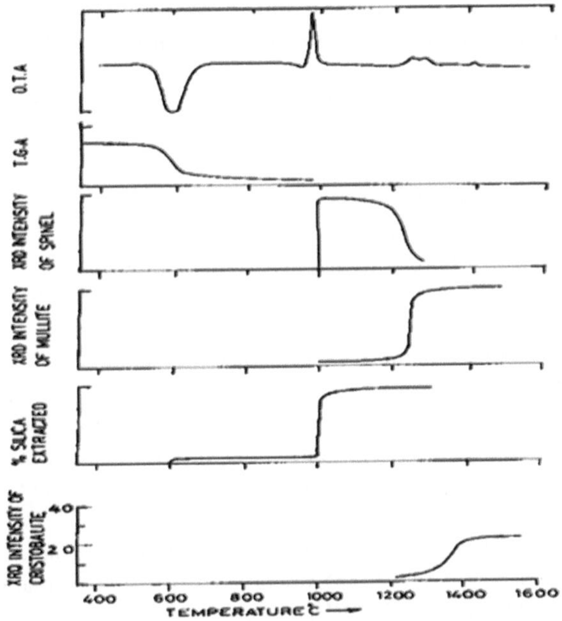

Fig. 1 Correlated diagram of various events of decomposition of kaolinite

(ii). Development of spinel phase by QXRD, and electron diffraction, High resolution electron microscopy (HRTEM), Energy-filtering electron microscopy (FE-TEM) studies.

(iii). Formation of mullite (A) phase by X-ray study followed by alkali leaching of siliceous phase.

4. Growth of mullite is studied by QXRD, SEM and TEM studies.

(i). Crystallization of mullite (A) phase by DTA and alkali leaching.

(ii). Changes in lattice constants of mullite, its size, and strain by Rietvled analysis.

5. Growth of cristobalite by QXRD.

(i). Disappearance of siliceous phase (A).

(ii). Change in lattice constant (L.C), size, strain of cristobalite by Rietveld analysis.

C. Major Problems of K–M Reaction Series

(i). Identification of three phases.

The most important one is constitution and structural characteristics of metakaolinite. It shows amorphous pattern in powder X-ray technique which imparts a great difficulty in proper characterization. By electron diffraction, it shows a-b axis periodicity and absence of c-axis order. Structure evolution is difficult.

The second intermediate phase i.e. spinel is also a poorly crystallized material which makes a more problem in ascertaining its composition. The ultimate phase is mullite which is also ill crystallized at the initial stage up to second exotherm. Most of the d (nm) peaks are broad resulting to a great difficulty in precise lattice parameter calculations. Nature of mullite phase, whether it changes it's composition on gradual heat treatment becomes a burning problem. General microscopic study (SEM) and even TEM show a dark background till mullite grows properly. These give a threat in proper elucidation of reaction process.

(ii). Mechanism of transformation of metakaolin to mullite. There are few suggestions . According to John (1953) chains of Al ion in metakaolin transforms to edge sharing chains in mullite. There is other mechanism in view of spinel concept. Brindley and Nakahira explained the way spinel structure transforms to mullite structure. [110] axes of the spinel phase running parallel to mullite chains.
(iii). Cause behind exothermic reactions.Different types of reaction have been put forward to explain the reasons of the first exothermic reaction, second, and last exothermic reactions. Variation in interpretations are many fold and the main reasons are due to, e.g., changes of crystallinity, and impurities content of kaolinite; technique of thermal analysis procedure, choice of thermally treated or untreated kaolinite, etc.

(iv). Existence of interrelationships among three phases. To elucidate this behavior, single crystal X-ray technique, single crystal electron diffraction technique, and then high resolution electron microscopy were used. Some information regarding poorly crystallized reaction products and reaction mechanism are focused. Problems still exist in presence of mixture of phases due to incomplete conversion or in some cases due to presence of lots of amorphous phases. This result to inconvenience in estimating the composition of spinel phase in EF-TEM microscopy.

(v). Mechanism of dehydroxylation of kaolinite leading to formation of meta-kaolinite and its final dehydroxylation process are not fully clear.

In the following chapters detail characterization of three phase and the reasons of exothermic reactions are put forward on the basis of the problems discussed above and the literature reviewed chapter wise from 2 to 18 in Part 1 of this book.

Chapter 19
Meta Kaolinite Phase

19.1 Introduction

The nature of metakaolin and its structural characteristics are being dealt in this chapter. The first question is: what is occurring on dehydroxylation and what happens to the Si–O–Al linkages present between two layers constituting kaolinite? There are two probabilities. It may happen that kaolinite partially or fully decomposes with disruptions of Si–O–Al bonding during dehydroxylation and liberate-free silica (A) or it may remain as an alumino-silicate (A) compound which decomposes at the vicinity at 980 °C exotherm. Various researchers conducted different chemical dissolution experiments in order to reveal indirectly the chemical linkages existing if any between tetrahedral silica sheet and dehydroxylated alumina layer. Mainly chemical techniques were employed. Since simple X-ray study exhibited amorphous pattern. Thus, it was speculated that metakaolinite be an amorphous mixtures of silica and alumina or amorphous compound. Acid dissolution study showed that alumina layer in metakaolinite either present in highly distorted state or it becomes amorphous. On the other hand, sodium carbonate extraction study showed that silica and alumina packets are not present in free state. Accordingly, two views were prevailed in the olden days.

19.2 Effect of Dehydroxylation on Two Layers of Kaolinite

Mellor and Holdcroft (1911), Rinnie (1924), Hyslop and Rooksby (1928), Tamman and Pape (1923), Keppler (1925), Krause and Wohner (1932), Newman and Kober (1926) assumed that during dehydroxylation SiO_2 and Al_2O_3 packets of the original kaolinite molecule moves apart and forms a homogeneous mixture of the two amorphous substances and metakaolinite becomes amorphous in nature. On the contrary, Clarke (1895), Vesterbcrg (1925), Salmang (1933), Budnikov (1935), Hyslop (1944), Insley and Ewell (1935) believed that kaolin does not decompose into free oxides on the other hand, it behaves as a chemical compound. Subsequently,

A. K. Chakraborty, *Phase Transformation of Kaolinite Clay*,
DOI: 10.1007/978-81-322-1154-9_19,

alkali extraction study of metakaolinite by Chakraborty and Ghosh (1978b) either in cold or in hot condition (Fig. 17.1a) substantiates the chemical bondage theory of metakaolinite. In addition, crystallization sequence study of HF treated metakaolinite of different degree presented in Chap. 11 proves that assumed protection of alumina layer on SiO_2 (A) as conceived by Colegrave and Rigby (1952) is purely chemical in nature. I.R. study made by Freund (1967) for ascertaining Si–O–Al bonds in metakaolinite is found inconclusive. Normal Si–O vibrations appear in the region 1,100–1,000 cm^{-1} and normal Al^{IV}–0 <–> Al vibrations in Al_2O_3 lie below 900 $cm^{-1.}$ Question is: where the Si–O <–> Al^{iv} vibration for kaolinite occurs? According to Freund (1967), in Si–O–Al(IV) bridge, the oxygen bond to the tetrahedrally Si^{+4} is stronger than that to octahedral Al^{+3}. One should expect a lower Si–O–Al(IV) <–> 0 frequency then the normal Al–O–Al(IV) <–> 0 frequency in Al_2O_3. So Al(IV) <–> 0 valence vibrations from the Si–O–Al(IV) bridge may likely lie in the area between 800 and 700 cm^{-1}. Unfortunately, in this region other bands of kaolinite appear. Thus, it is difficult to make proper assignment by I.R. technique. By indirect I.R. study on acid treated metakaolinite, Pampuch (1966) demonstrated the existence of Si–O–Al linkages in it. From IR absorption spectra of heated kaolinite, he showed that the most characteristic and intense bands at at 1,190, 1,100, and 800 cm^{-1} for free SiO_2; 570 and 1,035 cm^{-1} for alumina did not appear in metakaolinite sample. Following the dissolution of alumina layer of metakaolinite by treatment with boiling HCl as per procedure of Gastuche and Fripiat (1963), he showed IR spectra (Fig. 19.1) which displayed the very strong band at 1,090–1,102 cm^{-1} accompanied by weaker bands at 1,190 and at 800 cm^{-1} respectively. The later band had in addition the typical narrow sharp contour which differs markedly from the broad shape of 810 cm^{-1} band in spectrum of untreated metakaolinite. These bands indicate free SiO_2 evolution by chemical treatment of metakaolinite and thus Pampuch suggested that kaolinite transforms to a new phase or structure during 1st stage of heating. Thus, all these results point to the continuance of Si–O–Al bondings from kaolinite to metakaolin.

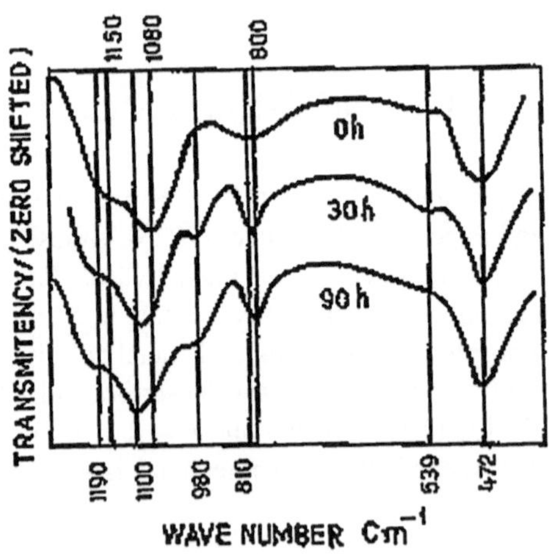

Fig. 19.1 Infrared Spectra of Zettlitz Kaolinite to 550 °C and Subsequently Treated with Hydrochloric acid for Various Period of Time (After Pampuch 1966). Reproduced with kind permission of Prof. Pampuch

19.3 Effect of Dehydroxylation on Tetrahedral Layer

Major changes involved during dehydroxylation of metakaolin are in its tetrahedral layer. Literature reveals the following observations.

(i) Metakaolinite shows the existence of hexagonal morphology.

By electron-microscopic study, Eitel et al. (1939), Eitel and Kedesdy (1943), Comeforo et al. (1948), Comer, Koenig and Lyons (1956), Sonuparlak et al. (1987), and other showed the persistence of kaolinite relicts after the dehydroxylation of kaolinite.

(ii) Metakaolinite shows a-b axis periodicity and absence of c-axis order as noted in the following studies.

(a) *X-ray powder diffraction analysis* Using monochromatic radiation in vacuum by Tscheischwili et al. (1939) noted (hko) reflections in dehydrated kaolinite. They first proposed the crystal structure of metakaolinite with the assumption that it dehydroxylated completely and second all Al atoms assumed tetrahedral co-ordination. Based on the observation of the persistence of hko reflection from kaolinites heated above dehydration temperature, they suggested a model characterized by the original Si–O network of kaolinite remaining almost unchanged. The original octahedral network of kaolinite only changed during dehydroxylation into edge sharing tetrahedral Al–O chains. Accordingly, he proposed a tentative structure of metakaolinite with tetrahedral silica sheet remained intact. It is based on monoclinic cell with $\boldsymbol{a} = 0.1514$ nm; $\boldsymbol{b} = 0.890$ nm, $\boldsymbol{c} = 0.68$ nm, and $\beta = 100.2$. Use of original cell dimensions of kaolinite $\boldsymbol{a}$ and $\boldsymbol{b}$ parameters resulted in Si–O and Al–O bond lengths to be 0.159 nm and 0.185 nm respectively.

(b) *Single crystal X-ray diffraction analysis* Brindley and Hunter (1955) noted hexagonal pattern of spots for nacrite flakes heated to 670 °C with no reflections in c-axis is noted in rotational diagram. Subsequently, Brindley and Nakahira (1959) showed (hko) reflection spots and absence of sports due to (001) reflections in kaolinite flakes (Fig. 5.3). With this observation and of previous density data of metakaolinite and assuming full and homogeneous dehydroxylation, they presented a new structure by modifying the structure proposed by Tscheischwili et al. (1939). They considered that the lattice parameters, $\boldsymbol{a}$ and $\boldsymbol{b}$ of kaolinite remain more or less unchanged in metakaolin. Regarding $\boldsymbol{c}$ axis dimension, they referred to the density data of Ricke and Mouve (1942) of Zettlitz kaolinite heated at 800 °C which is less than that of original kaolinite. They first suggested that ($\boldsymbol{c}$-layer) spacing must have diminished to $7.15–7.15 \times (13.95–1.52) = 6.3$ nm. The metakaolinite structural model consists of a two dimensional network of alternatively corner and edge shared aluminum oxygen tetrahedra superimposed on a silicon oxygen layer of identical in symmetry and lattice parameter to that of natural kaolinite. The structure of metakaolinite is based on triclinic cell of kaolinite proposed by Brindley and Robinson (1946) $\boldsymbol{a} = 0.514$ nm, $\boldsymbol{b} = 0.893$ nm, $\boldsymbol{c} = 0.63$ nm, $\alpha = 91.8°$, $\beta = 104.5°$ and $\gamma = 90°$.

(c) *Electron diffraction study* Roy et al. (1955) as shown in Fig. 6.1 and Comer (1961), Tsuzuki and Nagasawa (1969), McConnell and Fleet (1970) showed the existence of (hko) spots in kaolinites heated after dehydroxylation peak temperature. These results definitely signify that kaolinite still exhibiting two-dimensional (2D) order in a–b plane after endotherm.

These studies, e.g., microscopic and electron/X-ray diffraction analysis as mentioned above confirm that metakaolinite as a pseudomorph of kaolinite. Only exception is that it does not possess c-axis order. In kaolinite, the composite SiO_2 and Al_2O_3 layer in the ratio of 1:1 accomplishes sufficient ordering in the third dimension necessary for acquiring c-axis periodicity. In structure of kaolinite, the projection on (001) of the Si–O network of one layer and hydroxyl sheet of the adjacent layer where oxygen and hydroxy ions are grouped in pairs. These pairing suggest that a hydrogen or –O–OH bond forms which although inherently weak but sufficiently strong enough to build up the composite $SiO_2–Al_2O_3$ layer in c-direction. Newnhan (1961) is of the opinion that elevated hydroxyl and oxygen units in successive layers are involved in sharing the common hydrogen. The hydrogen is represented as continuously jumping backward and forward between the two oxygens, although in fact, a "hybrid" structure is likely formed.

$$--O^{-}H^{+} \dashrightarrow --O^{-} <---> --O^{-} \dashrightarrow \; -H^{+}--O^{-}$$

The resulting alternating resonance structure behaves as a rather weak bond and is known as a hydrogen bond. During dehydration of gibbsite layer of original kaolinite, hydrogen bonds connecting the two layers are disrupted. Since, it is essentially an interaction between portion of one hydroxyl group with the second hydroxyl group connected to the aluminum atom in the octahedral position of kaolinite structure.

Thus, as a result of dehydroxylation, bondings between two successive layers disappear and consequently the c-axis periodicity also vanishes.

```
HO    OH            O-
  \   /             I
-- Al --    ---> -- Al --   + H2O
  /   \             I
```

19.4 Effect of Dehydroxylation on Octahedral Layer

The other major change occurs in alumina sheet of kaolinite during its dehydroxylation is in the change of co-ordination number of aluminum with decrease of neighboring OH groups.

19.4.1 Coordination of Aluminum Shifts on Dehydroxylation

There are as many as four methods those detect the conversion of AlO(OH) octahedra to AlO tetrahedra.

(a) By *I.R. study* the following researchers namely Stubican and Roy (1961), Gastuche et al. (1962), Dekeyser (1963d), Pampuch (1966), Freund (1967), Tsuzuki and Nagasawa (1969), and Percival et al. (1974) showed the disappearance of bands due to Al(IV)–OH vibrations and with appearance of a Al(IV)–O band. This indicates that kaolinite changes from 6 to 4 in metakaolinite. Figure 7.2 of Dekeyser (1963) showed the gradual disappearance of four bands of kaolinite during dehydroxylation process.

(b) By *XRF study*, investigators namely Brindley and Mckinstry (1961), Tsuzuki and Nagasawa (1969), Udagawa et al. (1969), Leonard (1977), and Bulens et al. (1978) noted changes in the Al K_{α} wave length shifts of heated kaolinites in comparison with standard and confirmed shifting of Al(VI) in kaolin to Al(IV) in metakaolin.

(c) By *RED technique*, Lemaitre et al. (1975), Leonard (1977) showed that Al(VI)–O in kaolinite changes to Al(IV)–O in metakaolinite (Fig. 9.2). At 900 °C they noted both AlO_6 and AlO_4 groups. According to them, it might be possible that some AlO_6 is AlO_4 with long bond lengths of order of 1.92 Å instead of regular Al–O distance of 1.77 Å. Thus, all Al in metakaolinite is not strictly AlO_4 as shown by Gastuche et al. (1962). A comparison is to be cited on the XRF method of measuring coordination number of Al in metakaolinite from old verses new data. In earlier days, it was conceived that measured wave length shift is directly related to coordination number and accordingly it was compared with known calibrating standard materials. Recent concept is that measured wave length shift is probably related to average Al–O bond distance. Accordingly, newer procedure was adopted by Bulens et al. (1978) who used gibbsite and Na–X Zeolite as sixfold and fourfold standard end members and then interplotted a straight line out of which coordination number of Al in metakaolinite was found out. Their claims seem more reasonable. According to *Leonard* the value around 59 % for Al(IV)–O observed at 600 and 800 °C is purely artificial and only characterizes the unusual situation of Al atoms fully tetrahedrally coordinate (2OH of gibbsite-like layer are actually and completely removed, atleast at 800 °C) in an enlarged oxygen environment (1.92 Å) in classical Al(IV)–O distance. As pointed out by Wardle and

Brindely (1971), coordination number and cation–anion distances contribute together to the experimentally observed value. They pointed out that there was no sense in this particular situation to look after complementary 40 % Al(IV) content. At least at 800°, there are no more Al(VI) configuration, except that the distance b Al–O looks alike.

According to Freund (1960) the octahedral Al groups of kaolinite can lose hydroxyl water but still retain their octahedral bond lengths and configuration. He showed that on consequence of dehydration some lattice defects arise in octahedral layer. The concentration of lattice defects rise to the extent of 20 vol % which are fixed in the lattice removal of O and OH groups at the endotherm. At 970 °C exotherm, these vacancies are eliminated. It was conceived that in this eliminating process, the total lattice is not transformed into a new one, only the lattice vacancies having maximum energies are destroyed.

(d) *Electron density mapping* from X-ray study

Using dehydroxylated dickite, a polymorph of kaolinite, Iwai et al. (1971) made electron density map from 2D Fourier projections. On comparing two projections, they showed that the structural skeleton of the Dickite is nearly preserved. Electron densities at 535 °C at the points corresponding to OH(1) and OH(2) become thinner than the original and the oxygen maxima O(1) widens out anisotropically away from the Si atom. Only Al octahedra changes to tetrahedra with high degree of deformation/distortion. They may expand parallel to (001) and get compressed in a group vertical to the base plane and arrange themselves in a chain with a common edge (110). The structure of metakaolinite is essentially a 2D one.

(e) By ^{27}*Al MASNMR study*, Brown et al. (1985) Watanabe et al. (1987), Sanz et al. (1988) showed significant reduction of AlO_6 groups of kaolinite after dehydroxylation.

Meinhold et al. (1985) depicted the changes in relative $Al0_6/Al0_4$ ratio in gradually heat treated kaolinite (Fig. 13.1). As like above researchers, they also showed the existence of two peak in spectrum appeared between 65 and 70 ppm. These are due to development of tetrahedral Al during dehydroxylation of kaolinite. Now what happens to the octahedral sheet of kaolinite during dehydroxylation in general? The a-b axis dimension of octahedral sheet ($\boldsymbol{a}$ = 0.505 nm, $\boldsymbol{b}$ = 0.862 nm) is less than that of tetrahedral configuration ($\boldsymbol{a}$ = 0.514 nm, $\boldsymbol{b}$ = 0.893 nm). During dehydroxylation, Al(IV)–O is formed out of Al(VI)–O. Al(IV)–O distance is 0.176 nm (average) and is less than Al(VI)–O distance (0.194 nm). Then it is expected that the dimension of the resultant sheet should be reduced and not increased.

19.4.2 Change in Symmetry of Silica Sheet

Two different observations regarding the changes in symmetry of silica sheet are noted during the dehydroxylation of kaolinite.

(i) Pampuch (1966) first showed that tetrahedral Si–O sheet of metakaolinite takes up a higher symmetry after dehydroxylation. A profound change of spectrum in the range of internal stretching of Si–O–Si and Si–O vibrations in between 690 and 1,150 cm^{-1} occurs. They noted the appearance of two absorption bands at 1,150 and 1,080 cm^{-1} with concomitant disappearance of three bands namely 792, 750, and 692 cm^{-1} only on dehydroxylation. With these observations, he proposed a higher symmetry of Si–O layer in metakaolinite and assumed that Si–O layer in metakaolinite now assumes the open hexagonal packing of tetrahedra which are more stable at high temperature.

(ii) Contemporaly Brindley and Gibbon (1968) however showed ***b*** parameter of kaolinite increased by 2.2 % after dehydroxylation. According to them, this value could be accounted if it is assumed that Si–O bond length becomes normal value, i.e., tetrahedral sheet becomes hexagonal from originally ditrigonal in kaolinite.

19.4.3 Existence of OH Groups

Other important observation noted by various authors who showed the existence of a few –OH groups in metakaolinite. Pampuch (1958) noted the continued presence of 3,620–3,623 cm^{-1} bond in metakaolinite obtained from 550 to 600 °C and also from metakaolinite formed during heating H^+ form of kaolinite at 500–600 °C. (Fig. 19.2).

Considering open-hexagonal packing of silicon–oxygen tetrahedra, He proposed a structure of metakaolinite. This model consists of a chain of alternatively corner and edge shared aluminum-oxygen-hydroxyl tetrahedra superimposed on the silicon–oxygen layer having the open hexagonal packing of tetrahedra. This model differs from the model of Brindley and Nakahira by having a chain of aluminum-oxygen-hydroxyl tetrahedra instead of a continuous network of aluminum-oxygen tetrahedra. The presence of –OH ions and distributions of oxygen ions are the features which are also different from *Brindley and Nakahira and Tscheischwili model*. The structure of it is based on orthorhombic cell symmetry with parameters as follows $\boldsymbol{a} = 0.514$ nm, $\boldsymbol{b} = 0.893$ nm, and $\boldsymbol{c} = 0.706$ nm. However, it is very difficult to generate this structure from kaolinite as per Mackenzie et al. (1985). On further heat treatment *Pampuch* suggested a mechanism of further dehydroxylation which leads to separation of γ-Al_2O_3. Subsequently, Stubican (1959), Dekeyser (1963), and Perceival et al. (1974) also confirmed the persistence of weak bond of some residual OH groups in metakaolinite. For example, Dekeyser (1963e) noted the gradual disappearance of 4 OH bands with the start of dehydroxylation. After elimination of endothermic peak two wide absorption bands are still noted. These bands disappear completely at and before 980 °C exotherm as noted by Kubicki and Grim (1959) in their TGA analysis. Dekeyser (1963d, e) also correlated elimination of OH bands situated at 3,620 cm^{-1} with DTGA analysis. He noted by I.R. absorption measurement that

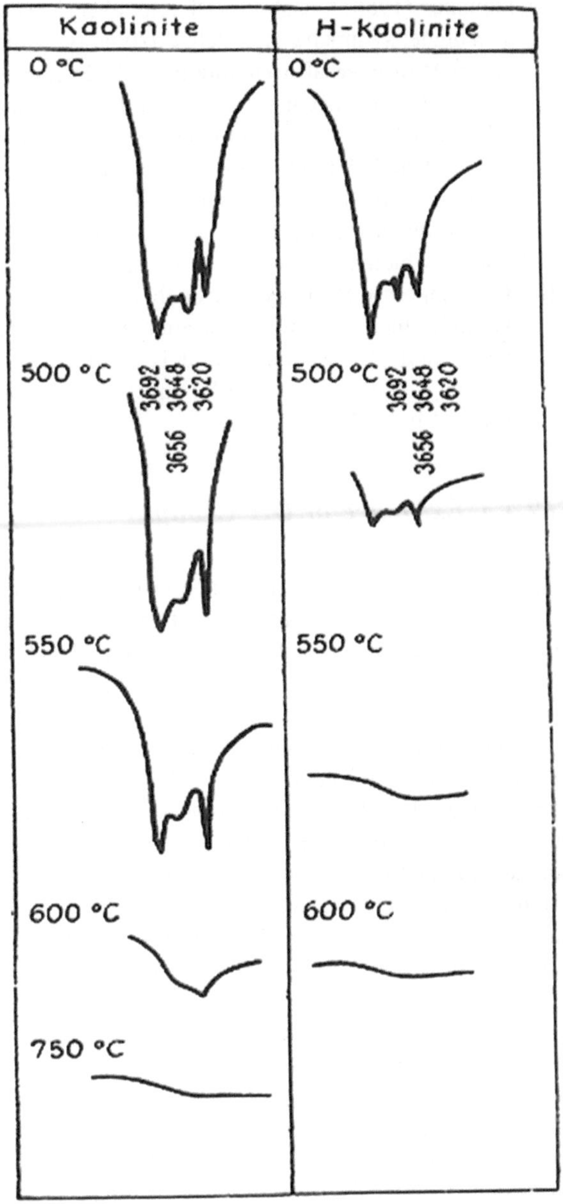

Fig. 19.2 IR—spectra of samples of kaolinite (*left*) and H- Kaolinite (*right*) heated to various temperatures (Spectral range 3,550–3,750 cm^{-1}) (After Pampuch 1958)

the residual OH groups situated at 3,620 cm^{-1} are progressively eliminated on further heating by gradual diminution of I.R. peak to a band and finally a loss of about 0.4 % water suddenly at 950 °C by rapid heating as observed in thermo gravimetric differential analysis. The temperature 950 °C lies precisely at the beginning of the exothermic reaction of the kaolinite when the sample is rapidly heated following the standard conditions of DTA. The elimination of the band due to –OH groups has also been noted by IR analysis. In the spectrograph of 980 °C heated sample, i.e., after completion of exothermal the band disappeared.

The presence of hydroxyl groups in metakaolinite structure is also noted by Redfern (1987) by TGA, Bergaya et al. (1996). Using transmission electron microscopy (TEM), and Heide and Földvari (2006) by evolved gas analysis and Rocha and Klinowski (1990) by NMR by TA analysis Trusilewicz et al. (2012) noted a weight loss up to the 1,000 °C (0.84 %). Thereafter, no hydroxyls left from the thermal activation.

Thus, it can be concluded that kaolinite losses its hydroxyl groups in two major steps. One, at the main endothermic peak. So the empirical formulae of metakaolinite will be partially hydrated alumino-silicate, e.g., Al_2O_3. $2SiO_2$ xH_2O instead of simple anhydrous $Al_2O_32SiO_2$ designated by several authors. Insley and Ewell (1935) showed the relationship between extent of heating, i.e., related to dehydroxylation and solubilization of Al_2O_3 component of metakaolinite by acid. Colegrave and Rigby (1952) noted when metakaolinite was subjected to the action of aqueous sodium carbonate solution both before and after treatment with mineral acid that an insoluble residue still left unattacked which indicate that a portion of alumina and silica could not solubilize under condition of leaching. This residue may be the unattacked portion of Al_2O_3 $2SiO_2$ XH_2O. Dekeyser (1963) claimed that residual hydroxyl groups are essential for metakaolinite structure. He and his group showed that on heating Zettlitz kaolinite at 650 °C/24 h, it did not contain not more than 1 % H_2O, but this small quantity of water escapes only with difficulty by prolong heating from 650 to 900 °C. The decrease in the amount of water is accompanied by an increase of specific gravity of the residue.

19.5 Comparison of the Predicted Metakaolinite Structures

Structural models of metakaolinite presented so far have wide variations. Table 19.1 and Fig. 19.3 show chronological developments of its structure shown by various authors.

However, it is to be reestablished on the basis of following information and accordingly it should confirm the following essential experimental facts.

(i) *X-ray/electron diffraction data*

   (a) It has residual X-ray order in a and b directions of parent kaolinite. So the model is likely to be based on a and b cell dimensions and which should nearly essentially be the same as in kaolinite.
   (b) Sufficient explanations are to be given to show that the proposed model does not exhibit c-axis periodicity.

(ii) *TGA/IR data* Dehydroxylated kaolinite contains the persistence of 10–12 % hydroxyl groups or ~1 % H_2O. Explanation should be given the regrading the essential role of those hydroxyl for stability of metakaolinite structure.

Table 19.1 Structural data of metakaolinite suggested by different authors

1	2	3
Brindley and Robinson (1948) (a) Kaolinite structure based on triclinic cell with following parameters. a = 0.515 nm b =0.895 nm c = 0.739 nm α = 91.8° β = 110.4 -105° γ = 90° (b) Bond distances Si-O = 0.162 nm Al(vi)-O = 0.194 nm	Tscheischwili et al (1939) (a) Structure of metakaolinite based on monoclinic cell with following parameters. a = 0.514 nm b = 0.890 nm c = 0.680 nm α = β = 100.2° γ = (b) Bond distances Si-O = 0.159 nm Al(iv)-O = 0.185 nm (c)Anhydrous structure with no (OH) groups. Oxygen sequence is 6 – 4 – 4.	Brindley and Nakahira(1959) (a) Structure of metakaoli--nite based on triclinic cell with following parameters. a = 0.514 nm b = 0.893 nm c = 0.630 nm α = 91.8° β = 104.5° γ = 90° (b) Bond distances Si-O = 0.164 nm Al(iv)-O = 0.184 nm (c) Anhydrous structure with no (OH) groups. Oxygen sequence is . 6 – 6 – 2.
4	5	6
Pampuch (1966) (a) Structure of meta-kaolinite based on orthorhombic cell with following parameters. a = 0.514 nm b =0.893 nm c = 0.706 nm (b) Bond distances Si-O = 0.172 nm Al(iv)-O = 0.172 nm (c) Partially hydrous structure of contains 1/8 (OH) groups.	Brindley and Gibbon (1968) b = 0.9145 nm ± 0.035	Iwai (1971) (a) Structure of metadeckite-based on triclinic cell conta--ins flatter configuration of Si – O layers. a = 0.5396 nm b = 0.9379 nm c = 0.737 nm α = 91.9° β = 104.8° γ = 89.90° (b) Bond distances Bond lengths confirm RED values.
7	8	9
Leonard (1977) Structural feature of metakaolinite at 800°C	Mackenzie et al (1985) Structure of metakaolinite a = 0.5396 nm b = 0.9379 nm c = 0. 737 nm α = 91.6° β = 104.8° γ = 89.90°	Chakraborty Proposed structural view of metakaolinite (a) No fixed structure

(continued)

Table 19.1 (continued)

(b) Bond distances	(b) Bond distances	(b) Bond distances and bond angles may vary with temperature of heating.
Si – O = 0.165 nm		
Al(iv) – O = 0.187 nm	Al(iv) – O = 0.197 – 0.203 nm	
(c) Partially anhydrous structure.	(c) Partially anhydrous structure, contains 1/8 (OH) groups and conforms Pampuch's data.	(c) Mostly anhydrous structure, with varying number of (OH) groups.

(iii) *XRF data* Structure of metakaolinite should be such as to contain both 6 and 4 coordinated Al.

(iv) *RED data* RED data indicates that some 6 coordinated Al in metakaolinite be in fact 4 coordinated in the large bond lengths. Bond lengths of Al(IV)–O, Si–O, Al(VI)–O in metakaolinite structure should be consistent with RED result shown by Leonard.

(v) *^{29}Si MAS NMR data* Structure should explain the occurrence of broad ^{29}Si MASNMR peak in the range of −99 to −102 ppm. The nature and position of the peak suggest the presence of a range of Si–O–Si(Al) bond lengths.

(vi) *^{27}Al MAS NMR data* Metakaolinite shows octahedral Al peak at −0.6 ppm and two tetrahedral Al peaks one at 65–66 ppm and the other at 34 to ppm. The proposed structure should be consistent with all those three peaks of less magnitude. The structure should also explain the loss of 90 % total Al signal during the dehydroxylation process of kaolinite.

(vii) *Electron density mapping* It suggests an increase of ***a*** and ***b*** parameters of metakaolinite to the extent of 5 %.

(viii) *Orientational relationship among phases* Structure of metakaolinite should possess an inter-relationship between parent kaolinite in one hand and its transformed spinel phase on the other hand. As kaolinite is rehydrated easily from metakaolinite so the proposed structure should easily be regenerated also smoothly to its parent form.

Sanz et al. (1988) assigned the 30 ppm intensity peak in ^{27}Al MASNMR to all penta coordinated. Accordingly, they proposed a different structure. The two Al share two oxygens corresponding to two OH Groups (03 and 04) and each Al is coordinated to three other tetrahedral oxygens, two tetrahedral apical oxygens (01 and 02), and oxygen of the next adjacent layer (05).

Mackenzie et al. (1985) critically analyzed the structure of it in detail which was proposed by various earlier researchers namely Tscheischwili et al. (1939), Brindley and Nakahira (1959), Pampuch (1966), Iwai et al. (1971), and Leonard (1977). On the bases of some newer findings from MASNMR study, they finally

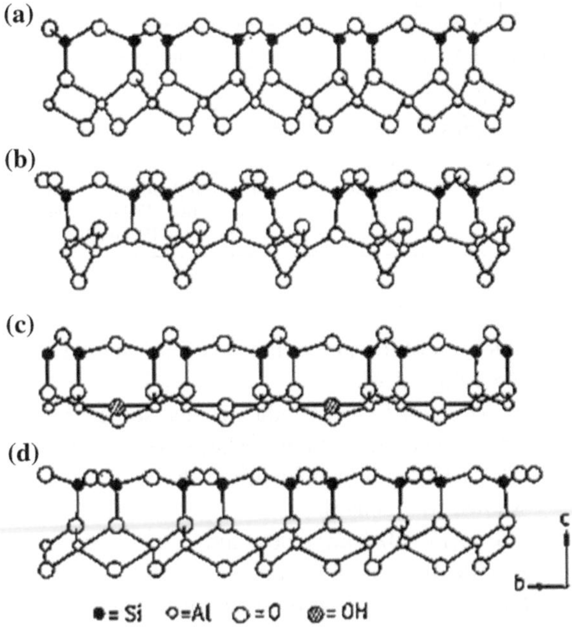

Fig. 19.3 Computer generated views along a axis for previously proposed metakaolinite structures. Suggested by **a** Structure of Tscheischwili et al. **b** Structure Brindley and Nakahira, **c** Structure of Pampuch, **d** Structure of Iwai et al. (After Mackenzie et al. 1985). Reproduced with kind permission of American Ceramic Society

presented a computer generated diagram for the proposed metakaolinite structure (Fig. 19.4).

It consists of a hydrous regions of distorted Al–O tetrahedra may be of Iwai type and other part contains randomly distributed isolated residual hydroxyl associated with Al–O configurations of regular octahedral and tetrahedral symmetry.

Carty and Senapati (1998) opined that the controversies such as the coordination of aluminum (more specifically, the ratio of fourfold to sixfold coordinated aluminum), the exact dimensions of the *c* parameter, and the amount of hydroxyl groups in the A1-0 layers continue to be far from resolved, and further characterization and structural modeling is necessary.

19.6 Newer Concept in Variability of Metakaolinite Structure

It has been noted that during dehydroxylation of kaolinite, some physical properties are changing concurrently. It is further observed that changes are still continuing during heating in the temperature between 1st endotherm to 1st exotherm.

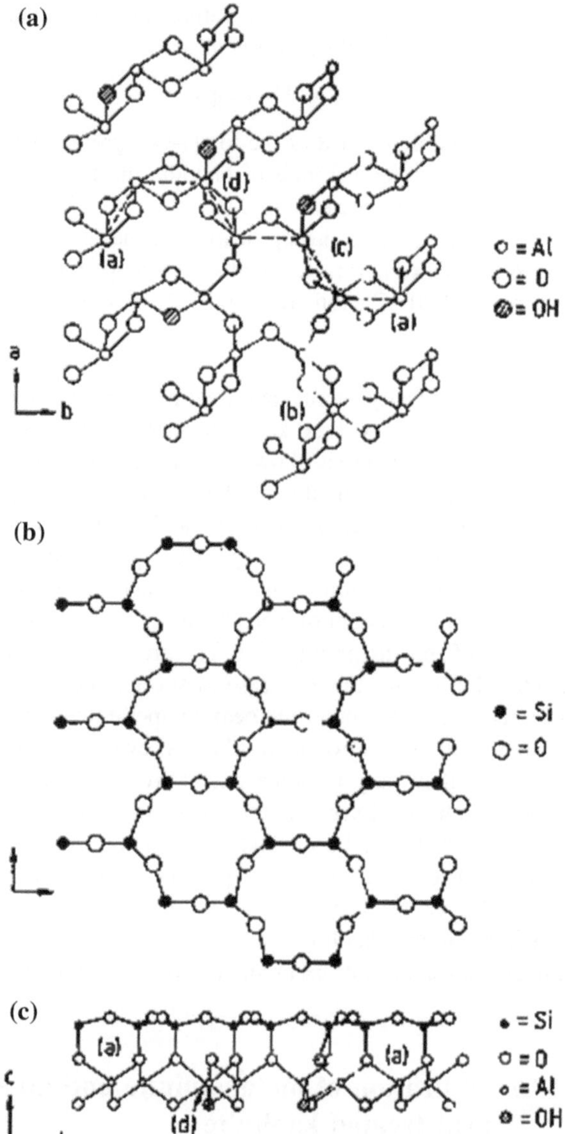

Fig. 19.4 Computer generated view of typical region of structure of metakaolinnite (After Mackenzie et al. 1985). Reproduced with kind permission of American Ceramic Society

(i) Expansion/contraction

At temperature of dehydroxylation, a gradual contraction is still followed up to the beginning of 1st exotherm (Fig. 3.4).

(ii) Surface area

Surface area changes with extent of dehydroxylation. It generally increases with rapid loss of structure at the time of dehydroxylation. Thereafter, it decreases gradually with increase of temperature (Lemaitre et al. 1975; McConnell and Fleet 1970). Then suddenly decreases at the exothermic peak temperature.

(iii) Concentration of –OH groups

It was shown that ~ 1/8 of the total –OH groups present in original kaolinite was removed at the temperature of endotherm. The remaining ~ 12 % of total –OH groups present in metakaolinite structure are eliminated in two ways. First portion is gradually removed during heating in the temperature range between 1st endotherm to 1st exotherm. Residual –OH groups are removed at the endothermic dip just before 980 °C exotherm (Fig. 7.2) as shown by Dekwser (1963) and others.

(iv) Density

Changes in density versus heat treatment of kaolinite have been shown to consist of three features. At first, kaolinite shows rapid decrease in density during dehydroxylation. Second, it shows a gradual increase in density with slow rise of temperature of heat treatment. Third, increase in density occurs rapidly near the occurrence of 1st exotherm (Ricke and Mauve 1942; Dekeyser 1963).

Thus, the changes of four physical properties are interrelated. Second, after formation of metakaolin it dehydroxylated in two ways.

Decrease in amount of water/hydroxyl out of metakaolin is accompanied by an increase of specific gravity of it. The gradual increase in density followed by slow contraction or by slow decrease in surface area of the heated kaolinites between endothermic and exothermic peak temperatures, suggests to assume that although heated kaolinites between two thermal events are designated as metakaolinite but in reality they vary to some extent from physical point of view. Obviously, it should not have a fixed structure as suggested by earlier researchers. Instead, it will change and it will be dependent upon the degree of dehydroxylation, i.e., the amount of residual OH groups present as demonstrated by Pampuch (Fig. 19.5).

As this value changes with progressive rise of temperature it is obvious that structure of metakaolinite is not a fixed one as proposed. Accordingly, c-axis dimensions could not be predicted from continuously changing density data.

19.7 Utilization of metakaolinite and alkali-leached heat-treated kaolinite

1. It is used for nuclear waste absorbing material.
2. It is used for production of porcelain clay filter. Since it shows high surface area and good pollutant catching capacity.
3. Metakaolin is used as cement mortar and has cementitious value. Since it is an amorphous alumino-silicate, which reacts aggressively with calcium hydroxide to form compounds with cementitious value. It is effectively used at 5–15 % replacement of cement by weight. It will be favorable for increasing strength

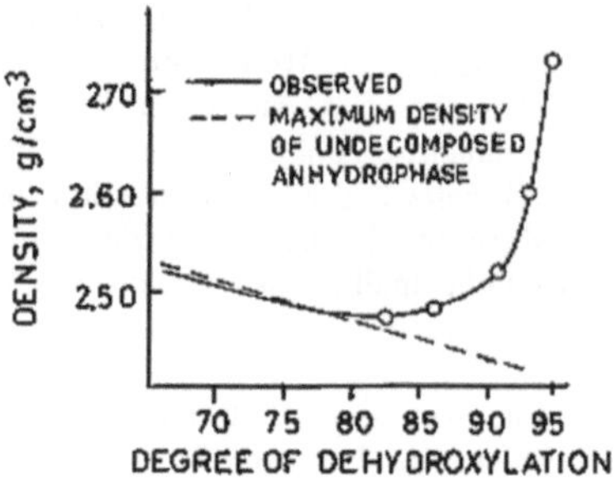

Fig. 19.5 Observed densities of Zettlitz kaolin according to Ricke and Mauve (1942) and maximum theoretical densities of unchanged metakaolin structure plotted against different degrees of dehydroxylation (After Pampuch 1966). Reproduced with kind permission of Prof. Pampuch

and reduce permeability. Meanwhile, it will also effectively reduce efflorescence and degradations caused by alkali-silica reaction (ASR) in concrete.

4. It is used for geopolymer formation which is used to make green concrete.
5. It is used as catalyst and is applied to petroleum refining because it has high matrix activity, strong antipollution capacity of heavy metal, excellent catalytic activity and selectivity.

19.8 Summary

Alkali dissolution study and direct/indirect IR analysis show no silica evolution in metakaolinite structure. This indicates that composite sheet comprising silica and alumina layer remains intact in metakaolinite. Obviously, X-ray, microscopy, and electron diffraction study interprets the existence of 2D orders with possibly change in symmetry and increase in ***b*** parameter in metakaolinite. Most of the octahedral coordination number of Al in metakaolinite becomes tetrahedral as observed in IR, XRF, RED, and MAS NMR, only a little extent of AlO_6 groups are present. Existence of small amount of Al–O (OH) groups is evident by IR and MAS NMR studies. These residual –OH groups are removed at the endothermic dip just before 1st exotherm as noted in DTA/DTG analysis. Some of the physical parameters, e.g., expansion/contraction, surface area/density of metakaolinite are changing concurrently in the temperature interval between 1st dehydroxylation to 1st exotherm. In addition, concentration of –OH groups in metakaolinite in this region also decreases which indicates that metakaolinite may not possess a fixed composition. The structure of it likely varies with temperature.

References

F. Bergaya, P. Dion, J.–F. Alcover, C. Clinard, D. Tchoubar, TEM study of kaolinite thermal decomposition by controlled-rate analysis . J. Mater. Sci. 31, 5069–5075 (1996)

G.W. Brindley, K. Hunter, Thermal reactions of nacrite and the formation of metakaolin, γ-alumina, and mullite. Mineral. Mag. **30**(288), 574–584 (1955)

G.W. Brindley, M. Nakahira, The Kaolinite-Mullite reaction series: I, II, and III,. J. Am. Ceram. Soc. **42**(7), 311–314 (1959)

G.W. Brindley, H.A. McKinstry, The Kaolinite-Mullite reaction series: IV, the coordination of aluminum. J. Am. Ceram. Soc. **44**(10), 506–507 (1961)

G.W. Brindley and K. Robinson, Structure of Kaolinite. Miner. Mag. **27**, 242–253 (1946)

I.W.M. Brown, K.J.D. MacKenzie, M.E. Bowden, R.H. Meinhold, Outstanding problems in the kaolinite–mullite reaction sequence investigated by ^{29}Si and ^{27}Al solid-state nuclear magnetic resonance: II, high-temperature transformations of metakaolinite. J. Am. Ceram. Soc. **68**(6), 298–301 (1985)

P.P. Budnikov, Ber. Dtsch. Keram.Ges. **16**, 349 (1935)

M. Bulens, A. Leonard, B. Delmon, spectroscopic investigations of the Kaolinite–Mullite reaction sequence. J. Am. Ceram. Soc. **61**(1–2), 81–84 (1978)

A. K. Chakraborty, D. K. Ghosh, Re-examination of the kaolinite to mullite reaction series. J. Am. Ceram. Soc. **61**(3–4), 170–173 (1978b)

F.W. Clarke, Constitution of the silicates. L7. S. Geol.Surury Bull. **125**(7–109), 32 (1895)

J.E. Comeforo, R.B. Fischer, W.F. Bradley, Mullitization of Kaolinite. J. Am. Ceram. Soc. **31**(9), 254–259 (1948)

J.J. Comer, New electron-optical data on the Kaolinite–Mullite transformation. J. Am. Ceram. Soc. **44**(11), 561–563 (1961)

J.J. Comer, J.H. Koenig, S.C. Lyons, What are ceramic materials really like? "ceram. Ind .**67**(4), 125,148,150 No. 6, 96 (1956)

E.B. Colegrave and G. R. Rigby, The Decomposition of kaolinite by heat. Trans. Brit. Ceram. Soc. **51**(6), 355–367(1952)

W.M. Carty and U. Senapati, Porcelain-raw materials, processing, phase evolution, and mechanical behavior. J. Am. Ceram. Soc. **81**(1), 3–20 (1998)

E.B. Colegrave and G. R. Rigby, The Decomposition of kaolinite by heat. Trans. Brit. Ceram. Soc. **51**(6), 355367(1952)

W.M. Carty and U. Senapati, Porcelain-raw materials, processing, phase evolution, and mechanical behavior. J. Am. Ceram. Soc. **81**(1), 320 (1998)

W.L. De Keyser, R. Wollast and L. De Laet, Contribution to the study of OH groups in kaolin minerals, Intl. Clay Conf., Pergamon Press. 75–86 (1963d)

W.L. De Keyser, Note concerning the exotherm reaction of Kaolinite and formation of spinel phase preceding that of Mullite. In International clay conference Pergamon press, pp. 91–96 (1963e)

W. Eitel, H. Kedesdy, Elektronen-Mikroskopie und Eeugung silikatischer Metaphasen: IV, Der Metakaolin (Electron Microscopy and Diffraction of Silicate Metaphases: IV, Metakaoliu). Abhandl. preuss. Akad. Wiss. Math.-nattir w. KI. **5**, 37–45 (1943)

W. Eitel, H.O. Miiller, O.E. Radczewski, Ubermikroskopische Untersuchungen an Tonminerdlien (Ultramicroscopic Exaniination of Clay Minerals). Ber. deut. keram. Ges. **20**(4), 165–180 (1939)

F. Freund, Explanation of exothermal reaction of Kaolinite as a 'Reaction of the Active State'. Ber. Deut. Keram. Ges. **37**(51), 209–218 (1960)

F. Freund, Kaolinite-metakaolinite, a model of a solid with Extremely high lattice defect concentrationsl. Ber. Dtsch. Keram. Ges. **44**(I), 5–13 (1967)

F. Freund, Infrared spectra of kaolinite, metakaolinite, and Al–Si spinel. Ber. Deut. Keram. Ges. **44**(181), 392–397 (1967)

M.C. Gastuche, F.Toussaint, J.J. Fripiat, R. Touilleaux, M. Van Meersche, Study of intermediate stages in the kaolin-metakaolin transformation. Clay Minerals Bull. **5**(29), 227 (1962)

M.C. Gastuche, J.J. Fripiat, Acid dissolution techniques applied to the determination of the structure of clay and controlled by physical methods. *Science of Ceramics*, ed. by G.H. Stewart (Academic Press, London,1963) p. 96

K. Heide and M. Fö ldvari, High temperature mass spectrometric gas release studies of kaolinite $Al_2[Si_2O_5(OH)_4]$ decomposition. Thermochim. Acta **446**, 106–12 (2006)

J.F. Hyslop, Decomposition of clay by heat. Trans. Brit. Ceranz. Soc. **43**(3), 49–51 (1944)

J.F. Hyslop, H.B. Rooksby, Further note on crystalline break up of kaolin. Trans. Br. Ceram. Soc. **27**(4), 299–302 (1928)

I.H. Insley, R.H. Ewell, Thermal behavior of the kaolin minerals. J. Res. Natl. Bur. Stand. **14**(S), 615–627 (1935)

S. Iwai, M. Tagai, T. Shimamnne, Procedure for Dickite structure modification by dehydration. Acta Cqstallogr. Sect. B: Struct. Crystallogr.Crysr. Chem. **27**, 248–250 (1971)

G. Keppeer, Sprechsaal **58**, 614 (1925)

O. Krause, H. Wohner, Uber Die Vorange Beim Bremmen Technischer Tone, Ber.Dtsch. Keram. Ges. **13**, 485 (1932)

G. Kubicki, R.E. Grim, A new method for thermal dehydration studies of clay Minerals. Min. Mag. **32**, 53 (1959)

J. Lemaitre, M. Bullens, B. Delmon, Influence of mineralizers on the 950°C exothermic reaction of metakaolinite, in *Proceedings of the International Clay Conference*, ed. by S. W. Bailey (Applied Publishing Ltd., Wilmette, IL, Mexico City, Mexico, 1975), p. 539–44

A.J. Leonard, Structural analysis of the transition phases in the Kaolinite–Mullite thermal sequence. J. Am. Ceram. Soc. **60**(1–2), 37–43 (1977)

K.J.D. MacKenzie, I.W.M. Brown, R.H. Meinhold, M.E. Bowden, Outstanding problems in the Kaolinite–Mullite reaction sequence investigated by 29Si and 27Al solid-state nuclear magnetic resonance: I, Metakaolinite. J. Am. Ceram. Soc. **68**(6), 293–97 (1985)

J.D.C. McConnell, S.G. Fleet, Electron optical study of the thermal decomposition of kaolinite. Clay Miner. **8**, 279–290 (1970)

V. Mellor, I.I. Holdcroft, Chemical constitution of the kaolinite molecule. Trans. Ceram. Sor. (Engl.) **10**, 94–120 (1911)

B. Neumann, S. Kober, Sprechsaal **59**, 607 (1926)

R.H. Meinhold, K.L.D. Mackenzie, I.W.M. Brown, Thermal reactions of kaolinite studied by solid state ^{27}Al and ^{29}Si NMR. J. Mater. Sci. Lett. **4**, 163–166 (1985)

R.E. Newnham, A refinement of Dickite structure ad remarks o polymorphism in kaolin minerals. Miner. Mag. 32, 683 (1961)

R. Pampuch, Infrared study of thermal transformations of kaolinite and the structure of metakaolin. Pol. Akad. Nauk. Oddzial Krakowie, Kom, Nauk, Mineral, Pr. Mineral, **6**, 53–70 (1966)

R. Pampuch, Mechanism of topotaxial thermal decomposition reactions of layer lattice silicates and hydroxides. in proceeding 9th conference silicate Ind., Budapest, 1958, pp. 144–151

H.J. Percival, J.F. Duncan, P.K. Foster, Interpretation of the Kaolinite–Mullite reaction sequence from infrared absorption spectra. J. Am. Ceram. Soc. **57**(2), 57–61 (1974)

S.A.T. Redfern, The Kinetics of dehydroxylation of kaolinite. Clay Miner. **22**, 447–456 (1987)

R. Rieke and L. Mauve, Zur Frage des Nachweises der mineralischen Bestandteile der Kaoline (Indications as to Mineral Constituents of Kaolin). Ber. deut. keram. Ges. **23**(141), 11–51(1942)

F. Rinne, Rontgrnographischc Diagnostik heini Rrennen von Kalkstcin, Dolomit,Kaolin. und Glimmer (X-Ray Study of Calcined Calcite, Dolomite, Kaolinite, and Mica). Z. Krist. **61**(1/2), 113 (1924)

J. Rocha, J. Klinowski, $_{29}$Si and $_{27}$Al Magic-Angle-Spinning NMR studies of the thermal transformation of kaolinite. Phys. Chem. Minerals **17**(2), 179–86 (1990)

R. Roy, D.M. Roy, E.E. Francis, New data on thermal decomposition of Kaolinite and Halloysite. J. Am. Ceram. Soc. **38**(6), 198–205 (1955)

H. Salmong, Physikalischen und cheinischen Grundlagcn der Keramik (Physical and Chemical Principles of Ceramics), p. 73. Julius Springer, Berlin, 1933. 229 pp.; Ceram. Abstr. **13**(4), 103 (1934)

J. Sanz, A. Madani, J.M. Serratosa, J.S. Moya, S. Aza, 27Aluminum and 29Silicon-magic-angle spinning nuclear magnetic resonance study of the Kaolinite–Mullite transformation. J. Am. Ceram. Soc. 71(10), C-418–C-421 (1988)

B. Sonuparlak, M. Sarikaya, I.A. Aksay, Spinel phase formation during the 980 °C exothermic reaction in the kaolinite-to-mullite reaction Series, J. Am. Ceram. Soc. **70**(11), 837–842 (1987)

V.S. Stubican, R. Roy, Isomorphous substitution and infrared spectra of the layer lattice silicates. Amer. Mineral. **46**(1–2), 32–51 (1961)

V. Stubican, Residual hydroxyl groups in the metakaolin range. Min. Mag. 32, 38–52 (1959)

G. Tamman, W. Pape, Ilber Den Wasserverlust Des Kaolines and Seinverhat-en In Festen Zuden Carbonatem Und Oxyden Der Erdalkalien, in Z. Anorg. Allg. Chem. **127**, 43–68 (1923)

L. Trusilewicz, F. Ferná ndez-Martínez, V. Rahhal, R. Talero, TEM and SAED characterization of metakaolin pozzolanic activity. J. Am. Ceram. Soc. **95**(9), 2989–2996 (2012)

L. Tscheischwili, W. Biissern, W. Wevl, Uber den metakaolin (Metakaolin). Ber. deut. keranz. Ges. **20**(61), 249–276 (1939)

Y. Tsuzuki, K. Nagasawa, A. transitional stage to 980 °C exotherm of Kaolin Minerals. Clay Sci. **3**(5), 87–102 (1969)

S. Udagawa, T. Nakada, M. Nakahira, Molecular structure of allophane as revealed by its thermal transformation. pp. 151 in Proc. Int.Clay Conf., Vol.1, Editor-in-chief Lisa Heller bisa Heller by Israil University Press, Gerusalem (1969).

K.A. Vesterbcrg, Kaolin and its thermal changes. Arkh Krmi, Jlinrml. GcoE. **9**(1141), 26 (1925)

R. Wardle, G.W. Brindley, Dependence of wavelength of, AIKα radiation from Alumino–Silicates'on the A1-0 distance. Amer. 2 Mineral. **56**(111–121), 2123–2128 (1971)

T. Watanabe, H. Shimizu, K. Nagasawa, A. Masuda, H. Saito, ^{29}Silicon ^{27}Al-MAS/NMR study of the thermal transformations of kaolinite. Clay Miner **22**, 37–48 (1987)

Further Readings

H. Abe, Preparation of porous mulllite ceramics by leaching method. J. Ceram. Soc. Jap. **97**, 604–611 (1989)

H. Abe, Preparation of porous mulllite ceramics by leaching method, (Part 2 , Effect of transition metaloxide on crystal growth of mullite. J. Ceram. Soc. Jap. **98**, (1990)

H. Abe, Preparation of porous mulllite ceramics by leaching method. J. Ceram. Soc. Jap. **44**, 339–347 (1990)

H. Katsuki, S. Furuta, H. Ichinose, H. Nakao, Preparation & some properties of porous ceramic sheet composed of needle-like mullite. J. Ceram. Soc. Jap. **96**, 1081–1086 (1988)

H. Katsuki, High temperature properties of neede like mullite obtained from New Zealand kaolin. J. Ceram. Soc. Jap. **97**, 1521–1524 (1989)

H. Katsuki, D. Matsuda, Preparation, some properties & application of needle-like mullite obtained from kaolin minerals, Fortschritts Berich-eta DKG **7**, 122–129 (1992)

LU Yinping, L.I. Kaiqi, LIU Qinfu, Study of preparation of mullite from desilicated kaolin. J. Chinese Ceramic Soc. **32**(8) 1033–1035 (2004)

S.M. Naga, I.M. Bakr, Effect of etching treatment on mullite – based bodies. Interceram. **51**(6), 404–407 (2002)

K. Okada , H. Kawashima, Y. Saito, S. Hayashi, A. Yasumori, New preparation method of mesoporous γ – alumina by selective leaching of calcined kaolin Minerals, J. Mater. Chem. **5**, 1241–1244 (1995)

R.M. Torres Sanchez, S.L. Perez De Vargas, E. Soto, E.I. Basadela, Influence of kaolinite crystalline structure in the production of γ-Al_2O_3 by alkaline lixiviation. Materials etters **57**, 1167–1170 (2003)

S. Yariv , E. Mendelovici and R.Villalba, The Study of the interaction between cesium chloride and kaolinite by thermal methods

Chapter 20
Spinel Phase

20.1 Introduction

The problem regarding the composition of spinel phase still remains. The various tools used for direct and indirect characterizations of it are the followings:

(i) Powder X-ray was used for identification of spinel phase and its lattice parameter measurement.
(ii) Single crystal X-ray diffraction showing existence of structural order in kaolinite to metakaolinite to spinel transformation.
(iii) Electron diffraction showing the relationship between spinel and mullite transformation.
(iv) RED study of the whole pattern when the components and their coefficients were measured.
(v) Coordination number of Aluminum measurement by XRF, IR, and MAS NMR was practiced for characterization of spinel phase.
(vi) Chemical analysis of spinel phase by EDS in an analytical TEM was done after leaching of heated kaolinite by NaOH with the assumption that only spinel phase was physically isolated and all associated silica phase was removed.

Based on the findings made by various authors, the problem regarding the proper characterization of spinel phase whether it is simple γ-Al_2O_3 or Al–Si spinel and if it is Al–Si spinel then what should be the probable composition of it is discussed in this chapter.

20.2 Arise of Controversy

Among the various physicochemical techniques as shown in previous chapters, by simple X-ray diffraction studies, researchers put forward two different opinions regarding the composition of it. As such two types of 980 °C exothermic reactions of kaolinite are prevailed.

A. K. Chakraborty, *Phase Transformation of Kaolinite Clay*,
DOI: 10.1007/978-81-322-1154-9_20,

20.2.1 First School

X-ray powder diffraction photographs of kaolinite heated in the temperature range 900–1050 °C at varying heat treatment schedule (Fig. 20.1a) consist of three board and diffuse reflections at 0.139, 0.197, and 0.238 nm, respectively. On comparison, these reflections are found to correspond to (400), (440), and (311) planes of γ-Al_2O_3 (Fig. 20.1b) and to ASTM data (Table 20.1).

Accordingly, Tammann and Pape (1923), MacVay and Thomson (1928), Hislop and Rooksby (1928a, b), Budnikov and Khiz (1929), Zwetsch (1934), Insley and Ewell (1935), Jay (1939), Tschischwili et al. (1939), Hyslop (1944), Colegrave and Rigby (1952), Richardson and Wilde (1952), Glass (1954), Vaughn (1955),

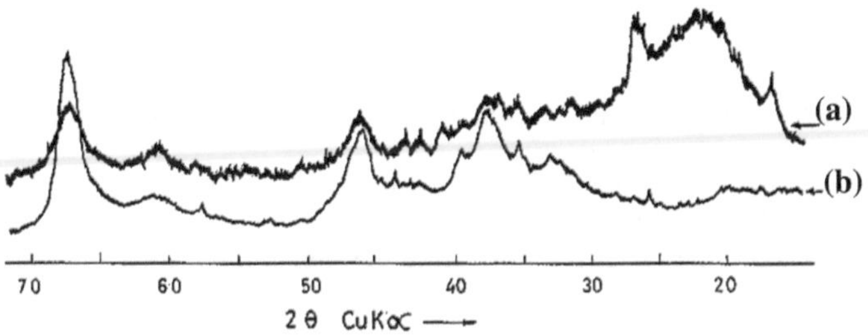

Fig. 20.1 XRD pattern of 980 °C heated **a** Rajmohol kaolinite and **b** Standard γ-Al_2O_3

Table 20.1 d(nm) Spacings of Bragg reflection peaks of cubic spinel phase obtained from different sources compared with standard pattern of γ-Al_2O_3 (After Chakraborty)

γ-Al_2O_3				Al–Si spinel derived out of different sources				
1 Ref-				Kaolinite (powder) 2 Ref-	Kaolinite (single crystal) 3 Ref-		Naturally occurring 4 Ref-	Halloysite 5 Ref-
d(nm)	I/I_0	hkl	L.C.	d(nm) I/I_0	d(nm) I/I_0	L.C.	d(nm) I/I_0	d(nm) I/I_0
0.455	10	111		–				
0.278	15	220		–				
0.238	35	311		0.238 m				
0.228	20	222		–				
0.197	100	400	7.908	0.197 s	0.197 s	7.804	0.197 s	0.19673 s Ring
0.152	10	511		–				
0.139	90	440	7.906	0.139 vs	0.139 vs	7.885	0.139 s	0.13919 vs Ring
0.114	5	444		–	0.114 w	7.804		0.1141 w
0.103	3	731		–				
0.098	5	800		–				
0.088	5	840						
0.080	5	884						

s strong, *vs* very strong, *w* weak, *vwb* very weak broad, *vvw* very very weak
1 JCPDS; *2*, *3* Brindley and Nakahira, *4* Lotgering (1959), *5* Changling (1985)

Lundin (1958), and Schieltz and Soliman (1966) with the X-ray evidence conjectured the spinel phase as pure Al spinel (γ-Al_2O_3). Based on electron diffraction studies, Eitel and Kedesdy (1943), Radczewski (1953), and Roy et al. (1955) noted the diffraction spots of this spinel phase as γ-Al_2O_3.

Sonuparlak et al. (1987) heat treated Georgia kaolinite for 1*d* and showed microdiffraction pattern (Fig. 6.4) in their high-resolution electron microscopic study, which confirms the fcc structure of spinel phase. Thus, first school believed the formation of γ-Al_2O_3 spinel at the first exotherm as per Eq. 12.4.

$$\text{Metakaolinite} \xrightarrow[\text{Exotherm}]{\text{At} \sim 980\,^{\circ}\text{C}} \gamma\text{-Al}_2\text{O}_3 + 2\text{SiO}_2\,(\text{A}) \tag{12.4}$$

20.2.2 Second School

By single crystal X-ray diffraction pattern of kaolintie flakes, (Fig. 5.3) Brindley and Nakahira (1959) first showed the existence of orientational relationship among kaolinitemetakaolinite and spinel phases. It was shown that b-axis of kaolinite is parallel to b-axis of metakaolinite and is subsequently parallel to face diagonal (110) of spinel phase. To explain the structural continuity among the three phases, they conjectured the incorporation of silicon in this spinel phase as per the Eq. 20.1.

$$\begin{aligned} \text{Metakaolinite} \xrightarrow[\text{Exotherm}]{\text{At} \sim 980^{\circ}\text{C}} & \; 2\text{Al}_2\text{O}_3 \cdot 3\text{SiO}_2 \\ & \text{Al}-\text{Si spinel} \\ & + \; 2\text{SiO}_2\,(\text{A}) \\ & (\sim 14\ \text{wt\%}) \end{aligned} \tag{20.1}$$

Only after this hypothesis was published by Brindley and Nakahira, the controversy i.e., whether it is simple γ-Al_2O_3 (Al-spinel) or Al–Si spinel arose. The size of the cubic spinel phase was shown by Comer (1961) in electron micrograph to be 12.5 nm. Similar value was also shown subsequently by Okada et al. and Srikrishna et al. by their electron microscopic studies. Sonuparlak et al. were however successful in developing larger grain of spinel phase about 100 nm. Schieltz and Soliman on the other hand tried to grow bigger size of spinel phase even on continued heating of kaolinite at exothermic peak temperature. It was not achieved however.

Comer (1961) showed electron diffraction pattern of this cubic spinel phase consisting of two oriented rings due to (400) and (440) reflections analogous to peaks observed generally in powder X-ray diffraction pattern. Besides, he also showed the orientational relationship between spinel phase and mullite. It is shown that c-axis of mullite runs parallel to (110) axis of spinel phase (Fig. 6.2). Thus, the entire axial relationship from kaolinite to metakaolinite to spinel phase and lastly from spinel phase to mullite is established by the single crystal studies jointly by Brindley and Nakahira and Comer.

Brindley and Nakahira (1959) showed the lattice constant of spinel phase was 7.886 Å out of single crystal kaolinite heated to 1050 °C. Srikrishna et al. (1990) noted lattice parameter value of spinel crystallities ~7.88 Å out of heat treated alkali leached kaolinite (Fig. 18.6).

Table 20.2 as summarized below shows that crystal size of spinel as observed by various researchers is small. Moreover, crystallinity of it is very poor (Fig. 20.1), three diffracted peaks are only observed, these are eventually very broad and weak in nature and are very difficult to measure peak maximum for lattice parameter measurement. Thus, proper identification of it is rather a challenging task upon mineralogist, crystallographers.

According to Bradley and Grim (1951), the spinel phase which appears first after decomposition of metakaolinite shows notably diffuse X-ray pattern and probably more true rather than having any particular composition, the phase is merely an irregularly constituted assemblage of small cations, some octahedrally coordinated and other tetrahedrally coordinated, with the oxygen packing actually approaching crystalline regularity. The character of the spinel type mixed oxides is thus in marked contrast to that of γ-Al_2O_3, a phase which has formed time to time been described as a spinel like structure, γ-Al_2O_3 also is probably some mixed arrangement of tetrahedrally and octahedrally coordinated aluminum ions.

Subsequent researchers by using other techniques namely: Alkali leaching, IR, XRF, RED, Leaching, HREM, analytical TEM, EDS, MASNMR, etc., are attempted to explain either of two views of spinel characterization.

Following experimental techniques have been undertaken by various researchers to reveal the composition of controversial spine phase.

Chakraborty and Ghosh (1978a) predicted that spinel phase is really a silicon bearing spinel other than the composition $2Al_2O_3{\cdot}3SiO_2$ suggested by Brindley and Nakahira (1959). Composition of it may only differ and adds to recent controversy. It may be analogous to the composition of 3:2 mullite i.e., $[Si_{4.92}Al_{3.06}][Al_{11.69}\ []_{4.31}]O_{32}$. On the contrary, it may contain up to 18 mol (~10 wt) % silica i.e., $[Si_2\ Al_6][Al_{12.67}\ []_{3.33}]O_{32}$ as per Okada and Otsuka (1986) and, Sonuparlak et al. (1987) (Table 20.3).

Table 20.2 Size of spinel phase observed under microscope by various researchers

Researchers	Size of spinel crystal (nm) observed by TEM
Mcconnell and Fleet	5–10
Okada et al.	5–8
Sonuparlak et al.	5–8
Srikrishna et al.	10–50
McConville et al.	5–10

Table 20.3 Composition of spinel phase proposed/determined by various authors

Researchers	Characterization method
1. Earlier researchers	X-ray identification and comparison with JCPDS Data
2. Brindley and Nakahira	Single crystal X-ray
3. Comer	Electron diffraction
4. Leonard	RED and XRF
5. Chakraborty and Ghosh	Alkali Leaching to estimate amount of free silica(A)
6. Percival et al.	I R with Alkali leaching
7. Okada et al.	EDS with Alkali leaching
8. Moya et al.	Alkali leaching
9. Sonuparlak et al.	HREM, Alkai leaching and EDS
10. Majumder and Mukherjee	Lattice energy calculation
11. Srikrishna et al.	Alkai leaching and EDS
12. Mackenzie, Hartman and Okada	MAS NMR
13. Lee et al.	EF-TEM

20.3 Thermodynamic Approach

Due to different experimental observations by various physicochemical studies of heated kaolinites, it is felt necessary to interpret the probable nature of spinel phase based on some thermodynamic considerations.

20.3.1 Free Energy and Heat of Formation

Many authors made an insight in the thermal reaction occurring when kaolinite is heated by considering the changes in free energy values, Vaughn (1955) calculated the change in free energy at $\Delta F_{298\ K}$ and thereafter, Schieltz and Soliman (1966) also calculated $\Delta F_{1250\ K}$ for some theoretical transformation of metakaolinite based on published free energy data. Later on, Chakraborty et al. (1986) recalculated the same out of free energy of formation and heat of formation data of different substances (Table 12.1) for at least 13 possible ways of decomposition of metakaolinite vide Table 12.2. It shows that a decrease in free energy occurs for 13 equations and this indicates the possibility of all such reactions in forward direction. That means theoretically different reaction products, e.g., γ-Al_2O_3 and SiO_2(A), mullite and cristobalite may form when kaolinite is heated at high temperature in accordance to the stated equations. It is ascertained that out of those reactions, the Eq. 12.8 shows maximum decrease in free energy. Therefore, Eq. 12.8 should be the final transformation reaction of metakaolinite. Schieltz and Soliman also noted that $-\Delta G$ (change in free energy) for same Eq. 12.8 shows maximum in the temperature range between 298 and 2100 K.

On comparing the free energy of formations of the two reactions namely 12.13 and 12.8, it shows from the Table 12.2 that $-\Delta F$ for Eq. 12.8 (-447.65 kJ/mol) is lesser than that of the Eq. 12.13 (-290.55 kJ/mol). When a greater free energy change is observed in a single step process, it is theoretically expected that metakaolinite may directly transform to mullite and cristobalite as per Eq. 12.8 instead of it decomposes first into γ-Al_2O_3 and SiO_2 in Step-I and recombination thereof in Step-II at high temperature according to Fig. 20.2.

It is highly probable that there may not be any need of γ-Al_2O_3 formation when ultimate transformation is directed toward mullitization, since $-\Delta F_1 \ll -\Delta F_3$, obviously formation of any γ-Al_2O_3 intermediary phase type spinel may be ruled out as per the above thermodynamic reasoning.

20.3.2 Lattice Energy Data

Quite analogous is the case in which the lattice energy concept is applied to metakaolinite transformation. The lowest value of lattice energy indicates the most stable reaction product. Accordingly, by comparing lattice energy values (Table 12.4) of different compounds, mullite should be the theoretical reaction product of kaolinite at 980 °C. But in practice, the cubic spinel phase is found as the major phase and weakly crystalline mullite as only a minor phase in the stability range between 950 and 1,100 °C of heated kaolinite. Where lies this discrepancy? In this connection, this can be solved or explained if *Chakraborty and Ghosh*'s old suggestions that cubic spinel phase is an alumino silicate of composition analogous to that of orthorhombic mullite or a polymorph of mullite other than simple Al-spinel (Al_2O_3) is accepted. In support of the view above, the thermodynamic calculations of the present study show that lattice energy of Al–Si spinel of mullite composition is slightly higher than orthorhombic mullite and, hence the Al–Si spinel is the most feasible 980 °C phase (Table 12.4). This view is reexamined by Chakraborty *using* free energy changes of Eqs. 12.13 and 12.8 vide Table 12.2 with other equations.

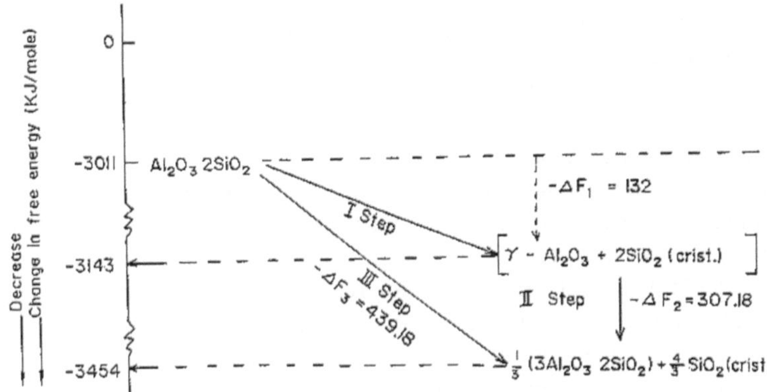

Fig. 20.2 Decrease in free energy of three steps of transformation of metakaolinite

20.4 Comparative Mullitization Behaviors of Kaolinite Versus Synthetic Mixture

It was felt that characterization of 980 °C spinel phase could be done indirectly by comparing the growth characteristics of mullite formations in natural kaolinite mineral with that of synthetic Al_2O_3/SiO_2 mixture.

Wahl et al. (1964) studied the mullitization behavior of various mixtures of silica and alumina forms e.g., diaspore—cristobalite, gibbsite—silicic acid, corundum—cristobaite, etc. They showed that initial form of the alumina and silica has a large influence on mullite development. Johnson and Pask (1982) noted interdiffusion of aluminum and silicon atoms between corundum and quartz grains and develops liquid phase. Mullite grows by a solution precipitation process.

Saruhan et al. (1994), Mullet and Schneider (1994), and Schmucker et al. (1994) showed that mullite formation process is a multiple-step process in various forms of silica–alumina mixtures e.g., quartz-corundum, cristobalite-corundum, and silica glass-corundum. However, the behavior of mulitization in kaolinite is out and out different.

The sequence of transformation of kaolinite was shown earlier by Okada et al. (1986), Udagawa et al. (1969), and Slaughter and Keller (1959). The later authors showed the formation of mullite and cristobalite phases schematically at each successive temperature of heating some kaolin clays of sedimentary origin. They plotted quantity of different phases which are approximately proportional to the height of the phase bands over a range of temperature.

Chakraborty and Ghosh (1991) showed that the nature and course of mullitization of natural clay are different from those of a synthetic mixture in the following two ways: (i) Corundum does not form before mullitization; (ii) mullitization takes place very suddenly in the temperature region 1,150–1,250 °C. In the synthetic mixture, crystallization of the reactants are predominant in the temperature range where much mullite forms from kaolinite. Substantial mullite formation occurs in this case at elevated temperature (Fig. 14.5). The different behaviors of mullite formation in the two cases are shown schematically in Fig. 20.3.

In case of kaolinite, a continuous increase in mullite occurs out of crystallization from residual mullite(A) phase at 1,050 °C to onward temperature. At ~1,250 °C a rapidity of mullite formation from Al–Si spinel also occurs and adds to general growth of mullite.

The mode of crystallization of the component oxides and mullite formation thereof from the synthetic mixture as shown in the schematic diagram (Fig. 14.6) indicated that γ-Al_2O_3 (formed at the intermediate stage) does not react with SiO_2 (A) to form mullite at a temperature between 1,100 and 1,250 °C. In a very similar way, the reaction between γ-Al_2O_3 and likely aluminosilicate phase (A) may not be feasible in the same temperature range. Instead the transition of γ-Al_2O_3 to α-Al_2O_3 which is the most obvious reaction product takes place and this is experimentally verified.

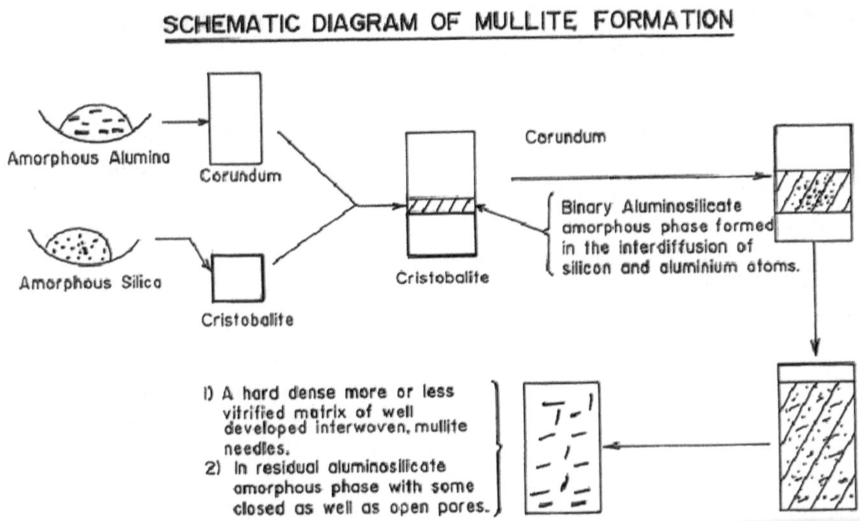

Fig. 20.3 Schematic course of mullitization of artificial γ-Al_2O_3 and SiO_2 mixture (After Chakraborty and Ghosh 1991). Reprinted by permission of the American Ceramic Society

Thus, the result concludes that spinel derived from kaolinite behaves differently in mullitization reaction then the reaction behavior of synthetic mixture of γ-Al_2O_3/SiO_2. Moreover, the XRD patterns of 980 °C heated kaolinite is different from that of pure γ-Al_2O_3 phase, so far as the missing of the 0.455, 0.278, and 0.228 nm Bragg diffraction peaks etc., and relative intensity ratio of 440 and 400 reflections of the cubic spinel phase are concerned.

20.5 Theoretical Concept of Al–Si Spinel Formation

The formation of Al–Si spinel from an alumina silicate connected structure (as like kaolinite) may be possible from the view of isomorphous substitution of Si^{+4} for Al^{+3} and vice versa, according to Bragg (1937) and Wells (1950). This phenomenon is well-known in various alumino-silicate structures. A question simultaneously arises, what will be the resultant effect of the substitution of Al^{+3} by Si^{+4} in cubic γ-Al_2O_3 structure in order that Al–Si spinel may form? Whatever be the change would take place, it depends upon the followings:

(i) The nature of the species substituting each other i.e., their ionic sizes.
(ii) Coordination number of substituting species.
(iii) The extent of such substitution.
(iv) The original structure of γ-Al_2O_3, its packing, vacancies, etc., is to be taken into consideration.

The sizes of Si and Al are close e.g., $Si^{+4} = 0.040$ and $Al^{+3} = 0.050$ nm, respectively. The cation–oxygen bond distances differ only by 10 % approximately e.g., Si–O = 0.162 , Al^{IV}–O = 0.178 nm, respectively. The structure of γ-Al_2O_3 is face centered cubic (spinel type) e.g.,

$$8\ \mathrm{Al}\ [\mathrm{Al}_{13.33}\ []_{2.66}]\ \mathrm{O}_{32}$$

Eight tetrahedral sites of the spinel structure are occupied by 8 Al atoms, rest Al atoms are present in 16 octahedral sites, and 2.66 sites are left vacant in the arrangement of 32 oxygen atoms of the cubic lattice.

If Al–Si spinel forms, then Si^{+4} will occupy into the tetrahedral sites. Simultaneously, Al^{+3} will shift from tetrahedral sites to octahedral sites (since Al possesses dual coordination). The number of vacancies will also change. Consequently, coordination number of Al will change. Since three exist a small difference in atomic sizes of the two cations and size of Si^{+4} is less Al^{+3} then, it is expected that Al–Si spinel if forms, its unit cell size will be little bit smaller than that of γ-Al_2O_3. But actual change cannot be predicted theoretically since during substitution a slight push–pull adjustment will take place within the structure and the final atomic positions will be established. Therefore, one cannot expect a large difference in cubic unit cell size although the difference in bond distance is of the order of ~10 %. Inspite of this, researchers expected that spinel phase would be characterized by measurement of its unit cell size and then comparison with γ-Al_2O_3.

20.6 Lattice Parameter Measurement

Glass (1954) first noted that the size of cubic spinel phase obtained from heating kaolinite was larger than that of γ-Al_2O_3. However, Brindley and Nakahira showed that there was a real difference in lattice parameter between the two phases and the size of Al–Si spinel (0.7886 nm) was slightly less than that of pure γ-Al_2O_3 (0.7906 nm). The accuracy of lattice parameter value of course depends upon the precise measurement of d nm lines. Secondly, the small difference in lattice parameter arises from very slight difference in d nm values. This difference is observed in third decimal places. This is in expectation of theoretical conception as discussed. The exact measurement up to the accuracy of third decimal places of d nm values of the cubic phase, however, seems to be very difficult due to the following drawbacks and these may lead to erroneous results in practice.

(i) Number of *d* nm peaks of the cubic phase are very limited. In power photocopy only two to three lines are observed namely: 0.139, 0.197, and 0.238 nm. Brindley and Nakahira noted 0.1138 nm peak in addition to others.
(ii) The diffraction peaks at 0.139 and 0.197 nm are broad and diffuse. From these broad lines or peaks, the peak maximum is very difficult to note precisely.

According to Yamada and Kimura (1962), "X-ray peaks of spinel phase crystal are considerably wide and hence it becomes very difficult to calculate the exact value of lattice parameter". Thus, by lattice parameter measurement it is unlikely to characterize the spinel phase properly.

20.7 Density Measurements

Lemaitre et al. (1975) compared the measured densities of heated kaolinites with calculated theoretical densities for various transformations of metakaolinite. Results rule out the possibility of formation of Al–Si spinel as per their Path A proposed by Brindley and Nakahira (1959). For other two types of transformations namely, Path B and Path C, they choose quite large proportions of mullite formation which are unlike to occur in practice. As such prediction of spinel phase as γ-Al_2O_3 by them seems unattainable.

20.8 Main Distinguishing Parameter for Two Types of Spinel

The spinel phase whether it is γ-Al_2O_3 or Al–Si spinel will differ first of all to a great extent in the chemical composition as given below:

Material	Composition	Unit cell formula
Al–Spinel	Al_2O_3 (as per Leonard)	8 $Al[Al_{13.33}\ []_{2.66}]O_{32}$
Al–Si Spinel	$2Al_2O_3{\cdot}3SiO_2$ (Brindley and Nakahira)	8 $Si[Al_{10.66}\ []_{5.33}]O_{32}$

Thus, it is seen from above that γ-Al_2O_3 contains only Al as cation where as besides Al atoms, some amounts of Si-atoms are present in tetrahedral part of Al–Si spinal structure. Obviously, there may be two approaches for characterization of 980 °C spinel phase.

20.8.1 First Approach

The presence or absence of SiO_2 may be one of the clues for proper characterization of cubic phase. Conversely, the amount of silica to be liberated on to the decomposition of metakaolinite will vary.

$$Al_2O_3 \cdot 2SiO_2 \rightarrow Al_2O_3 + 2SiO_2 \tag{12.4}$$

In the above equation, 2 molecules of SiO_2 will be liberated per mole of conversion of metakaolinite. On the other hand, if the cubic phase be Al–Si spinel, the amount of exolved SiO_2 will be different. So it is thought that 980 °C heat treated spinel could be characterized indirectly by determining the amount of silica liberated on heating kaolinite at different temperatures.

20.8.2 Second Approach

This includes fitting of experimental RED curve with that of the theoretical decomposition products of kaolinite at first exotherm.

20.8.3 Third Approach

This approach includes the measurement of change in coordination number of Al in heated kaolinite. Since the relative proportion of Al i.e., Al(IV) and Al(VI) contents and likely the ratio of their coordination numbers would change on substitution of Si^{+4} and Al^{+3}. Thus, I.R; XRF; and NMR studies would be helpful for characterization of spinel phase that the different researchers explore for solving γ-Al_2O_3 versus spinel controversy.

20.9 First Approach of Characterizing Spinel Phase

20.9.1 Estimation of Siliceous Phase

First step to deal with the problem is to understand the quantity of silica exolved out during the decomposition of kaolinite. By alkali leaching study, Chakraborty and Ghosh (1978b) showed the quantitative value of silica (A) liberated on heating kaolinite at different temperatures. They showed that metakaolinite liberates siliceous (A) phase (SiO_2 and Al_2O_3 to the extent of 35–38 and 5 %) only once at 980 °C exotherm with the crystallization of poorly crystallized spinel and mullite phases. So the amount of bonded SiO_2 present in heated kaolinite is $\sim$18 wt% (Fig. 17.1). Verification of silica data by similar alkali leaching study followed by chemical analysis was done by Moya et al. (1985) and Rincon et al. (1986). *Former author* reported the alkali leaching study on heated Halloysite rock by using 10 % NaOH solution at 900 °C with continuous stirring for periods between 5 and 50 min. The amount of silica extracted at each temperature is shown in Fig. 20.4. It is a monotonically increasing function, with an inflection point (maximum slope) of the SiO_2 extraction curve with the temperature T_m of the DTA

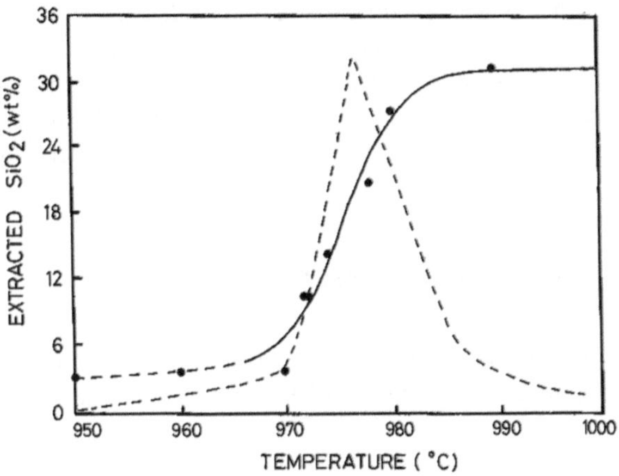

Fig. 20.4 Extracted silica (wt% with respect to metakaolin) by alkaline attack. The *dashed line* represents the exotherm at 980 °C (After Moya et al. 1985). Reprinted by permission of Springer

peak and approaches a symptomatically the value corresponding to a final Al_2O_3/SiO_2 molar ratio of 1:5 in the solid phase, which corresponds to mullite composition. Thus, their data indicate that 980 °C exothermic peak is associated with silica liberation.

20.9.2 *Characterization of Alkali Leached Residue*

The chemical composition of the leached residue is thus analogous to the composition of 3:2 mullite. The leached residue still shows the same two poorly crystalline phases in XRD analysis. With these data, Chakraborty and Ghosh (1978a) concluded that 980 °C spinel phase is not simple γ-Al_2O_3, but Al–Si spinel of similar composition to 3:2 mullite or $Si_2Al_6O_{11}$ instead of $2Al_2O_3{\cdot}3SiO_2$ (Si_3 Al_4O_{12}) as conjectured by Brindley and Nakahira. The estimated value of SiO_2 (A) fits well with the SiO_2 part shown in RHS of the theoretical transformation Eq. 12.8 given by Schieltz and Soleman which showed maximum $-\Delta G$ value as per thermodynamic calculations for free energies. Leached residue on further heat treatment develops well-crystallized mullite with complete disappearance of spinel phase. Chakraborty and Ghosh (1991) describe this mullitization process may be a case of polymorphic transformation of Al–Si spinel to mullite other than solid-state reaction between γ-Al_2O_3 spinel and SiO_2(A) to form mullite.

The unit cell formulae of mullite-like composition spinel as follows:

$[Si_{4.9}\ Al_{3.10}]$	$[Al_{11.7}\ \square_{4.3}]$	O_{32}
Tetrahedral site	Octahedral site	Oxygen
IV	VI	
8A	16B	O_{32}

This spinel containing both Al(VI) and Al(IV) which is unlike to proposed spinel by Brindley and Nakahira. Since it contains only Al(IV) and has some vacancies in octahedral site which is less likely to occur as discussed by IR and XRF evidences. Contrary, the spinel presented above is feasible from the standpoint of crystal chemistry.

Growth curve of mullite out of heating leached residue is shown in Fig. 20.5, which shows a sudden structural transformation of spinel to mullite during heat treatment at the second exotherm at 1,250 °C .

20.9.3 Detection and Estimation of Other 980 °C Phases by QXRD

Tszuki et al. (1969), Okada et al. (1986), *and later on* Chakraborty and Ghosh (1991) estimated semi quantitatively the two weakly crystalline phases namely spinel and mullite formed during first exotherm by usual X-ray diffraction technique using standards. Quantity of spinel formed shows variations. These may be due to the nature of kaolinite and their source or origin. The value of spinel lies in the range 20–38 wt% and the amount of weakly crystalline mullite is approximately 5 wt%.

Chakraborty and Ghosh (1991) detected and estimated of mullite (A) phase indirectly. They pointed out the existence of amorphous band at 0.22 nm out of alkali leached residues beyond 30 min of leaching (Fig. 18.1). This indicates that the decomposition product of kaolinite at 980 °C contains amorphous aluminosilicate phase besides siliceous (A) phase. Weight percent value of it has been calculated indirectly by subtracting the values of crystalline phases from 100.

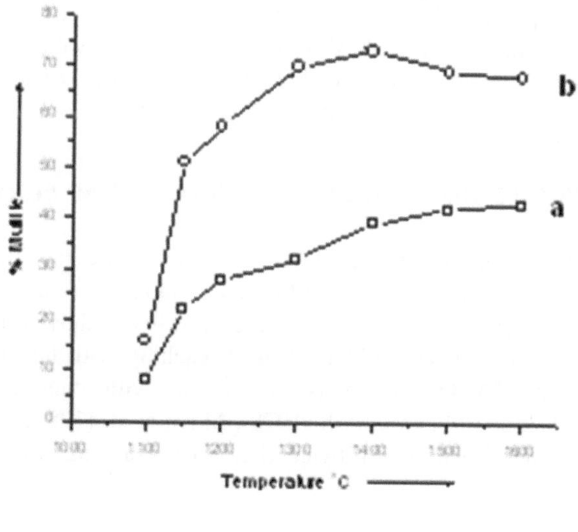

Fig. 20.5 Growth curve of mullite formed during heating of alkali leached residue. **a** Mullitization from M.P. Kaolinite, **b** mullitization from alkali leached heated kaolin

It is observed that an amount of this residual alumino-silicate phase associated with Al–Si spinel and mullite phase is of the order of 30–40 %. Considering the estimated data of Al–Si spinel, mullite shown by Tszuki et al. (1969), Okada et al. (1986), Chakraborty and Ghosh (1991), and Mackenzie et al. (1987) by QXRD technique, the 980 °C transformation is expressed as follows:

$$\begin{aligned} \text{Metakaolinite} \xrightarrow[\text{Exotherm}]{980^\circ} & \text{ i) Al−Si spinel phase, 20−30 \%} \\ & \text{ ii) Mullite phase, 5 \%} \\ & \text{ iii) Mullite(A) phase, 30−40 \%} \\ & \text{ iv) SiO}_2\text{(A) phase, 36 \%} \end{aligned} \tag{20.2}$$

Rincon et al. (1986) collected alkali leached mass out of Halloysite heated to 1,000 °C and called as premullite. The overall chemical analysis (wt%) is the following : Al_2O_3–72.20, SiO_2–26.92, CaO–0.2, Fe_2O_3–0.15, Na_2O–0.05, MgO–0.02, and K_2O–0.02. This powder was isopressed at 200 p.a. and fired at 1,570 °C for 16 h. The TEM-EDX analysis of the specimen shows different kinds of crystals. Glass or amorphous phase present at the junction point shows composition similar to mullite in composition (25 wt% SiO_2 and 75 wt% Al_2O_3) possibly with some impurities. This result suggests us to believe that composition of Al–Si spinel and mullite (A) phase is of the same composition of 3:2 mullite.

20.9.4 Comparative Analysis of Leached Residue by Analytical Method Versus EDS Method

20.9.4.1 Analytical Method

The major objective of alkali leaching technique is to demarcate between SiO_2 rich (A) phase and bonded SiO_2 bearing phases present in 980 °C heat treated kaolinite. Chakraborty and Ghosh (1978a) showed the value of SiO_2 content in silica rich phase is 35–37 % and likely the composition of leached reside which contains spinel is similar to mullite-like composition.

20.9.4.2 Direct EDS/TEM Analysis of Alkali Leached Residue

The compositional analysis of spinel phase has been directly made by EDS technique in analytical TEM after isolating it from surrounding amorphous phases by a leaching process. Okada et al. (1986a) showed the chemical composition of the spinel phase (obtained after leaching with hot alkali for once only) around $2SiO_2{\cdot}3Al_2O_3$ which is compatible with chemical analysis data given by Chakraborty and Ghosh (1978), Moya et al. (1985), and Rincon et al. (1986). On further leaching for twice (i.e., long leaching time) Okada et al. showed that composition approaching to 8 wt% SiO_2 as determined by TEM which corresponds approximately to composition SiO_2 $6Al_2O_3$ $[Si_{16}$ $Al_{10.2}]$ O_{32} spinel other than simple γ-Al_2O_3 spinel of Leonard's model. Contrary to this data of Okada et al. and Somuparlak et al., by using similar EDX analysis Srikrishna et al. (1990) showed chemical analysis data of residual spinel phase to be close to that of 3:2 mullite composition (Fig. 17.4).

20.9.4.3 Comparison of Analytical Versus EDS Data

Jantzen (1990) raised question on the quantitative data of SiO_2 to note the dissolution behavior of heated kaolinite data of Chakraborty and Ghosh. She pointed out that both P- and Q-type zeolites are formed during leaching heated kaolinite at water bath and in boiling condition. As the SiO_2 content in the two zeolites are varied and secondly a slow dissolution of SiO_2 took place with extended period of leaching she raised doubt about the estimated value of Chakraborty and Ghosh (1978).

Although the TEM/EDS data for direct compositional analysis of alkali leached residue shown by Okada et al. and Sonuparlak et al. (1987) lead to a value of 10 wt% SiO_2 containing spinel phase, some question to be raised on the procedures chosen by them. For example,

(i) Percival et al. (1974) made alkali extraction of heated kaolinite at direct boiling condition instead of leaching at boiling water bath condition as chosen by Chakraborty and Ghosh (1978a)

(ii) Okada et al. (1986) leached the heated kaolinite once with alkali and then did XRD analysis of the leached residue but assumed that extraction was not complete and as such they leached twice; i.e., a long leaching time was given instead of 40–50 min as specified by Chakraborty and Ghosh. Question is: why they leached for an extended period of time?

(iii) Sonuparlak et al. (1987) leached heated Georgia kaolinite in direct boiling condition as well as in prolonged period. Question is: why they increased the temperature and time of leaching? This procedure is unlike to that standardized by Chakraborty and Ghosh. On scrutinizing their data, following observations are made:

  (i) Percival et al., Okada et al., and Sounparlak et al. did not present the XRD data of leached residue obtained at several stages during drastic leaching process.

  (ii) They did not report the formation of any new crystalline phases by possible reaction of heated kaolinite with NaOH at boiling condition. Thus, IR spectrum of leached residue as shown by Percival et al. which showed absorption bands due to AlO_4 and AlO_6 groups from alkali leached residues and formation of corundum on subsequent heating are explained to be due to newly formed aluminous phase developed during extensive leaching process indicated by Chakraborty and Ghosh (1978a, 1989).

(iii) The microdiffraction patterns of the alkali leached residue shown by Aksay et al. (1989) would be due to liberated aluminous phase and not due to residual Al–Si spinel phase. They did not submit any supportive X-ray data of leached residue. Moreover, the presence of spinel and liberation of newly formed aluminous phase could not be differentiated probably by TEM. Furthermore, the formation of x- or k-alumina phases during exhaustive leaching process may be responsible for displaying only Al peaks in their EDS study (Chakraborty and Ghosh 1989).
(iv) Extraction condition used by *Jantzan* was as drastic as adopted by Sonuparlak et al. and likely the comparison of leaching behavior of C—glass with heated kaolinite may be unrealistic.

It is really essential to ascertain the correct end point of alkali—free silica reaction for obtaining quantitative data of siliceous phase content of 980 °C kaolinite.

Chakraborty and Ghosh (1992) showed how a slight variation in leaching condition leads to followings:

(i) A great variation in the attack of heated kaolinite which causes dissolution of Al_2O_3 component of aluminosilicate phases.
(ii) Formation of sodium containing hydrated aluminosilicates. He compared the dissolution behavior of SiO_2 rich (A) phase by four leaching process (Fig. 17.4) and finally modified the earlier leaching condition and followed a mild and even a less mild leaching procedure. Since selective leaching of SiO_2 rich (A) phase vis–a–vis enrichment and comparatively less attack of spinel phase and aluminosilicate phase to be as low as practicable. These are the main aim of the leaching process. Any type of branch/side reaction e.g., zeolite formation during solubilization of free SiO_2 out of 980 °C heated kaolinite should be minimized as low as possible. To achieve this, the temperature of extraction was first reduced from direct boiling condition to water bath condition and secondly the concentration of NaOH is reduced from 10 to 5 wt%. Instead of leaching in static condition, Chakraborty and Ghosh (1992) finally applied stirring system into the leaching bath and succeeded in removing SiO_2 rich (A) phase only without altering spinel phase (Fig. 17.6).

20.10 Second Approach of Characterizing Spinel Phase: Structural Analysis of Three Spinel Models by RED Technique

Leonard et al. (1979) drew experimental electron distribution (RED) profile of Zettlitz kaolinite heated at 900° and 970 °C, extending from 0 to 6 Å and compared them with the theoretical RED profiles for the 980 °C transformation as per the model of Chakraborty's mullite like Al–Si spinel normalized to one stoichiometric unit.

$$(Al_2O_3)_{0.333}(SiO_2)_{0.666} = 0.361\,[Si_{0.615}Al_{0.385}]^{IV}[Al_{1.462}[\,]_{0.518}]^{VI}$$

Chakraborty's mullite composition Al–Si spinel model (20.3)

+

$0.445\ SiO_2$

(i) Cation–anion band 1.4–2.1 Å RED domain. Table 20.4 provides a comparison between the experimental RED components and the ones obtained for the corresponding three theoretical hypotheses under discussion.

The electron density values are calculated according to the methodology described earlier by Leonard (1977). The results of cation–anion band corresponding to three theoretical models (described here) lead to discrepancies with respect to 900 °C sample experimental profile is shown in Table 20.5. The corresponding discrepancies with respect to 970 °C sample RED amounts to 1.5, 72.2, 41.3, and 17.7 %, respectively. Thus, integrated electron density of Chakraborty's mullite-like composition spinel is in half way position between two Brindley's composition and Leonard's composition.

(ii) 2.5–6 Å RED domain. The analyses of 2.5–6 Å of samples are given by Leonard et al. (1979). Figure 20.5 reproduces these experimental profiles (A) and the superimposed theoretical curves for both Leonard's γ-Al_2O_3 spinel (B) and Chakraborty's mullite, like composition Al–Si spinel (C) with the respective suited amount of free SiO_2 according to the formulae of equivalent composition. The agreement with experimental values decreases from Leonard's model to Brindley passing through Chakraborty. With this comparison, they argue for more favorable γ-Al_2O_3 model and closer to Chakraborty's siliceous phase content. But they raised doubt about the role played by the remainder part of SiO_2 content.

As shown earlier (Fig. 20.6), RED curve B (Leonard's model) in the region Al^{VI}–O passes slightly above the experimental profile, whereas curve C Chakraborty's model) in the same region passes above curve B; i.e., curve C is still

Table 20.4 Comparison between the experimental RED distribution of components (Cation–Anion) for the 900 °C sample and the corresponding ones for the three involved structural models (Brindley, Chakraborty, Leonard). After Leonard et al. (1979)

Bond	Bond length	Experimental best-fitting results	Brindley composition	Chakraborty composition	Leonard composition
Si–O	1.62 λ	2180 e^2/λ	483 e^2/λ	1264 e^2/λ	1903 e^2/λ
(Al, SI)^{IV}–O	1.77 λ	835 e^2/λ	1420 e^2/λ	988 e^2/λ	642 e^2/λ
Al^{VI}–O	1.94 λ	1207 e^2/λ	2568 e^2/λ	2033 e^2/λ	1603 e^2/λ
Integrated electron density (e^2) (1.38–2.1 λ)		1156	1238	1176	1125
Discrepancy (RSD(λ))		2.5	63.1	34.9	15.9

Table 20.5 Summarizes in a more quantitative approach how four spinel models, ie γ–Al_2O_3 (Leonard), $[Si_{0.613}Al_{0.385}]^{IV}$ $[Al_{1.462}\square_{0.538}]^{VI}$ O_4 (Chakraborty and Ghosh), Si^{IV} $[Al_{1.333}\square_{0.667}]^{VI}$ O_4 (Brindley and Nakahira), $Si_{0.5}$ Al_2O_4 (Magnesion type (Mg → Si)) are compared with experimental values for five selected inter atomic distances of γ–Al_2O_3 and cristobalite respectively. (After Leonard et al.1979). Reprinted by permission of the American Ceramic Society.

	Characteristic γ-spiel contributions at r(Å)					Characteristic β-spiel contributions at r(Å)					Integrated electron density.e2 (0-6 Å)	Rel.std. deviation (%) (0-6 Å)
Model	2.79	3.28	3.44	4.84	5.59	2.52	3.09	3.88	5.04	5.27		
Experimetal 900°C	3371	5180	4319	8766	11621	1474	5113	4761	10562	9259	26647	
Leonard	94	108	104	115	94	134	79	110	153	165	28,812	39.0
Chakraborty and Ghosh	114	138	135	128	98	125	91	89	147	142	30,118	34.1
Brindley and Nakahira	139	180	180	145	113	114	107	63	140	115	32,324	43.8
Magnesian type	101	80	102	129	78	125	69	90	145	141	26,619	35.3

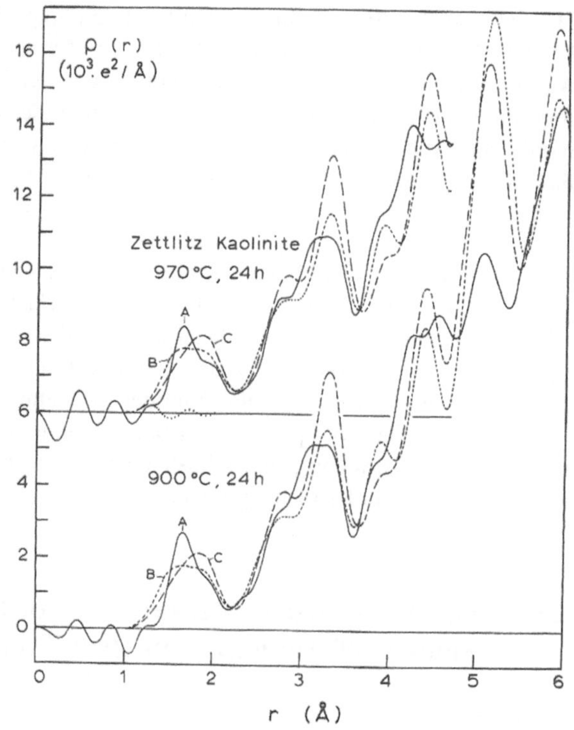

Fig. 20.6 Superposition of experimental RED profiles of kaolinite treated at 900 and 970 °C (*solid line*, **a**) and theoretical *RED* curves corresponding to Leonard's hypothesis (*fine dashed line*, **b**) and Chakraborty and Ghosh's model (*dashed line*, **c**, After Leonard et al. 1979). Reprinted by permission of the American Ceramic Society

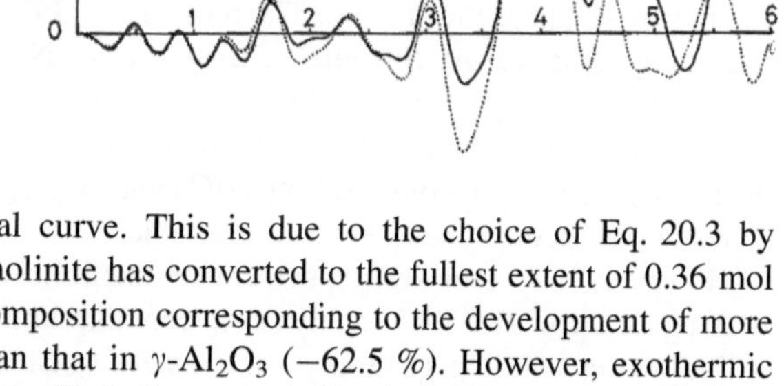

Fig. 20.7 Electron density map for silica contribution above 0.45 nm (After Leonard et al. 1980)

higher than the experimental curve. This is due to the choice of Eq. 20.3 by Leonard et al., where metakaolinite has converted to the fullest extent of 0.36 mol of Al–Si spinel of mullite composition corresponding to the development of more Al^{VI}–O content (−79 %) than that in γ-Al_2O_3 (−62.5 %). However, exothermic reaction products according to Chakraborty and Ghosh (1991) consists of about 0.1 mol of Al–Si Spinel of mullite-like composition (called as cubic mullite), 0.2 mol of binary aluminosilicate of mullite-like composition (called as mullite (A)) and 0.03 mol of ill crystallized mullite as in Eq. 20.2 but unlike Eq. 20.3. With these data, it is now possible to predict the imperfect matching of the model of Chakraborty's Si–Al spinel of mullite-like composition with the experimental RED profile. When 0.1 mol of Al–Si spinel is formed, the expected sharp rise of Al(VI)–O content at 980 °C should not be appreciable. On the other hand, the formation of 0.2 mol of mullite (A) indicates that the Al(IV)–O component will be more than that expected in Eq. 20.3. This explains why the experimental profile of the Al(IV)–O and Si–O global region is higher than in the theoretical model (Eq. 20.2) of Leanard et al. For verification, it is now necessary to perform the RED in the light of a quaternary mixture using the coefficients of the products formed during the 980 °C exothermic reaction of kaolinite.

Regarding verification of silica data given by Chakraborty and Ghosh, Leonard et al. (1979) selected β-cristobalite structure as the nearest ideal crystalline model, which reveals silica contribution above 4.5 Å. Figure 20.7 shows that dashed line (mullite composition Al–Si spinel) model fits rather correctly with experimental 900/24 h data. Thus, 3/4 part of total SiO_2 content corresponding to Eq. 12.8 and measured by Chakraborty and Ghosh by alkali leaching study is confirmed by RED as the right order of magnitude.

20.11 Third Approach: Measurement of Coordination Number of Aluminum in Heat Treated Kaolinite

20.11.1 I.R.

Structural change of metakaolinite to spinel phase was noted by Pampuch (1966), Freund (1967), and Percival et al. (1974) by their I.R. spectral observation. All of them noted that transformation of metakaolinite to spinel occurs by the appearance of a new spectral band at 580 cm^{-1} due to AlO_6 groups with the simultaneous disappearance of broad band at 810 cm^{-1} due to AlO_4 groups in metakaolinite. With this observation, Pampuch first conjectured the development of this band is due to γ-Al_2O_3 phase formation. The most recent I.R. spectra are given by Percival et al. In hypothetical Al–Si spinel of Brindley and Nakahira AlO_4 groups should be absent. Accordingly, it is argued that as AlO_4 are still detected in I.R. spectrum of kaolinite heated to about 1,000 °C then Brindley's spinel should be less probable than γ-Al_2O_3. Since it contains both AlO_4 and AlO_6 groups, formation of the latter phase is on the other hand is more probable.

However, they took I.R. spectrum of kaolinite heated to 950 °C for a long soaking of 24 h. On XRD analysis, it contains both spinel and mullite as crystalline phases. Mullite generally contains AlO_6, SiO_4, and isolated AlO_4 groups; the AlO_4 groups of mullite gave absorption bands in the range 650–900 cm^{-1}. The bands at 740 and 830 cm^{-1} were assigned to mullite. The question then arises as to how, in presence of weakly crystalline mullite, the γ-Al_2O_3 phase could be differentiated from Al–Si spinel by I.R. spectrography? Since AlO_4 and AlO_6 groups are present in mullite. The AlO_4 identified might have originated from either

γ-Al_2O_3 or mullite or both. Formation of mullite along with the spinel phase during exothermic reaction was reported by several workers on the basis of X-ray and electron diffraction studies. Chakraborty and Ghosh further showed the presence of aluminosilicate (A) phase in addition to both the above crystalline phases. The former phase also contains AlO_4 groups. Therefore, in presence of mullite and alumino-silicate (A) phase, the spinel phase could not be conclusively identified as γ-Al_2O_3 by I.R. spectrophotometer.

20.11.2 XRF

A complementary semiquantitative check on the structural analysis of intermediate phases was made by the systematic determination of the relative amount of Al in four coordination by XRF technique during thermal evolution of kaolinite. By X-ray fluorescence, Leonard (1977) showed that the amount of AlO_4 present in Zettlitz kaolinite heated at 980 °C is 38 ± 10 %. This value does not agree with Chakraborty's 21 % (0.385/1.847) AlO_4 in mullite-like Al–Si spinel model. Again the values of (25–50 %) Al(VI) and (75–50 %) Al(IV) shown by Udagawa et al. (1969) in the range of temperature 900–1,000 °C disagree with 20.8 % Al(IV) for Al–Si spinel of mullite composition. Leonard is of the opinion that instead of observing a regular increase in Al(IV) from 800 to 1,000 °C as shown by Udagawa, for the favorable Al–Si spinel phase, the increase of Al(VI) must be steeper and would give rise to a pronounced maximum in its proposed Al(VI) content to the extent of 79 %. Apparently, the measured AlO_4 data shown by Leonard in Table 8.1 (~38 %) agree with that of γ-Al_2O_3 according to Leonard's γ-Al_2O_3 spinel model (Eq. (9.1)). In reality, some excess Al_2O_3 is always present in the mullite(A) phase. This appears inconsistent with the γ-Al_2O_3 model Chakraborty and Ghosh (1979). However, the measured AlO_4 should always be greater than 21 % because of the association of mullite (A) phase as the fourth component in the product of kaolinite heated at 980 °C. The total amount of AlO_4 calculated by Chakraborty and Ghosh (1991) from kaolinite heated at 980 °C (Table 20.6) is approximately 37.1 %.

This value is in close agreement with the XRF value for Zettlitz kaolinite heated at 900 °C for 24 h as shown by Leonard (1977). This provides an answer to the pertinent question regarding the validity of Chakraborty's mullite-like Al–Si

Table 20.6 Calculated amount of AlO_4 in Zettlitz Kaolinite heated at 980 °C (After Chakraborty and Ghosh 1991). Reprinted by permission of the American Ceramic Society

Phases	AlO(IV)		AlO(IV) in kaolinite heated at 980 °C (%)	
	Amount (%)	Content (%)	Chakraborty and Ghosh (1991)	Leonard (1977)
1. Al–Si spinel	34	7.14	37.1	38 ± 10
2. Mullite (A) phase	31	17.05		
3. Mullite (weakly crystalline)	Trace			
Total		24.19		

spinel model as raised by Leonard et al. (1979). Moreover, the expected sharp rise in AlO_6 content during metakaolinite to Al–Si spinel transformation at 980 °C should not be observed because of the development of mullite (A) phase.

20.11.3 MAS NMR

Brown et al. (1985) compared ^{27}Al NMR spectrum of pure γ-Al_2O_3 with those of kaolinite heat treated at 970 °C and 1050 °C, respectively. γ-Al_2O_3 sample (obtained by heating pseudo bohemite at 830 °C/0.5 h) shows broad pattern in X-ray study and contains a prominent octahedral resonance at 5 ppm, whereas a small tetrahedral peak at 76 ppm. Kaolinite sample heated to 970 °C contained well-defined quantity of cubic spinel phase, the spectrum shows higher proportion of tetrahedral peak than that in pure γ-Al_2O_3. Therefore, these two spectrums are not identical (Fig. 20.8). So NMR evidence is not consistent with γ-Al_2O_3 spinel. When kaolinite heated to 1,050 °C, where cubic phase is reduced and mullite is advanced, the spectrum shows NMR increase of AlO_4 peak at the cost of decrease of octahedral Al peak.

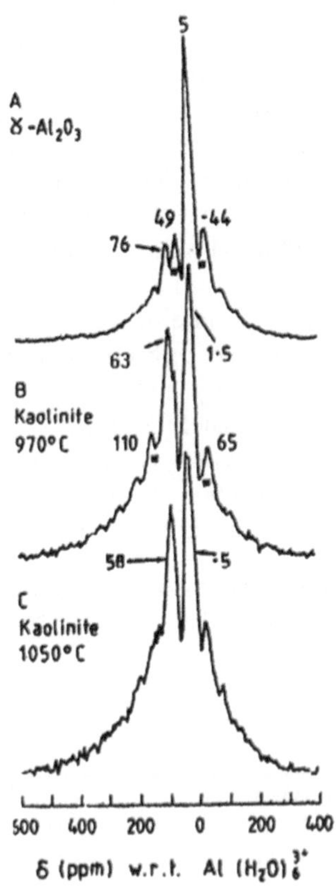

Fig. 20.8 Aluminum-27MR spectra of γ-Al_2O_3 and heated kaolinite containing a spinel phase (Brown et al. 1985). Reprinted by permission of the American Ceramic Society

Thus, (i) by quantitative detecting the presence of both AlO_4 and AlO_6 groups by I R; (ii) by determining their quantitative value by XRF and even (iii) by summing up of XRD intensifies of heated kaolinite by RDF functions it is difficult to conclude that the intermediate cubic phase in K–M reaction series as γ-Al_2O_3. On the contrary, it is more feasible to be silicon bearing spinel phase.

20.11.4 Effect of Grinding on Spinel Formation

The area of the exothermic peak as observed by Sánchez-Soto et al. (2000) became almost constant or slightly smaller as grinding time increased (Fig. 20.9), and it is independent of the state of order (KGa-1) or disorder (KGa-2) of the kaolinite crystal used for grinding experiments. Thus, grinding did not seem to favor the metakaolinite-to-spinel phase transformation. Low-crystalline mullite and Al–Si spinel phase were identified as the main crystalline phases formed after heating both ground kaolinite samples.

For longer grinding times, the transformation of amorphous phase to mullite occurred at relatively lower temperatures than for the original kaolinites, especially when the starting material was a disordered kaolinite, KGa-2. Thus, grinding has neither influenced nor discarded the formation of Aluminum–Silicon spinel from metakaolinite. So formation of Al–Si spinel is an essential step and not as an accidental step prior to mullite formation in K–M reaction series.

This study concludes that spinel is not al a metastable phase in K–M reaction series. Therefore, conjecture of Comeforo et al. (1948) and John (1953) and that spinel is a transitory phase and metakaolinite directly converts to mullite does not hold true. Al–Si spinel formation is thus a necessary step in K–M reaction series.

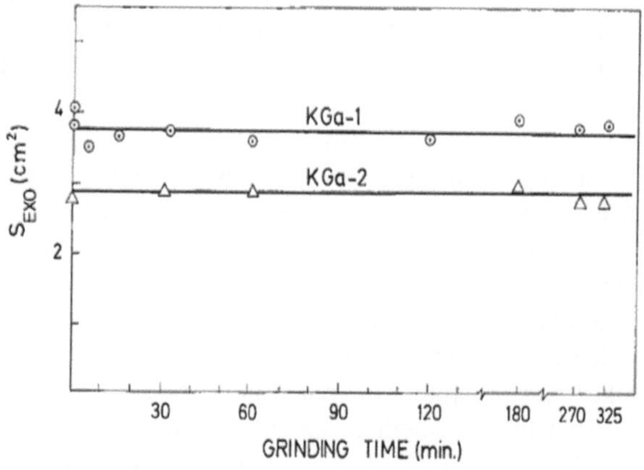

Fig. 20.9 Plot of exothermic DTA peak area evolution (*S*EXO), as a function of grinding time, for samples KGa-1 and KGa-2 (area is normalized per gram of sample (After Sánchez-Soto et al. 2000)). Reprinted by permission of the American Ceramic Society

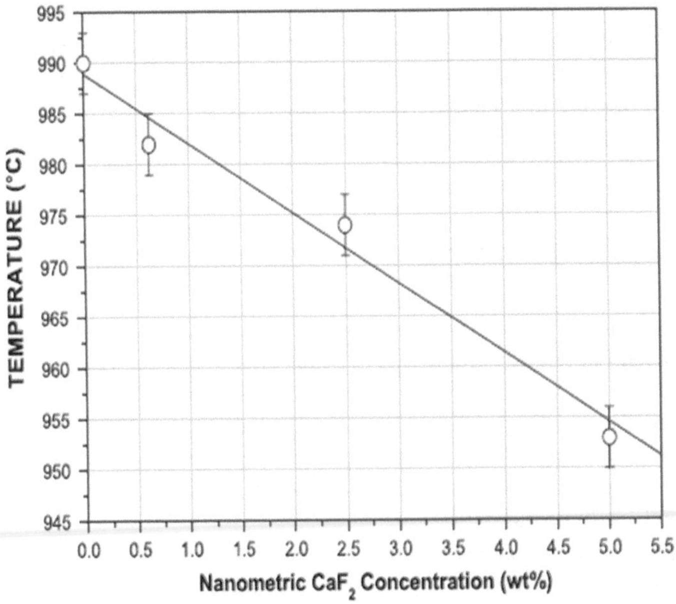

Fig. 20.10 Relationship between transformation of metakaolinite to cubic Al–Si spine phase and nanometric CaF_2 concentration (After Lomeli et al. 2009). Reprinted by permission of the American Ceramic Society

20.11.5 Mineralizing Effect

Lomeli et al. (2009) noted a negligible down word shift of 980 °C exotherm even with addition of substantial quantity of fluorspar (Fig. 3.8). With this observation, they concluded that the long-range order of the mineralized kaolinite collapses at the same temperature as that of conventional kaolinite and the micrometric CaF_2 particles do not notably promote the formation of the spinel phase.

Even with addition of nanometric CaF_2, the temperature of the exothermic peaks (occurring at 989 °C) was slightly shifted to a lower value as the amount of it was increased (Fig. 20.10).

Larger amount of several fluxes and oxides e.g., Fe_2O_3, Na_2CO_3/NaCl, CaO, AlF_3, and H_3BO_3 reduces and or totally screen off 980 °C exotherm of kaoinite. These results indicate that above-mentioned oxides react with metakaolin during dynamic and static heating processes. The gradual reduction of exothermic peak intensity is related to more and more interaction of metakaolinite with quantity of oxide added. Only the residual unreacted or unconverted residual metakaoinite takes part in exothermic reaction process.

That the intermediate spinel formation is an essential step, is again substantiated by the studies of external oxide addition on DTA exothermic reaction process of kaolinite.

20.12 Utilization Alkali Leached Heat Treated Kaolinite

It is effectively used for production of ceramic filters and also for membranes (See References).

20.13 Summary

There were two school of thoughts regarding composition of spinel phase. It may be simple γ-Al_2O_3 due to Leonard or Al–Si spinel of the composition $2Al_2O_3{\cdot}3SiO_2$ earlier due to Brindley and Nakahira.

Calculation of free energy and heat of formation of some possible transformation of metakaolinite indicate that segregation followed by recombination theory seems to be untenable theoretically. On the other hand, Chakraborty's suggestion regarding spinel analogous to 3:2 composition of mullite seems feasible from standpoint of lattice energy calculation and crystal chemistry point of view. Its unit cell formulae is given $[Si_{4.9}\ Al_{3.1}]\ [Al_{11.7}\ []_{4.3}]\ O_{32.}$

This spinel containing both Al(VI) and AL(IV) groups unlike to proposed by Brindley and Nakahira (1959). For that reason, mullitization behaviors of kaolinite and artificial mixture of alumina/silica show differently. This theory has been verified by alkali extraction study of 980 °C heated product of kaolinite. Chemical analysis of leached residue confirms the composition analogous to $3Al_2O_3{\cdot}2SiO_2$ which verify theoretical concept. QXRD analysis suggests the existence of aluminosilicate(A) phase of the same composition as Al–Si spinel which is verified by TEM study of heated alkali leached mass. Thus, Eq. 20.2 may corresponds the first exothermic reaction of metakaolinite decomposition.

Verification of this Chakraborty's mullite like Al–Si spinel model was attempted by Leonard et al. by RED technique. Complementary study by the determination of the relative proportion of AlO_6 and AlO_4 in heated kaolinite was done by XRF. The total amount of AlO_4 calculated from kaolinite heated to 980 °C is 37.1 % approximately, which confirms the validity of the above 980 °C reaction.

Sánchez-Soto et al. (2000) noted that first exothermic peak remains almost constant even on degradation of structure of kaolinite on prolong grinding. Thus, grinding neither favor nor eliminate the metakaolinite-to-spinel phase transformation. It is an essential step in K–M reaction series. It is also verified by DTA experimentation study with oxide addition.

Some externally added oxides e.g., Fe_2O_3, Na_2CO_3/NaCl, CaO, AlF_3, and H_3BO_3 to metakaolin powder gradually reduce the intensity of first exotherm, which is due to solid-state reaction of fluxing materials with metakaolinite during gradual heating process otherwise occurrence of DTA exotherm could not be put off. Lomeli et al. (2009) noted the long-range order of the micrometric CaF_2 added kaolinite collapses at the same temperature as that of kaolinite without any added mineralizer.

References

I.A. Aksay, M. Sarikaya, B. Sonuparlak, Reply to the comment on spinel formation during 980 °C exothermic reaction in the kaolinite–mullite reaction series. J. Am. Ceram. Soc. **72**(8), 1571 (1989)

W.F. Bradley, R.E. Grim, High-temperature thermal effects of clay and related materials. Am. Miner. **36**(3/4), 182–201 (1951)

W. L. Bragg, *Atomic Structure of Minerals*. Ithaca, (Cornell Univ. Press, N.Y.1937) p. 33, 144, 257

G.W. Brindley, M. Nakahira, The kaolinite–mullite reaction series: I, II, and III,. J. Am. Ceram. Soc. **42**(7), 311–324 (1959)

I.W.M. Brown, K.J.D. MacKenzie, M.E. Bowden, R.H. Meinhold, Outstanding problems in the kaolinite–mullite reaction sequence investigated by 29Si and 27Al solid-state nuclear magnetic resonance: II, high-temperature transformations of metakaolinite. J. Am. Ceram. Soc. **68**(6), 298–301 (1985)

P. B. Budnikov and B. A. Khizh, Mullitizationof refractory grog, Ber. Detsch. Keram. Ges. **10**(10), 445–448 (1929)

A.K. Chakraborty and D.K. Ghosh, Supplementary alkali extraction studies of 980°C heated kaolinite by X – ray diffraction analysis, J. Mater. Sci., **27**(8), 2075–2082 (1992)

A.K. Chakraborty, D.K. Ghosh, Comment on the interpretation of the kaolinite–mullite reaction sequence from infra-red absorption spectra. J. Am. Ceram. Soc. **61**(1–2), 90–91 (1978a)

A.K.Chakraborty and D.K.Ghosh, Re - examination of the kaolinite to mullite reaction series, J. Am. Ceram. Soc., **61**(3–4), 170–173 (1978b)

A.K. Chakraborty, D.K. Ghosh, Comment on structural analysis of the transition phases in kaolinite–mullite thermal sequence. J. Am. Ceram. Soc. **62**(9–10), 529 (1979)

A.K. Chakraborty and D.K. Ghosh, Comment on spinel formation during 980°C exothermic reaction in the kaolinite – mullite reaction series, J. Am. Ceram. Soc., **72**(8), 1569–1570 (1989)

A.K. Chakraborty, D.K. Ghosh, Kaolinite–mullite reaction series. The development and significance of a binary aluminosilicate phase. J. Am. Ceram. Soc. **74**(6), 1401–1406 (1991)

A.K. Chakraborty, D.K. Ghosh, P. Kundu, Comment on structural characterization of the spinel phase in the kaolin–mullite research series through lattice energy. J. Am. Ceram. Soc. **68**(8), 200–201 (1986)

L. Changling, Discovery of Al–Si spinel nature. Yogo Kyokai Shi. **13**(2), 190–197 (1985)

E.B. Colegrave, G.R. Rigby, The decomposition of kaolinite by heat. Trans. Brit. Ceram. Soc. **51**(6), 355–367 (1952)

J.E. Comeforo, R.B. Fischer, W.F. Bradley, Mullitization of kaolinite. J. Am. Ceram. Soc. **31**(9), 254–259 (1948)

J.J. Comer, New electron-optical data on the kaolinite–mullite transformation. J. Am. Ceram. Soc. **44**(11), 561–563 (1961)

W. Eitel, H. Kedesdy, Elektronen-Mikroskopie und Eeugung silikatischer Metaphasen: IV, Der Metakaolin (Electron Microscopy and Diffraction of Silicate Metaphases: IV, Metakaoliu), Abhandl. preuss. Akad. Wiss. Math.-nattir w. KI., NO. 5, pp. 37–45 (1943)

F. Freund, Kaolinite-metakaolinite, a model of a solid with extremely high lattice defect concentrations. Ber. Dtsch. Keram. Ges. **44**(I) 5–13 (1967)

H.D. Glass, High-temperature phases from kaolinite and halloysite. Am. Mineral. **39**, 193–207 (1954)

J.F. Hyslop, Decomposition of clay by heat. Trans. Brit. Ceranz. Soc. **43**(3), 49–51 (1944)

J.F. Hyslop, H.B. Rooksby, Effect of heat on crystalline break up of kaolin. Trans. Br. Ceram. Soc. **27**(4), 93–96 (1928a)

J.F. Hyslop, H.B. Rooksby, Further note on crystalline break up of kaolin. Trans. Br. Ceram. Soc. **27**(4), 299–302 (1928b)

I.H. Insley, R.H. Ewell, Thermal behavior of the kaolin minerals. J. Res. Natl. Bur. Stand **14**(S), 615–627 (1935)

C.M. Jantzen, Formation of zeolite during caustic dissolution of fibre glass: implications for studies of kaolinite to mullite transformation. J. Am. Ceram. Soc. **73**(12), 3708–3711 (1990)

A.H. Jay, An X-ray study of alumino-silicate refractories. Trans. Brit. Ceram. Soc. **35**, 455–463 (1939)

W.D. Johns, High-temperature phase changes in kaolinites. Miner. Mag. **30**(222), 186–198 (1953)

S.M. Johnson, J.A. Pask, Role of impurities on formation of mullite from kaolinite and AI, O,-SiO_2 mixtures. Am. Ceram. Soc. Bull. **61**, 838–842 (1982)

A.J. Leonard, Structural analysis of the transition phases in the kaolinite–mullite thermal sequence. J. Am. Ceram. Soc. **60**(1–2), 37–43 (1977)

A.J. Leonard, M.J. Genet, J. Lemaitre, M. Bulens, Reply to the Comments On structural analysis of transition phases in kaolinite - mullite thermal sequence, J. Am. Ceram. Soc., 9–10, 529–531, (1979)

A.J. Leonard, Personal commuication (1980)

J. Lemaitre, M. Bullens, B. Delmon, in *Influence of Mineralizers on the 950 °C Exothermic Reaction of Metakaolinite*, ed. by S.W. Bailey. Proceedings of the International Clay Conference (Mexico City, Mexico, July 1975). Applied Publishing Ltd., Wilmette, IL, pp. 539–544 (1975)

F.K. Lotgering, J. Inorganic Nucl. Chem (in J.D. Bernal; Topotaxi. Schw. Archiv. Jharb., 26, 69) **9**, 113 (1959)

M. Lomeli, L. Maria Flores-Velez, I. Esparza, R. Torres, O. Domínguez, Catalytic effect of CaF_2 nanoparticles on sintering behavior of kaolin-based materials. J. Am. Ceram. Soc. **92**(7), 1526–1533 (2009)

S.T. Lundin, Studies on triaxial white ware bodies. Almqnist Wiksall, Stockholm, Sweden. **32**,103 (1959)

K.J.D. MacKenzie , J.S. Hartman and K. Okada, MAS NMR evidence for the presence of silicon in the alumina spinel from thermally transformed kaolinite, J. Am. Ceram. Soc. **79**(11), 2980–2982 (1987)

T.N. Mc Vay, C.I. Thompson, X-ray investigation of effect of heat on China clays. J. Am. Ceram. Soc. **11**(11), 839–841 (1928)

J.S. Moya, C.J. Serna, J.E. Iglesias, On the formation of mullite from Kandaties. J. Mater. Sci. **20**, 32–36 (1985)

B. Mullet and H. Schneider, Reaction sintering of slip cast Quartz plus α- Al_2O_3, in *Formation of Mullite from Kaolinite and Related Minerals*. ed. by H. Schneider, K. Okada and J. Pask, Mullite and Mullite Ceramics, (John Wiley & Sons Ltd, Chichester, 1994) p. 154

K. Okada, N. Ostuka, J. Ossaka, Characterization of spinel phase formed in the kaolin–mullite thermal sequence. J. Am. Ceram. Soc. **69**(10), C-251–C-253 (1986)

R. Pampuch, Infrared study of thermal transformations of kaolinite and the structure of metakaolin. Pol. Akad. Nauk. Oddzial Krakowie, Kom, Nauk, Mineral, *Pr. Mineral,* **6**, 53–70 (1966)

H.J. Percival, J.F. Duncan, P.K. Foster, Interpretation of the kaolinite–mullite reaction sequence from infrared absorption spectra. J. Am. Ceram. Soc. **57**(2), 57–61 (1974)

O. E. Radczewski, Electron optical investigations of the decomposition of clay minerals. Tonind. Ztg. **77**, 291 (1953)

H.M. Richardson, F.G. Wilde, X-ray study of the crystalline phases that occur in fired clays. Trans. Brit. Ceram. Soc. **51**(7), 367–400 (1952)

J.M.A. Rincon, G. Thomas, J.S. Moya, Micro structural study of sintered mullet. J. Am. Ceram. Soc. **69**(2), 29–31 (1986)

R. Roy, D.M. Roy, E.E. Francis, New data on thermal decomposition of kaolinite and halloysite. J. Am. Ceram. Soc. **38**(6), 198–205 (1955)

N.C. Schieltz, M.R. Soleman, in *Thermodynamics of Various High Temperature Transformation of Kaolinite*, ed. by E. Ingerson. Proceedings of 13th National Conference on Clays, Madison, Wisconsin, Pergamon press, Monograph, No. 25, pp. 419–425 (1966)

P.J. Sánchez-Soto, M. del Carmen, J. de Haro, L.A. Pérez-Maqueda, I. Varona, J.L. Pérez-Rodríguez, Effects of dry grinding on the structural changes of kaolinite powders. J. Am. Ceram. Soc. **83**(7), 1649–1657 (2000)

B. Saruhan, W. Albers and H. Schneider, Reaction sintering of Quartz, Cristobalite and SiO_2 glass with α-Al_2O_3 to mullite. in *Formation of Mullite from Kaolinite and Related Minerals*. ed. by H. Schneider, K. Okada and J. Pask, Mullite and Mullite Ceramics, (John Wiley & Sons Ltd, Chichester, 1994) p. 154

M. Schmucker, W. Albers and H. Schneider, Mullite formation by reaction sintering of Quartz and $\alpha - Al_2O_3$ – A TEM Study. J. Euro. Ceram. Soc. **14**, 511–15 (1994)

M. Slaughter, W.D. Keller, High temperature transformation from impure kaolin clays. Am. Ceram. Soc. Bull. **38**, 703–707 (1959)

B. Sonuparlak, M. Sarikaya, I.A. Aksay, Spinel phase formation during the 980 °C exothermic reaction in the kaolinite-to-mullite reaction series. J. Am. Ceram. Soc. **70**(11), 837–842 (1987)

K. Srikrishna, G. Thomas, R. Martinez, M.P. Corral, S. Aza, J.S. Moya, Kaolinite–mullite reaction series: a TEM study. J. Mater. Sci. **25**, 607–612 (1990)

G. Tamman, W. Pape, Ilber Den Wasserverlust Des Kaolines & Seinverhat-en In Festen Zuden Carbonatem Und Oxyden Der Erdalkalien. Z. Anorg. Allg. Chem. **127**, 43–68 (1923)

L. Tscheischwili, W. Biissern, W. Wevl, Uber den Metakaolin (Metakaolin), Ber. deut. keranz. Ges. **20**(61), 249–276 (1939)

Y. Tsuzuki, K. Nagasawa, A transitional stage to 980°C exotherm of kaolin minerals. Clay Sci. **3**(5), 87–102 (1969)

S. Udagawa, T. Nakada, M. Nakahira, in *Molecular Structure of Allophane as Revealed by Its Thermal Transformation*, in Proceedings of International Clay Conference, vol. 1, Editor-in-chief Lisa Heller bisa Heller by Israil University Press, Gerusalem, pp. 151 (1969)

F. Vaughan, Energy changes when kaolin minerals are heated. Clay Mineral Bull. **2**(13) 265–274 (1955)

F.M. Wahl, R. E. Grim, High temperature DTA and XRD studies of reactions, in 12th National Conference on Clays and Clay Minerals, p. 69–81 (1964)

A.F. Wells, *Structural Inorganic Chemistry*. 2nd edn. (Oxford, Claredon Press, 1950) p. 571

H. Yamada and S. Kimura, Studies on coprecipitates of alumina and silica gels and their transformation at higher temperature. J. Ceram. Assn. Japan. **70**(3), 65–71 (1962)

A. Zwetsch, Rontgenunter suchungen in der keramik. Ber. Dtsch. Ges. **15**(6), (1934)

Chapter 21
Mullite Phase

21.1 Introduction on the Formation of Mullite

Sequential phase development that occurs on heating different kaolinites was done by other researchers, e.g., McCom et al. (1999), Gualtieri (1995), Viswabaskaran et al. (2002) based mainly on X-ray diffraction and TEM analysis data of heated kaolinites.

21.1.1 X-ray

At 900–1,000 °C, mullite crystallization ensues and at this stage, mullite is found to be weakly crystalline. All Bragg diffraction peaks characteristics of well-crystallized mullite are not fully observed in the diffraction pattern. Major peaks are broad, some are missing. Previous record shows that the characteristic lines, namely, 0.330 and 0.341 nm are merged instead of splitting as is found in case of mullite formed at elevated temperature. Comeforo et al. (1948) noted that in early stage of mullite crystal growth, different diffraction lines exhibit different line breadths. The first few lines in the powder diffraction diagrams may be indexed without ambiguity, the prism reflections are diffuse, flat pyramids are moderately sharp, and the basal reflections are distinct. The asymmetry of mullite developed from kaolinite is apparent in the original negative but less clear in reproduction. So it may be called as mullite type phase instead of well-developed mullite. Recently, Chakraborty et al. (2003) indicated splitting of those peaks in case of English kaolinite heated to 1,000 °C (Fig. 21.1a, b).

At 1,100–1,250 °C, mullite development occurs predominantly at the cost of disappearance of broad X-ray pattern due to spinel phase. Intensity of mullite peak shows increasing from 1,100 °C. At about 1,200 °C and above increase in both intensity and crystallinity of mullite are evident. Splitting of 0.341 and 0.338 nm is apparent, other XRD lines are becoming well sharp and recognizable with standard pattern of mullite.

A. K. Chakraborty, *Phase Transformation of Kaolinite Clay*,
DOI: 10.1007/978-81-322-1154-9_21, © Springer India 2014

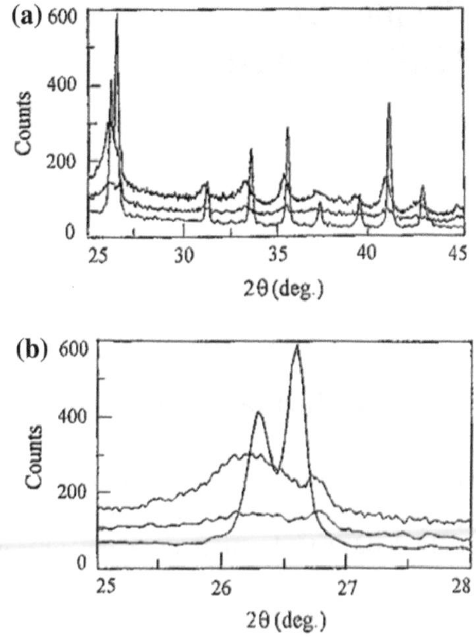

Fig. 21.1 **a** 120/210 reflection pair for English kaolinite fired at different temperatures. *Bottom curve* fired at 1,550 °C; *middle curve* fired at 1,150 °C; *top curve* fired at 1,050 °C. **b** The said reflection pair in 25–28° 2θ range

21.1.2 TEM

On transmission electron microscopy, two types of micrographs are revealed. Mullite formed from kaolinite by a reaction in the solid state during firing is called as primary mullite. This type of mullite exhibits a typical scaly appearance in the electron micrograph. It was postulated that the scale appearance results from the original sheet structure of clay mineral. This shape seems to be preserved during firing to high temperatures even when kaolinite is completely converted to mullite at high temperature (Schuller and Kromer 1975).

Other type of mullite which recrystallizes from the melt is called secondary mullite. They are characterized by a needle-like habit. These mullite needles have just the same X-ray diagram and thus the same crystal structure as the scaly mullite. The stability of the scaly mullite depends upon firing cycle. It changes its appearance and acquires the well-known needle like habit on continued firing and preferably in the presence of glassy phase. By scanning electron microscopy, Schuller and Kromer (1975) showed booklet like texture of Burda-Kaolin. After firing at 1,000 °C, kaolinite crystals are retained as idiomorphic crystal. On raising the firing temperature to 1,300 °C where mullitization was completed but shape of the crystal does not differ in any way from the appearance in the unified state.

This transformation of kaolinite to pseudo morphs with mullite structure is explained by them to be a reaction in purely solid state. The recrystallization to a needle-like habit may be considered either as a solid-state reaction or as a solution and crystallization from a melt produced by a small amount of fluxing oxide present in the clay.

21.2 Size of Mullite

The shape and size of mullite crystal increase on continued firing as noted in TEM diagrams of Comeforo et al. (1948), Comer (1960) and Dekeyser and other SEM photographs. Comeforo et al. (1948) showed the range of grain growth of mullite needles with increasing temperature. Thus, the overall nucleation of mullite and its growth are dependent on both time and temperature.

21.3 Views of Mullite Formation from Kaolinite

There are two views of mullite formation in kaolinite.

21.3.1 Solid-State Reaction of γ-Al_2O_3 and SiO_2 (A)

This reaction was proposed by first group. Researchers belong to same old groups and a few new groups to be called as first group for examples Okada et al., Percival et al., Sonuperlak et al., Mackenzie et al., Mazumdar and Mukherjee, etc. are of the opinion that γ-Al_2O_3 reacts with SiO_2 (A) to form mullite. This conjecture is based on the following independent studies namely

(a) Qualitative and quantitative studies of heated kaolinites by X-ray analysis.
(b) By prediction through theoretical calculation of free energy change of possible thermal transformation of metakaolinite.
(c) By I.R. Study.
(d) By RED.
(e) By thermodynamic calculation, Vaughn (1955), Mazumdar and Mukherjee (1986) tried to compare the measured heat of reaction value ($\Delta H_{1250\,K}$) with theoretically possible reactions and finally concluded that γ-Al_2O_3 as the phase formed at 980 °C and subsequently γ-Al_2O_3/SiO_2 reaction is responsible for 1,250 °C exotherm. Chakraborty et al. (1986) recalculated $\Delta H_{1250\,K}$ and $\Delta H_{1250\,K}$ values using $\Delta H_{298\,K}$ for γ-Al_2O_3 from different standard charts and is given in Table 12.3. It is shown that calculated enthalpy values vary to a great extent. Since standard values are different and secondly the exact co-efficient of the reaction products were not taken into calculation. As a result, on comparison of the measured value with theoretically calculated values for Eqs. 12.4, 12.7 and 12.16, it is very difficult to ascertain which reaction would actually occur at 980 °C. Similar is the case for mullitization reaction as proposed for Eqs. 12.2 and 12.11.

21.3.2 Polymorphic Transformation of Al–Si Spinel

Researchers belong to seconds group, e.g., Brindley and Nakahira (1959), Comer (1961), Von Gehlen (1962) are of the opinion that Al–Si spinel gradually releases SiO_2 on further heating and assume 3:2 mullite. Brindley and Nakahira (1959) gave a well integrated picture of the sequence of events from the initial layer structure of kaolinite, through a residual layer structures in metakaolin, followed by a cubic spinel type phase and finally a chain type structure in mullite as follows.

$$2[Al_2O_3.SiO_2.2H_2O] \underset{925\,^{\circ}C}{\overset{500\,^{\circ}C}{\longrightarrow}} \underset{\text{Metakaolinite}}{2Al_2O_3.4SiO_2} + 4H_2O \quad (21.1)$$

$$\underset{1100\,^{\circ}C}{\longrightarrow} \underset{\text{Spinel type phase}}{2Al_2O_3 \cdot SiO_2} + SiO_2 \quad (21.2)$$

$$\underset{>1400\,^{\circ}C}{\longrightarrow} \underset{\text{Mullite type phase}}{2Al_2O_3 \cdot 2SiO_2} + \underset{\text{Cristobalite}}{SiO_2} \quad (21.3)$$

$$\longrightarrow \underset{\text{Mullite}}{2Al_2O_3.1/3\ SiO_2} + \underset{\text{Cristobalite}}{2/3SiO_2} \quad (21.4)$$

In order to explain a consistent structural reaction sequence, they first proposed that metakaolinite continuously discards SiO_2(A) and secondly the intermediate spinel phase would be Al–Si type spinel in which 8 Si are occupied in eight tetrahedral sites, 10.2/3 Al ions are statistically filling 16 octahedral sites. This spinel on progressive heating forms mullite with further discard of SiO_2, which appears as cristobalite. In this way, they tried to explain how the continuity of the oxygen framework is retained and how cat ions are migrated and a portion of SiO_2 is discarded by diffusion of Si ions. However, this hypothesis becomes untenable in the light of measured data of exolution of SiO_2 rich (A) phase by Chakraborty and Ghosh (1978a, b). They showed that SiO_2 rich (A) phase forms only once on heating metakaolinite at 980 °C and no further evolution of said phase is occurred during mullitization process. Moya et al. confirmed the above observation and noted the persistence of second exothermic peak about 1,250 °C in DTA analysis of leached kaolinite. Similarly, expansion phenomenon is also noticed in DTMA analysis of alkali-treated residue by Chakraborty (1993). He noted both predominant mullitization by XRD along with increase in intensity of inflection in differential peak (Fig. 18.7). Contrary to two process of mullitization as predicted by previous researchers viz. (i) γ-Al_2O_3/SiO_2 solid-state reaction; (ii) gradual elimination of silica out of silica rich spinel phase, polymorphic transformation of Al–Si spinel is more likely responsible for 3:2 mullitization. Tentative arguments against those two processes of mullitization are as follows.

By free energy calculation, it is shown by Chakraborty that the theoretical possibility of direct mullite formation is more easier than decomposition of metakaolinite into γ-Al_2O_3 and SiO_2 (A) and recombination there of (Fig. 20.2). Accordingly, formation of any intermediate γ-Al_2O_3 phase in K-M reaction seems untenable.

21.3.3 *Comparison of Mullite Formation in Kaolinite Versus Mixture of γ-Al_2O_3 and SiO_2 (A)*

Chakraborty and Ghosh (1991b) monitored the mullitization behavior in synthetic Al_2O_3–SiO_2 mixture to that of kaolinite by XRD studies.

The results show that component oxides first crystallize into their respective high temperature crystalline forms. Figure 20.3b shows the schematic course of mullitization of artificial Al_2O_3–SiO_2 mixtures. Afterward, as the temperature increases, the intensities of the two newly formed crystals, namely, corundum and β-cristobalite decrease gradually with the formation of an aluminosilicate phase (A). Thereafter, mullite nucleation takes place and then steady crystallization is recorded in the growth curve (Fig. 14.5). Whereas mullite formation in case of kaolinite occurs very rapidly around 1,250 °C; a gradual slow rise of mullitization occurs up to elevated temperature. Thus, mechanisms of mullite formation in two cases are different.

Comeforo et al. (1948) concluded two stages other than normal three stages of thermal behavior of kaolinite by using R.C.A. Type B electron microscope in their mullitization study. First is the dcomposition of kaolinite to the formation of metakaolinite. Second is the collapse of metakaolinite into crystalline mullite nuclei. Figure 21.2 shows the retention of outlines of hexagonal kaolinite and deposition of mullite needles. Growth of mullite seems to occur along the hexagonal boundrary, which strongly suggests that long strings of alumina coordination octahedra of kaolinite are carried over directly to structure of mullite.

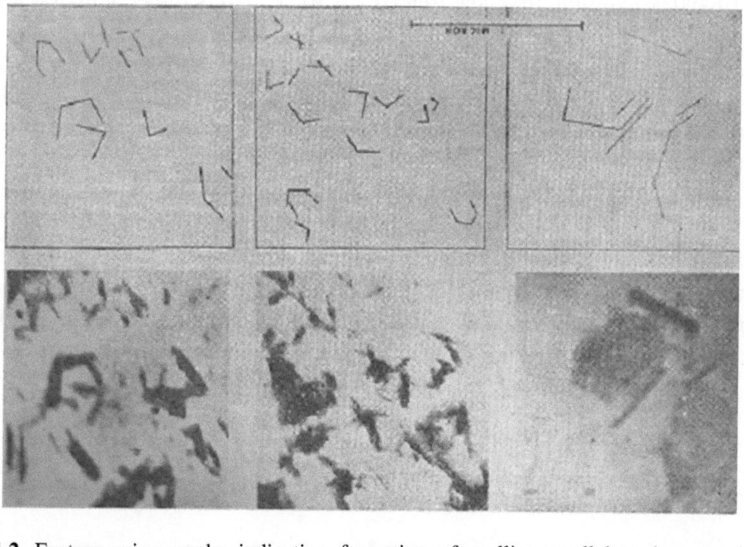

Fig. 21.2 Eectron micrographs indicating formation of mullite parallel to hexagonal grain boundaries of kaolinite; adjacent tracings of photographs outline the more pronounced instances of parallelism between mullite crystal formation and original hexagonal form (after Comeforo et al. 1948). Reprinted by permission of the American Ceramic Society

Johonson and Pask (1982) also proposes similar view. On comparing the mullite formation behavior and microstructure development at high temperature of kaolinite and an equivalent alumina-silica mixture consisting of α-alumina and SiO_2 (α-quartz), they noted that mullite formed in two cases show different crystal morphologies. In case of kaolinite, mullite are largely acicular with pointed ends and diamond shaped cross-sections. They further observed that the principal growth of the crystals is lengthwise in the c-direction.

In case of mixture, the mullite forms on the surfaces of the alumina grains and grows by interdiffusion of aluminum and silicon. Simultaneously, small mullite particles dissolves in the surrounding aluminosilicate liquid phase, leads to saturation and finally precipitates as larger crystals by the principle of Ostwald ripening. The growth rate is dependent upon the quantity of liquid phase generated, its composition and viscosity related to temperature. The microstructure shows mullite of rectangular shape with rounded ends and elongate with increase of firing times. Accordingly, they predicted that mechanism of mullitization in two cases is dissimilar.

Similar variation is observed when kaolinite is fumed with an acid prior to heating vide Eq. 11.1.

21.3.4 Comparison of Mullite Formation in Kaolinite and Fumed Kaolintie

When Bhandak kaolinite first heat treated at 980 °C and is fumed with a few drops of concentrated H_2SO_4 in a Pt crucible at 1,200 °C for 1 h, corundum forms along with mullite and β-cristobalite. When fuming is prolonged corundum and β-cristobalite develop, but not mullite. On continued heating, mullite forms gradually, whereas the amounts of corundum and β-cristobalite decreases. With this behavior, Chakraborty and Ghosh (1978a) suggest that the H_2SO_4 attacks the cubic phase and liberates free Al_2O_3 as $Al_2(SO4)_3$, which decomposes to γ-Al_2O_3 on heating. This γ-Al_2O_3 although it is associated with large amounts of free SiO_2, does not transform directly into mullite on heating, in contradiction to the suggestion of Percival et al. (1974) and by Percival (1978), that γ-Al_2O_3 and SiO_2 react to form mullite. Rather, γ-Al_2O_3 crystallizes further into α-Al_2O_3 and the reacts with cristobalite and/or SiO_2 (A) to form mullite gradually. This evidence proves that the cubic phase is Al–Si spinel rather than simple γ-Al_2O_3 as conjectured.

21.3.5 Peak Width and Semi Quantitative Analysis of Mullite

Chakraborty et al. (2003) calculated the changes in peak width of mullite and semi quantitative values of it formed at different stages of heating using PC APD software on the basis of 121 reflection (40.95° 2 θ Cu k_α) of it and are shown in Table 21.1.

Table 21.1 Changes in (121) peak width and intensity of mullite (area) formed during heating English Kaolinite

Parameters	Heat treatment temperature (°C)					
	1,050	1,150	1,250	1,350	1,450	1,550
FWHM	0.443	0.446	0.303	0.265	0.234	0.218
AREA	9.3	24.1	73.8	81.6	80.1	79.2

It shows a decreasing trend of FWHM value of mullite on progressive heat treatment. It also shows the formation of mullite starts at 1,050 °C, it increases at 1,150 °C considerably as noted by area under the (121) peak. It further increases suddenly to a significant quantity on heating at 1,250 °C. Increment continues to a smaller extent up to 1,350 °C and then remains stationary on further heating to 1,550 °C as shown in the plot (Fig. 21.3).

Regarding the nature of mullite produced, it is noted that at the lower temperature region (<1,200 °C), mullite so developed is of poorly crystallized. Since all XRD peaks pertaining to mullite are really broad. However, Fig. 21.1a, b shows splitting of peaks of Miller indices 120/210 at 0.3418/0.3383 nm for mullite formed out of kaolinite heated to 1,050 and 1,150 °C in reverse order in intensity in comparison to well-crystallized mullite appears on heating kaolinite at 1,550 °C. Such observations are not reported previously for kaolinite of other source. It may be due to choice of purer variety of English kaolinite, which contains only a trace of free quartz and may be due to precision X-ray recordings taken at low step size.

At 1,200 °C, splitting of peaks with Miller indices of 120/210, 240/420, 041/401, and 250/520 are noted in the diffractograms of heated clay. Moreover, the resolution is sufficiently low (Fig. 21.4a–d). With increase of calcinations temperature (>1,200 °C), a clearer angular separation becomes evident. So it is to be considered that two types of mullite are forming in the entire reaction series. First formed mullite at and below 1,200 °C may be pseudo orthorhombic. As the pattern shows splitting of 120/210 reflecting pair then Fig. 21.4a and b lattice constants

Fig. 21.3 Area under (121) of mullite formed on heating English Kaolinite versus temperature

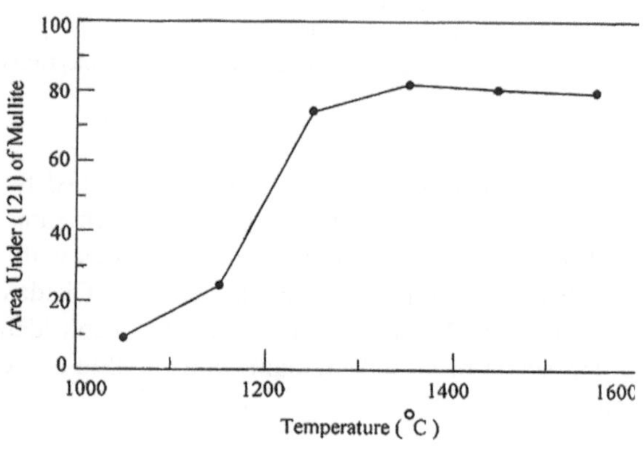

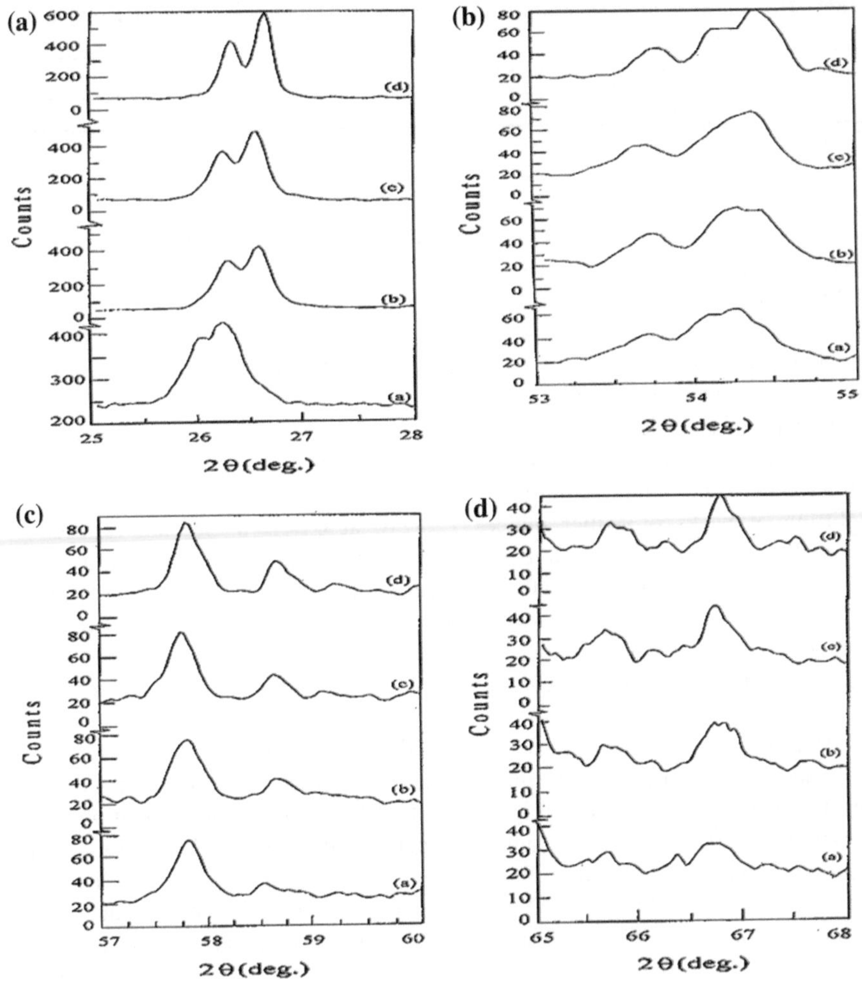

Fig. 21.4 **a** 120/210 reflection pairs for mullite formed on heating English Kaolinite fired at 1,250–1,550 °C. **b** 240/420 reflection pair of mullite formed on heating English Kaolinite at 1,250–1,550 °C. **c** 041/410 reflection pair of mullite formed on heating English Kaolinite at 1,250–1,550 °C. **d** 250/520 reflection pair of mullite formed on heating English Kaolinite at 1,250–1,550 °C

would not supposedly equal. Mullite developed thereafter on continued heating kaolinite at >1,200 °C may be purely orthorhombic form.

The percentage of mullite formation varies with the nature of clay chosen and their heat treatment temperature as shown by Okada et al. (1986), McGee (1966) in their (Figs. 14.2 and 14.4). It is noted that a considerable amount of vitreous phase to the extent of 40 wt% remains after mullite and cristobalite formations at high temperature.

21.4 Nature of Mullite

The X-ray pattern at the beginning of mullite formation at ~ 980 °C is weak and broad as shown in Fig. 20.1 and cited earlier (Comeforo et al, 1948). The mullite lattice evolves at this stage of heating is far from perfection. The mullite lattice is far from perfection. Brindley and Nakahira (1959) noted a gradual change in lattice constant at different stages of heating. At higher temperature, the crystallinity becomes more perfect. Both the cell volume and **a** parameter become steady on increasing heating conditions as shown by Brown (1985), Tomura et al. (1986), and Okada et al. (1986). The nature of mullite formed out of kaolinite was the most thought-provoking subject in earlier days. Brindley and Nakahira heat-treated single crystal kaolinite and measured the unit cell parameters of mullite formed at increasing temperatures by X-ray technique. During the course of transformation of kaolinite, they noted a change in the lattice constants (L.C.) and in cell volume of mullite. Significantly, the **a** parameter contracts with increase of temperature. Brown et al. (1985) measured the above parameters by X-ray powder diffraction using quartz as the angular calibration substance. They noted X-ray cell volume and X-ray **a** parameter decrease from the early stage of heating to high temperature. Okada et al. (1986) noted the similar fact of changes in lattice parameter of mullite formed out of the decomposition of Allophane, Halloysite, Kaolinite, and Dickite. They noted two-stepped formation curve of mullite which corresponds with the change in length of **a** axis. In the first step, when primary mullite was formed **a** axis length decreases with increase of firing temperature in the lower temperature region. In the second step, when secondary mullite was formed **a** axis remains constant with rise of temperature. Tomura et al. (1986) calculated the lattice parameter change of mullite in heat-treated spherical kaolinite. Changes in crystallite size and lattice strain of different compounds or during grinding those are generally made by X-ray line broadening analysis utilizing Williamson and Hall's equation (1953). Chakraborty and Das (2004) calculated the changes in crystallite size and lattice strain of mullite in addition to changes in L.C. values during phase transformation process of kaolinite type minerals by use of a computer program named Quasar based on *Rietveld* refinement.

21.4.1 Changes in L.C. Values of Mullite and Its Explanation

The above program was undertaken for the determination of quantitative phase abundance and for crystallite size/strain analysis-using X-ray diffraction data followed by Rietveld scale refining. L.C., crystallite size and strain are derived from scale factors, peak width and shape functions. First of all, diffraction data of heat-treated kaolinites were collected and analyzed for phases. Thereafter, quasar program was started as per stated procedure. XRD patterns are too weak in the temperature range of 1,000–1,200 °C to be analyzed by QUASAR. Quasar run of each heat-treated kaolinite sample was done, and the results are tabulated and are given in Table 21.2.

Table 21.2 Effect of heat treatments on strain and crystallite size and lattice constants of mullite developed in English kaolinite

Heat treatment temperature (°C)	Size (μm)	Strain (rms)	Lattice constant of mullite (nm)		
			a	b	c
1,250	455	1,271	0.75560	0.77010	0.2888
1,350	698	0.0604	0.75470	0.77030	0.2886
1,450	847	0.0346	0.75470	0.77010	0.2889
1,550	1,031	–	0.75490	0.76940	0.2889

The lattice constant values change with temperature of heat treatment. Among them, **a** axis value shortens drastically from 0.7556 nm at 1,250 °C to 0.7547 nm at 1,450 °C. A gradual decrease is noted for **b** and for **c** axis values with heating. The changes in lattice parameter values and sizes of mullite at different stages of heating of kaolinite calculated by applying quasar software are comparable with the data measured earlier by Brown et al. (1985), Okada et al. (1986), and Tomura et al. (1986). The mean micro strain is found to decrease proportionately with the reciprocal of the crystallite size during heating English kaolinite (Fig. 21.5). Changes in lattice parameter of mullite are found to be related to that of width of XRD peak of mullite (Table 21.1). As width decreases on heating, lattice constant values particularly **a** value likely decreases from higher side to a value close to that of 3:2 mullite.

This relationship suggest us to assume that mullite formed at lower temperature region is sufficiently weak in nature and on successive heat treatment its crystallinity gradually increases and becomes perfectly well crystallized at a temperature of the order of 1,550 °C. These results predict a change in the order of mullite occurring during heating English kaolinite in the entire transformation region.

21.4.2 Composition of Mullite

Crystal structure, chemical composition of primary and secondary mullite is a subject of much discussion. These two varieties may be identical as predicted by earlier authors Agrell and Smith (1960), Krishna Murthy and Hummel (1960),

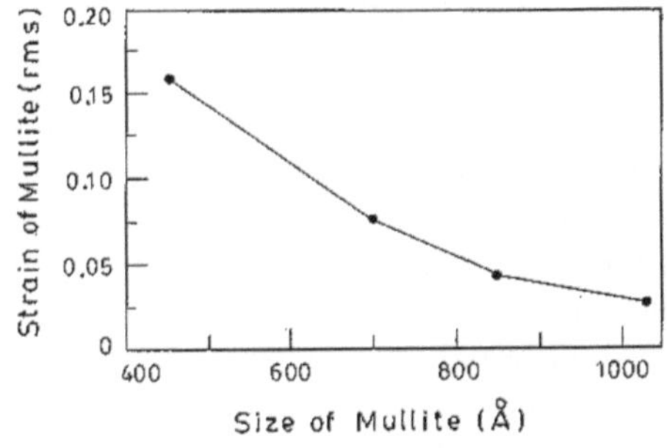

Fig. 21.5 Changes in micro strain with crystallite size of mullite formed at different stages of heating English Kaolinite

Durovic (1962). However, some authors express divergent views on the changes in L.C values with temperature.

(i) According to Brindley and Nakahira (1959) mullite formed at lower temperature may contains excess silica. They assumed the composition of this primary mullite is to be $2Al_2O_3.3SiO_2$, which liberates out silica on further heating. On the other hand, Chakraborty and Ghosh (1978b) by alkali leaching technique indicated experimentally that during heating kaolinite exolves siliceous phase only once and that occurred exclusively at the temperature of first exotherm to the extent of ~35–37 wt%. Thus, it seems that there is no scope of the presence of free silica in mullite lattice at lower temperature.
(ii) Recent views interprets that these two parameters are likely different to some extent. Brown et al. (1985) applied Cameron relationship (1977) of change in composition of mullite with **a** value to explain the constitution of initial mullite and phenomenon of changes of L.C. values. It is suggested that as the firing temperature was increased from 1,200 to 1,400 °C, the alumina content decreased from 61 to 59.8 mol.% and accordingly they believed that the mullite formed at lower temperature was considerably more silica deficient than 2:1 :: Al_2O_3:SiO_2. On prolonged heating at high temperatures, the silica content increased and tended toward 3:2 composition. After leaching out the free silica (A) content, this alkali leached heat-treated residue containing ~5 wt% primary mullite develops considerable quantity of orthorhombic mullite on further heat treatment. There is no free silica left in leached residue, there is no question of silica deficient primary mullite (2:1) silica deficient forms 3:2 orthorhombic variety as conceived by them.

Okada et al. (1991) reported two-step mullite formation during heating kaolinite versus firing temperature (Fig. 15.3). They predicted that the first step (lower temperature side) corresponds to tetragonal-like mullite, and the second one corresponds to ordinary 3:2 orthorhombic mullite. It is further reported that tetragonal-like mullite should be corresponds to the Al_2O_3 rich composition as estimated from the relation reported by Cameron (1977).

Question is: whether Cameron's relationship could be applied on the formation process of mullite during the decomposition of metakaolinite in the low temperature region, e.g., 1,000–1,200 °C as well as the temperature beyond 1,200 °C.

Cameron used various mullite samples for lattice parameter study, and these are all well-ordered mullite but differ only in composition, mostly are fused mullite, or mullite crystal obtained from electro cast brick, corhat refractory, forester mullite, sintered mullite, etc. Obviously, these mullite were fully developed at the start of the X-ray diffraction study. On the contrary, mullite is gradually developing from its incipient stage at ~1,000 °C on continued heating process of kaolinite. Here lies a difference between nature and character of the samples used for comparative L.C. measurements. Therefore, a question is still left whether conclusion (**a**—X or **a**—V plots) relationship drawn by Cameron on the basis of some highly crystallized mullite could be applied to the result obtained using mullite samples collected ongoing heating process of kaolinite. Increase in heat treatment of kaolinite decreases width and rises the sharpening of mullite peaks (Table 21.1). So the cause of changes of **a**, **b**, and **c** parameter values are more likely to be due to increase in order of crystallinity of mullite. Roy et al. indicated that composition of mullite may not be obtained from lattice parameter calculation due to the observation that even the mullite of fixed composition changes its L.C. values on heat treatment. In the present investigation, it is observed that mullite formed at ~1,200 °C is of low crystallite size and of highly strained (Fig. 21.5). Then, it is a matter of great question, how the composition of early formed mullite could be ascertained from the plot of L.C. versus alumina mol.% used by Cameron. It may be conjecture that lattice constant-alumina mol.% interrelationship is not applicable in the second stage of mullite formation processes occur at temperature >1,200 °C. Applicability of said relationship would be even more critical in the primary stage of mullite formation. Since the breadth of mullite peaks noted in heated kaolinite at ~1,050–1,150 °C are still broad (Table 21.2). The conjecture that this mullite as alumina rich may be of 2:1 variety by Okada et al. (1986), Schneider et al. (1988), the question is: if it is correct then why these peaks are so broad? Most probably, the change in **a** value is due to increase in the order of crystallinity of mullite. It is obvious that newly formed mullite at lower temperature would be broad and that it would require both time/temperature treatment to increase of its crystallinity. This study predicts that the composition of this early formed mullite (primary mullite) would be 3:2 as 120/210 peaks shows resolution (Fig. 21.3). Moreover, in view of increased growth rate as noted in (Fig. 21.2) in the temperature 1,050–1,150 °C and also due to said resolution behavior of mullite it may be conjecture that the type of mullite in this region is certainly different from the type of mullite formed at temperature >1,200 °C. These verify the concept of two stages of mullite formation as suggested by Okada et al. (1991).

21.5 Kinetics of Mullite Formation

Growth of mullite out of kaolinite shows nucleation control process. Okada et al. (1986) convincingly showed in Fig. 14.2 that transformation of spinel phase to mullite is interrelated. Since disappearance of spinel and development of mullite are concurrent in heated sample of kaolin polymorphs.

Dunkan et al. (1969) determined the kinetics of mullite growth from kaolinite and halloysite by quantitative X-ray diffraction analysis. The most probable rate-determining step is deduced by comparing with three theoretical rate equations with experimental results. Model of phase boundary control crystal growth is expressed mathematically as follows.

$$\alpha = 1 - \left(1 - \frac{k_f}{a}\right) \quad (21.5)$$

a radius of the spherical particle,
α fraction reacted at time,
k rate constant.

On comparing the plot of the above equation with experimental growth curve, a good agreement is noted up to 80 % conversion to mullite. However, the model does not accord with physical properties of the resultant mullite as shown by Lundin (1969) in his electron microscopy study. Comer (1960) in his electron optical study showed that mullite crystallizes from the reaction material in the form of numerous fine crystal oriented with respect to parent kaolinite. According to Kingery (1960), this observation supports nucleation control model. The mathematical equation corresponding the nucleation mechanism is as follows.

$$\text{In}(1 - \alpha) = k\,t \tag{21.6}$$

Accordingly, plot of ln $(1 - \alpha)$ against t shows a good agreement with experimental data (Fig. 21.6).

Latter on Mackenzie (1969) carried out kinetic study of a pure synthetic kaolinite by infrared frequency shift method for mullite analysis. Plots of percentage mullite formation derived from the IR measurement of aluminosilicate peak at 790–850 cm^{-1} versus firing time are identical in shape to those obtained for natural kaolinite or halloysite. Only the mullite formation temperatures are higher in synthetic kaolinite than in natural clays. The rate controlling step is similar to that natural kaolinite which involves a nucleation control process.

Tecilazie-Stevanovic et al. carried out the kinetic analysis of H-form and H-form of degraded Zettlitz kaolinite. Apparently, the mullitization degree is almost double in samples, which were degraded in structure. Mullitization process in freshly formed H-kaolinite follows nucleation kinetics model. However, in H-kaolinite with degraded structure, mullitization kinetics more corresponds to Janders diffusional model.

Bellotto (1994) studied the kinetic analysis of mullite formation for two kinds of clays (KGa-1 and KGa-2). He plotted ln $\{-\ln(1 - \alpha)\}$ against ln t and determined the slope (m) to get the idea about reaction mechanism. The mean m values obtained for the different heating rates are $m = 0.46$ for KGa-1 and $m = 1.1$ for KGa-2, respectively. He inferred that mullite nucleation followed by subsequent

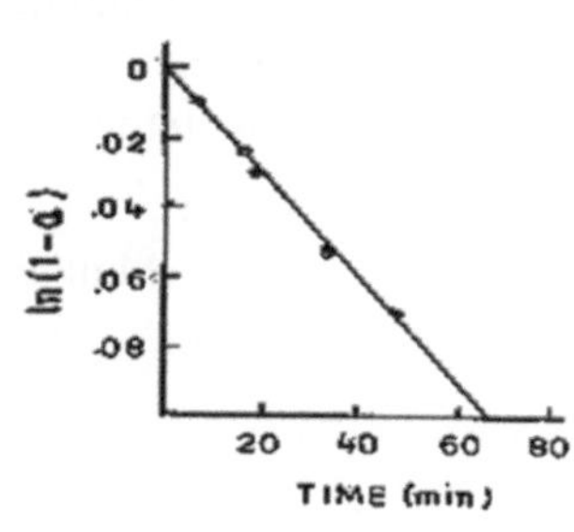

Fig. 21.6 Ln(1-α) versus firing time for kaolinite (after Duncan et al. 1969). Reprinted by permission of the American Ceramic Society

grain growth, which is diffusion controlled in case of KGa-1 whereas phase boundary controlled in case of KGa-2. But the question remains how mullitization takes place.

21.6 Newer Concept of Mullite Formation

Chakraborty and Ghosh (1991b) carried out the quantitative mullite formation of Bhandak kaolinite by X-ray study (Fig. 14.5). The overall growth curve in the temperature range from 1,000 to 1,600 °C consists of two parts.

(i) A sudden or steep increase of mullite formation from about 1,100–1,250 °C occurs followed by complete disappearance of the 0.197 nm and 0.139 nm Bragg reflection of spinel phase in the corresponding XRD pattern. It is most likely that transformation of spinel is interrelated to the development of mullite in the above temperature range. It may be concluded that only a crystallographic change might have occurred as per the equation below.

$$\underset{\substack{\text{Al-Si spinel}\\ \text{Cubic}\\ (25-30\ \text{wt\%})}}{3Al_2O_3.2SiO_2} \underset{1250\ ^\circ C}{\overset{1100\ ^\circ C}{\longrightarrow}} \underset{\substack{\text{Mullite}\\ \text{Orthorhombic}}}{3Al_2O_3.2SiO_2} \tag{21.7}$$

Chemical composition of both Al–Si spinel and mullite are likely similar but they are different in crystallography. Thus, this transformation may be polymorphic in nature.

Although Mcgee (1966) noted different growth curves for mullitization of their three different varieties of clays. In Missouri plastic fireclay after sudden rise mullite content the curve flattens. In Missouri flint fireclay, the curve after attaining a maximum value decreases with rise of temperature. In Pioneer kaolin, a sharp mullitization is noted at about 1,250 °C thereafter it slowly increases with continued firing which confirms the mullitization behavior of Rajmahal kaolinite noted during heating as shown by Chakraborty and Ghosh in Fig. 14.5.

(ii) A rather slow rise of mullite formation is indicated from 1,250 to 1,600 °C, i.e., after steep mullitization process. It is assumed that nucleation of mullite may occur in large residual part of mullite (A) phase generated during 980 °C exotherm. These mullite nuclei subsequently grow during further heat treatment. Nucleation and crystallization mechanism may be responsible for mullite formation in the second stage.

$$\text{Mullite (A) phase} \underset{1600\ ^\circ C}{\overset{1250\ ^\circ C-}{\longrightarrow}} \underset{\text{Mullite}}{3Al_2O_3.2SiO_2} \tag{21.8}$$

Leonard et al. (1979) not only confirmed the estimated value of silica (A) liberated by the decomposition of kaolinite as shown by Chakraborty and Ghosh but also agreed with the presence of the binary aluminosilicate phase.

It was noted earlier that alkali leached kaolinite that was heated at 980 °C formed mullite on further heating (Eq. 18.1). This equation shows that mullitization in kaolinite does not depend in any way on the associated 35 % free silica (A).

Brindley and Lemaitre (1987) are of the opinion that the development of mullite-like structure upon heat treatment of alkali-leached residue is a consequence of sodium contamination during the alkali leaching process. However, alkali-leached residue was washed and then tested for chloride ion to assure complete removal of adhered sodium ion for clarity. Moreover, the contention of Brindley and Lemaitre is not tenable since a mullite-like structure with about 2.4 mol.% Na_2O developed only with alumina as shown by Perotta and Yong (1974). By time-temperature-transformation (TTT) studies of well-crystallized kaolinite, Onike et al. (1986) plotted ln time versus wt% of mullite at temperatures ranging from 1,000 to 1,500 °C. These indicate that curves consist of two parts (Fig. 21.7). This result may conform the two paths of mullitization of kaolinite as shown by Chakraborty above.

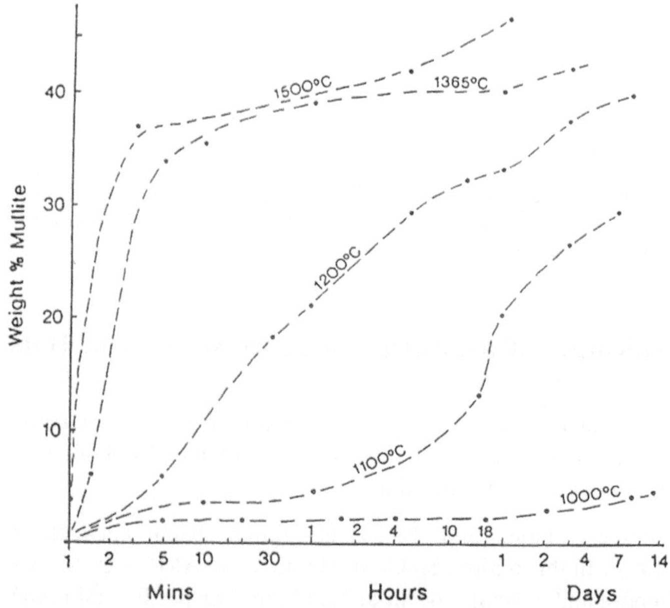

Fig. 21.7 Plot of in time versus wt% of mullite (after Onike et al. 1986). Reprinted by permission of the Trans. Tech Publication

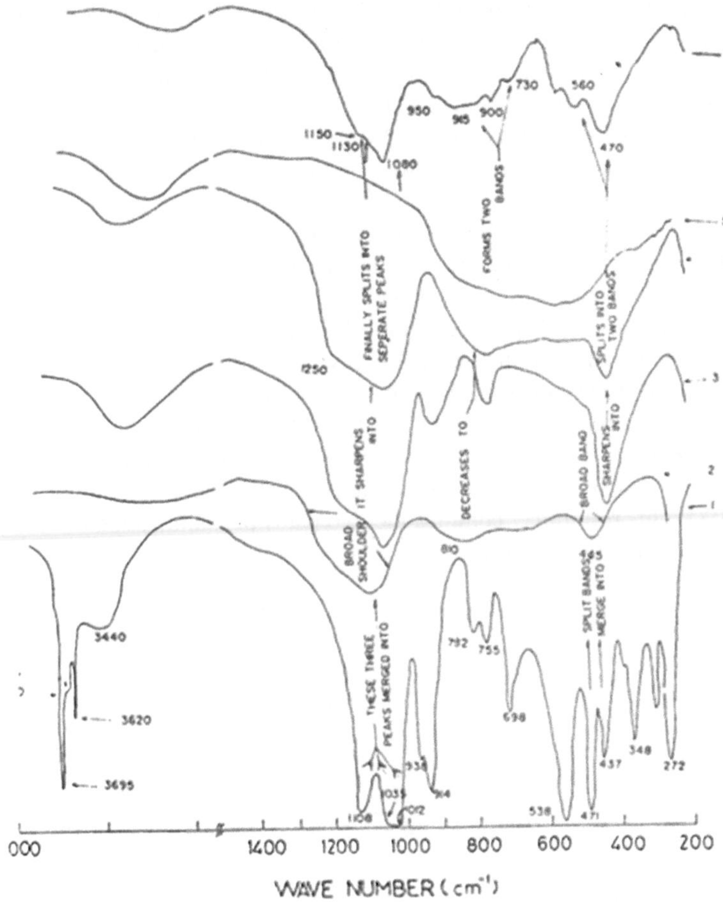

Fig. 21.8 I.R. studies of Rajmahal Kaolinite heated to different temperatures. *1* Raw clay, *2* Clay heated to 600 °C, *3* Amorphous silica, *4* Clay heated to 1,000 °C, *5* γ-Al_2O_3, *6* Clay heated to 1,300 °C. Reprinted by permission of Clay Science

21.7 Verification of Mullitization Processes by I.R. Data

To support metakaolinite—Al–Si spinel—mullite reaction mechanism the following wave number regions of the I.R. spectra shown by Chakraborty and Ghosh (1991a), in Fig. 21.8 are discussed as follows.

(i) Comparatively large and broad absorption band between 1,200 and 1,000 cm^{-1}. In this region, Si–O–Si vibration and Al(IV)–O–Al(IV) vibrations superimpose and generally occur at 1,075 cm^{-1} as per Freund (1967a, b). This broad absorption shoulder becomes more sharp from metakaolinite stage to

Al–Si spinel stage and thence to mullite stage where it splits into separate Si–O vibrations at 1,160 cm^{-1}, 1,130 cm^{-1} and Al(VI)-O vibration at 1,075 cm^{-1}. So due to mullitization, one can see the decrease in the broadband intensity with simultaneous formation of new Al(VI)-O absorption band.

(ii) Comparatively smaller broadband between 900 and 700 cm^{-1}. This band due to corner and edge sharing of $Al0_4$ groups characteristic of metakaolinite decreases in width during successive heating to higher temperatures. Two new bands appear, one at 725 cm^{-1} due to the change of Si–O–Al(IV) to Si–O–Al(VI) during Si–Al spinel formation and another at 900 cm^{-1} due to Al(VI)-O vibration in mullite. The remaining portion of $Al0_4$ changes its place and appears at 830 cm^{-1} for isolated $Al0_4$ groups of mullite.

(iii) Broadband between 600 and 400 cm^{-1}. In this range, 445 cm^{-1} due to $Al0_6$ groups because of formation of mullite and secondly another at 470 cm^{-1} due to SiO_2 groups because of the exolution of SiO_2 (A) during metakaolin to Al–Si spinel transformation persist.

It was already shown that mullitization takes place from Al–Si spinel as well as from mullite (A) phase on further heating. In both the transformations, the coordination number of Al changes. Now the question is: whether from the changes of $Al0_4$ and $Al0_6$ I.R. bands in successive heat-treated kaolinite, is it possible to demarcate from which of the two paths, mullitization is taking place? The basic information which one gathers from the spectroscopic studies is the disappearance of mostly $Al0_6$ groups in kaolinite to $Al0_4$ in metakaolinite and then partial development of $Al0_6$ during 980 °C reaction and then further decrease of the development $Al0_6$ with consequent increase of $Al0_4$ during mullite formation. From this observation, it is difficult to conclude from which path mullitization occurs. However, major mullite development takes place from mullite (A) phase. Obviously, the changes in the contours or shapes of the three absorption bands, e.g. 1,200–1,000 cm^{-1}, 900–700 cm^{-1}, and 600–400 cm^{-1} have been explained to be related to metakaolinite to mullite(A) to mullite reaction. This is analogous to the concept of chemical continuity of Si–O–Al bond present in kaolinite to metakaolinite to Al–Si Spinel + mullite(A) to orthorhombic mullite transformation process as follows.

$$\begin{array}{ccccc} \text{Kaolinite/} & \rightarrow & \text{A1-Si Spinel} & \rightarrow & \text{Mullite} \\ \text{Metakaoinite} & & + & & \\ & & \text{Mullite(A) phase} & & \\ \text{(having} & & \text{(into which mullite} & & \text{(persistence of} \\ \text{Si-O-Al} & & \text{nuclei originates)} & & \text{Si-O-Al} \\ \text{bonds)} & & & & \text{bonds)} \end{array} \tag{21.9}$$

The results of Udagawa et al., supports the above view. They showed that the ratios of $Al0_4/Al0_6$ of 970 and 1,250 °C heated kaolinite were found to be nearly close. This leads us to believe that the coordination state of Al analogous to the of mullite has already been attained in the matrix of 1,000 °C heated kaolinite and wherein nucleation has started, only slight heating or soaking is necessary for the mobility of atoms to have sufficient growth of the size of crystals to be revealed by XRD analysis. In some kaolinites, e.g., Zettlitz kaolinite referred by Leonard (1977), Rajmahal kaolinite referred by Chakraborty and Ghosh (1978) and MgO mineralized kaolinite by Bulens et al. (1978), the relative ratio of $Al0_4/Al0_6$ changes gradually and approaches toward the $Al0_4/Al0_6$ ratio of mullite with the development of more mullite as well as increase in its crystallinity. This indicates that Al sites in 1,000 °C heated kaolinites are reorganized on further heat treatment to develop more $Al0_4$ groups connected with remaining $Al0_6$ chains, which is necessary for building up of mullite structure. This type of transformation is related to the polymorphic change of Al–Si spinel (cubic) to mullite (orthorhombic) as shown earlier by Chakraborty and Ghosh (2003). The predominance of either of the two paths of metakaolinite's transformation may be dependent upon various factors as revealed from the analysis of different kaolinites studied by various workers, e.g., crystallinity of kaolinite, its particle size, inherent oxides present in the clay mineral.

As such these are indirectly related to the origin/source of clay mineral as mentioned in the appendix and in Chap. 5.

Tecilazic-Stevanovic et al. presented the following concept of mullite formation. According to the suggested mechanism, the donor and acceptor ranges are formed in metakaolinite structure and counter diffusion of both Al and Si ions occur. When mullite is formed from metakaolinite, the number of Si ions that must move is greater. The two-dimensional sheets of oxygen sub lattice of metakaolinite structure is changed on to three-dimensional array in mullite. Small cat ions and Si and Al fill the existing interstitial spaces in the structure of metakaolinite and in the spinel type phase structure. The diffusion of Si ions progress and cause the SiO_2 exclusion. Donor range is part of structure that is degraded and the acceptor range is a part in which mullite nuclei are formed.

Mechanism of Mullite Formation from Metakaolinite

Mechanism of Mullite Formation from Metakaolinite

"The acceptor range"		"The donor range"
$4(Al_2O_3\,2SiO_2)$	3Si →	$2(Al_2O_3\,2SiO_2)$
↓	4Al ←	↓
$2(3Al_2O_3\,2SiO_2) + SiO_2$		$7SiO_2$

21.8 Mechanism of Mullite Formation

The formation of mullite by heating diaspore, gibbsite with cristobalite, the silicic acid has been studied by Wahl et al. (1964). For example, diaspore first converts to corundum at relatively low temperature (~500 °C) and then reacts with cristobalite and contributes to mullite development at 1,200 °C, whereas gibbsite forms corundum at 1,100 °C and then reacts with cristobalite to form mullite at 1,440 °C. When silicic acid is used in place of cristobalite and mixed with diaspore, it first transforms to β-cristobalite at 900 °C and then there is mullite formation at 1,300 °C. When quartz is used and mixed with diaspore, it transforms to β-cristobalite at 1,200 °C and after that there is mullite formation at 1,325 °C. Thus, they concluded that the original structures of the alumina and silica component materials, which is turn control the transformation to corundum and β-cristobalite, are fundamental factors in limiting the ultimate development of mullite. Subsequently, they noted that on heating corundum with cristobalite, no mullite formation is noted even on heating at 1,375 °C with 72 h soaking. Therefore, the nucleation of mullite from a heated alumina-silica mixture is not related to temperature for the development of corundum and β-cristobalite from Al_2O_3 and SiO_2 component materials.

They also observed that when corundum/cristobalite mixtures are heated gradually, the intensities of corundum and cristobalite decrease. These results indicate that corundum and cristobalite react with each other and form an amorphous aluminisilicate phase at high temperature. By the combined use of X-ray fluorescence and optical and X-ray crystallo-graphic analysis, Dekeyser (1965) noted that reaction at the interface of contact of SiO_2 and Al_2O_3 pellets and showed that a glassy phase is formed when Al_2O_3 pellets and showed that a glassy phase is formed when Al_2O_3 diffuses into the SiO_2 zone. Staley and Brindley (1969) also indicate that the noncrystalline product is an important transitional phase in the reaction between corundum and cristobalite at 1,550 °C. Chakraborty and Ghosh (1991b) noted that the formation of amorphous aluminosilicate phase can be increased by adding a suitable flux to the synthetic mixture before heating. Results show that with a gradual increase of H_3BO_3, the percentage of mullite formed increases on heating an Al_2O_3–SiO_2 synthetic mixture even at 1,000 °C for 4 h. Therefore, the controlling factor in mullite formation in cases of artificial mixtures is the formation of an intermediate aluminosilicate phase (A) by the solid-state reaction of SiO_2 and Al_2O_3 components before nucleation of mullite. When formation of it has been facilitated by external use of fluxing material, mullite formation takes place at a comparatively lower temperature. Kaolinite forms intermediate aluminosilicate phase at 980 °C, which accounts for the early mullite development before the second exotherm. However, in the case of pure reactants, the formation of aluminosilicate occurs at a very high temperature. In the meanwhile, crystallization of the Al_2O_3 component to γ-Al_2O_3 and then to its stable modification, i.e., α-Al_2O_3 and similarly crystallization of the silica component to β-cristobalite take place preferentially at successively higher temperature and they react to form a binary aluminosilicate glassy phase. Finally, mullite nucleates in the glassy matrix since thermodynamically it is the stable state in the phase diagram of the binary Al_2O_3–SiO_2 system.

In the mullitization process of fumed heated kaolinite as studied by Chakraborty (1978a) some researchers still assumed that addition of sulphuric acid in fuming process might have disturbed the homogeneity of silica and spinel phases in heated kaolinite matrix and their arrangements. As a result, some quantity of corundum formation starts at the beginning of heating process and mullitization is deled. Of course, homogeneity of silica (A) and spinel phase ($\sim$10 nm in size) mixture is a major cause of earlier mullite development in metakaolinite. In this connection, a compilation of mullitization data out of various mixtures of two components ranging from micro-to-nano-to-molecular level carried out by some researchers is presented in Table 21.3 and discussed below.

1. In micro composite (corundum coated with hydrous silica), α-Al_2O_3 is found to remain stable as high as at $\sim$1,499 °C. Thereafter, a solid-state reaction between them ensues, which results to the exhibition of an endothermic peak (Fig. 21.9) in DTA as shown by Wang and Sacks (1996). The obvious crystalline phase beyond the endotherm is mullite(o).
2. Pure γ-Al_2O_3 exhibits exotherm during DTA studies at $\sim$1,269 °C with formation of α-Al_2O_3. When it was mixed with fume silica, Saito et al. (1998) showed that the intensity of the said peak decreases and it becomes broad. On the other hand, the formation of corundum is retarded vis-a-vis exothermic peak is increased to 1,422 °C with addition of TEOS as silica source (Fig. 21.10).

Table 21.3 Alumina–Silica components used to synthesize mullite showing mixing scale

State of silica–alumina components	Size	Level/degree of homogeneity
1. Oxide mixture (1) α-Al_2O_3/α-quartz	Micrometer range	Inhomogeneous, possibility of agglomeration
2. Oxide coated silica (2) α-Al_2O_3/TEOS	Alumina-0–18 μm (micro composite particles)	
3. γ-Al_2O_3/fume silica (3)		
4. γ-Al_2O_3/TEOS (3)		Partly homogeneous
5. Diphasic/Colloidal Al_2O_3–SiO_2 gel (4,5,6)	Nanometer range	More homogeneous
6. Monophasic/Polymeric Al_2O_3–SiO_2 gel (4.5,6)	Molecular range	Most homogeneous, develops chemical bondings (Si–O–Al) during heating at 600–900 °C
7. Metakaolin	Existence of Si–O–Al linkages (a two dimensional chain like structure)	Most homogeneous as like as synthetic polymeric gel (heated)

1. Johnson and Pask (1982), 2. Wang and Sacks (1996), 3. Saito et al. (1998), 4. Hoffman et al., 5. Yoldas (1992), 6. Chakraborty (1996a, b)

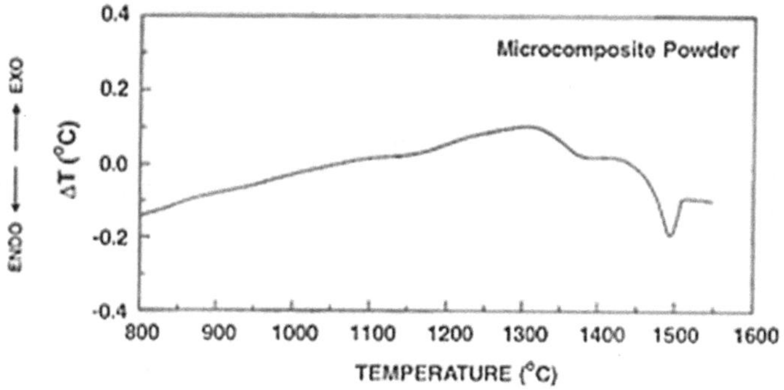

Fig. 21.9 DTA plot of micro composite powder showing endothermic peak (after Wang and Sacks 1996). Reprinted by permission of the American Ceramic Society

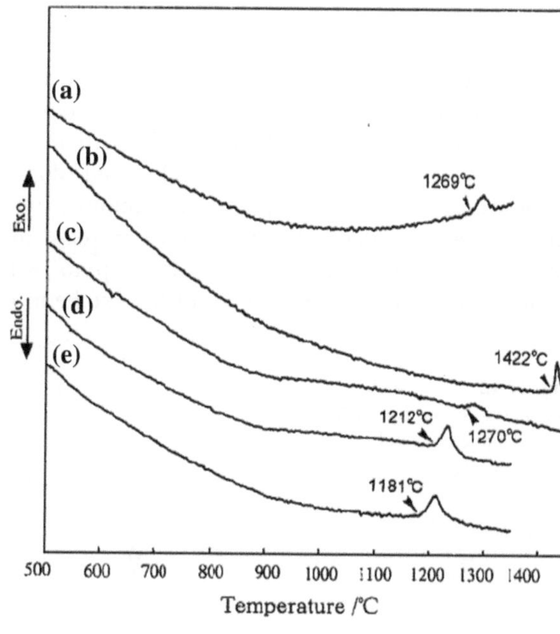

Fig. 21.10 DTA curves of γ-Al_2O_3 with 10 mol.% addition of various silica: no addition (*a*); TEOS (*b*); fumed-silica (*c*); quartz (*d*); and cristobalite (*e*) (after Saito et al. 1998). Reprinted by permission of the American Ceramic Society

The extent of inhibition is found to be related to amount of SiO_2 additive (TEOS) used (Fig. 21.11). The peak temperature is increased rapidly at the beginning and is attained constant at addition beyond 10 mol.%. Thus, phase transition of γ-Al_2O_3 to α-Al_2O_3 is retarded and is affected largely by TEOS impregnation process than with fume silica mixture. Even, retardation is more effective when coating of γ-Al_2O_3 with silica was done by CVD process as shown by Niwa et al. (1990).

They showed that a small quantity of it would suffice the purpose stabilization. Thus, the nature of suppression is due to physical properties of silica source, its structural difference and it might be due to reactivity of silica at the surface of γ-

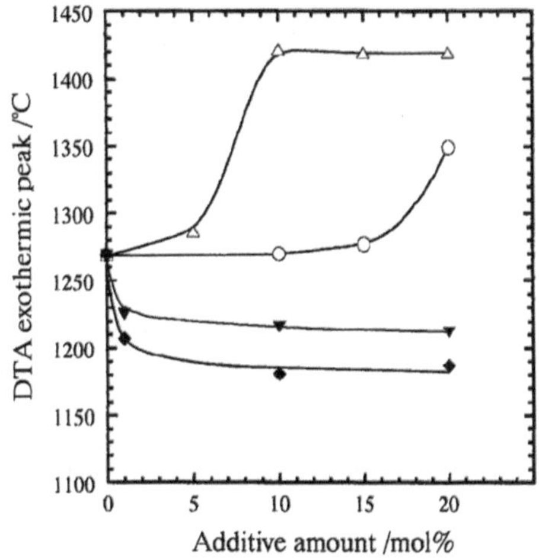

Fig. 21.11 Amount of SiO_2 additives versus DTA Exothermic Peak Temperature ((Δ TEOS (*open circle*) fumed silica (*filled down-pointing triangle*) quartz and (*filled diamond*) cristobalite) (after Saito et al 1998). Reprinted by permission of the American Ceramic Society

Al_2O_3 particles. According to Yoldas (1976), the elevation of the exothermic peak or the rise in temperature stability of α-Al_2O_3 phase is due to structural incorporation of silica into the surface of γ-Al_2O_3 particle. He synthesized alumina gel by hydrolysis of aluminum-sec butoxide with doping of various amounts of $Si(OC_2H_5)_4$. The undoped alumina gel exhibits exotherm at ~1,200 °C due to α-Al_2O_3 transformation. With the gradual increase in concentration of silica, the stabilization temperature also increases up to ~1,380 °C with 6 wt% silica addition thereafter the temperature of the exotherm rapidly falls off (Fig. 21.12).

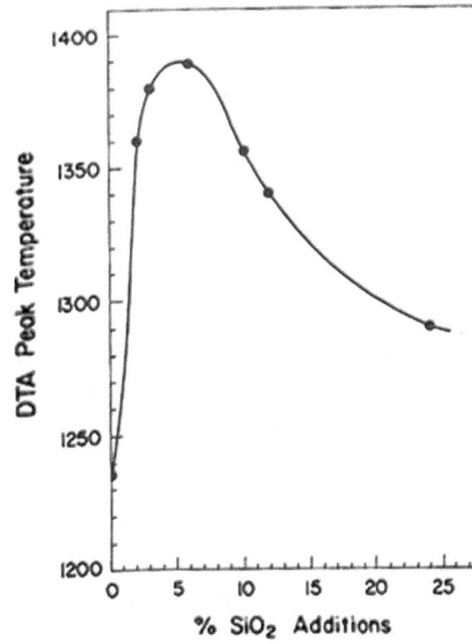

Fig. 21.12 Effect of SiO_2 concentration on the DTA Peak Temperature (after Yoldas 1976)

He contended that exothermic peak above 6 % silica is entirely a different origin and may be due to mullitization. XRD study is needed to verify this conjecture. On summarizing the above experimentations, the followings are emerging out.

(i) Mixing homogeneity definitely plays a major role in α-Al_2O_3 stabilization vis-à-vis mullitization. It gradually increases in compact mixtures down the serial number from 1 to 6 as given in Table 21.3.
(ii) The physical property (size and its surface area) and its structural aspect of silica phase play a vital role in development of corundum.
(iii) In case of solid–solid mixture, 20 mol.% of fume silica is needed to affect the crystallization of γ-Al_2O_3.
(iv) In impregnated mixture, 10 mol.% TEOS increases the temperature of stabilization.
(v) Whereas 6 % silica is enough for corundum stabilization when mixing is done by sol-gel process.
(vi) Corundum formation is absolutely ceases, i.e., its thermal stability vanishes out when silica and alumina components in the ratio (Al/Si = 3/1) are gelified as per the procedures of synthesis through diphasic and lastly by monophasic route. Phase transformations of diphasic according to Chakraborty (1996a) shows two exothermic peaks (Fig. 21.13) due to crystallization of mullite from mullite(A) and Al–Si Spinel, respectively.

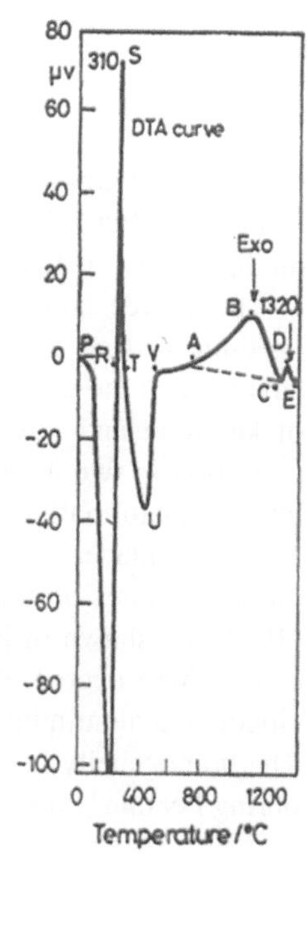

Fig. 21.13 DTA curve of diphasic mullite gel (after Chakraborty 1996a)

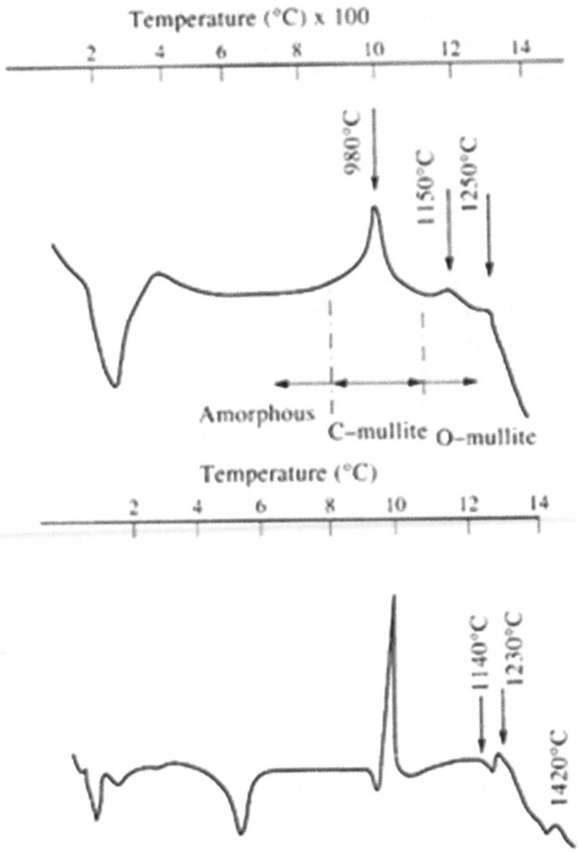

Fig. 21.14 DTA peak of polymeric mullite gel, G-152(1) compared with same of Bhandak Kaolinite (after Chakraborty 1996b)

Monophasic/polymeric gel according to Chakraborty (1996b) shows three exotherms (Fig. 21.14) due to consecutive crystallization of Al–Si Spinel and mullite by two paths.

These three exotherms are concurrent with the exhibition of three exotherms of kaolinite and no exotherm corresponding to corundum formation was observed anywhere in the DTA trace. Moreover, the nature of growth curves of mullite out of kaolinite and Al_2O_3–SiO_2 gel are quite similar, which suggest that mullite formation in two cases follow the same course and same mechanism. Therefore, phase transformation of kaolinite and polymeric gel are analogous (Chakraborty 1978c Chaklader ad Wheat). Similar view was also due to Hoffman et al. (1984). Ultra structures of colloidal, polymeric gels of two varieties according to Yoldas (1992) are shown in Fig. 21.15.

The high degree of homogeneity in polymeric gel and the bondings between silicon and aluminum leads to spontaneous crystallization of mullite at ~980 °C. The homogeneity is described as development of Si–O–Al network connectivity during heating process of polymeric gel before the occurrence of exotherm. Such

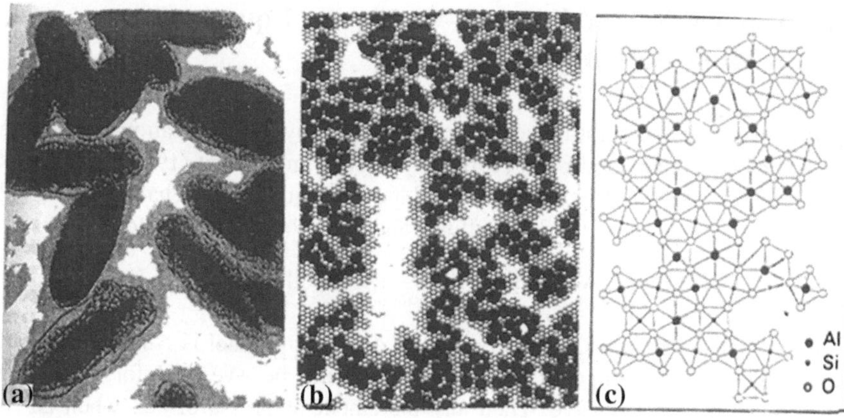

Fig. 21.15 Schematic representation of aluminosilicate ultra structure formed by reaction of $Si(OCH_3)_4$ with **a** Colloidal Alumina, **b** Aluminum Nitrate, and **c** Aluminum A oxide by Co polymerization Method (after Yoldas 1992)

linkages are inherently exist in chain like metakaolin structure (Fig. 19.4). As the former directly transform to mullite (t), the later also develops substantial quantity of mullite (A) during the exhibition of exotherm. Both precursors develop Al–Si spinel phase and absolute absence of γ-Al_2O_3.

In concluding the discussion, the principal factor which emerges in mullite formations.

1. First is the importance of Si–O–Al linkages

This chemical bodings are maintained throughout the entire reaction sequence of kaolinite. During the course of the reaction, it is only the excess SiO_2 which is exolved at 980 °C and the rest of SiO_2 is bonded to Al_2O_3 to form mullite (A) and weakly crystalline Al–Si spinel. These two phases, still containing Si–O–Al bonding, transform to mullite in the course of heating. Thus, chemical continuity is maintained throughout the reaction sequence. In the absence of any pre-linkages between component oxides, e.g., in artificial mixtures, a high temperature is necessary to grow the Si–O–Al network first and after that nucleation of mullite takes place. So, here lies the difference. This explains why the mullitization process is different in the two cases.

2. Second is the homogeneity of silica and alumina phases

Finally, it is concluded that if bonding is originally present between cations like Si to Al or Si to Mg via oxygen, then on heating such compounds, high-melting constituents, i.e., Al_2O_3 and MgO, will not crystallize out. The compounds on the other hand, will transform into other binary compounds, e.g., metakaolin to mullite and/or metakaolin to Al–Si spinel, talc to enstatite, pyrophyllite to mullite, etc.

21.9 Applicability of Clausius–Clapeyron Equation in Mullitization Process

Mullite formation temperature could be lower down by the application of pressure as noted by Carruthers and Scott (1968), Carruthers and Wheat (1965), and Weiss et al. (1970). Applicability of Clausius-Clapeyron equation in Chakraborty's 3:2 composition spinel model to 3:2 mullite transformation is given below.

$$\text{A1-Si spinel} \xrightarrow[\text{Transformation}]{} \text{Mullite} \quad (21.10)$$

In this transformation $C = 1$, $P = 2$ and density of spinel is >than density of mullite. Therefore, change in molar volume would be +ve. ΔH for the above transformation is surely −ve. Then any increment of pressure, i.e., when ΔP is +ve, T should be less than T_o. Therefore, increase of pressure lowers the transformation temperature. This is actually happens in practice.

21.10 Effect of Mineralizer on Mullite Formation

The formation of mullite from clays is determined by various factors, the most important of which are temperature, heating duration, and the presence of foreign mineralizing compounds. Some mineralizers can reduce the temperature of mullite formation considerably/or reduce both the time and temperature required for the reaction, i.e., accelerate mullitization. Some other catalyst would yield the formation of mullite by decomposition of cays. It is largely necessary to induce mullitization from industrial point of view considering the economics of time and fuel consumption with accomplishment of better mullite articles.

It is reported below that small amounts of chemical compounds such as carbonates of alkali and alkaline-earth, iron oxide, titanium dioxides, some fluorides, e.g., NaF, CaF_2, AlF_3, MgF_2, LiF, etc. enhance the formation of mullite.

Johnson and Pask (1982) has shown that CaO and K_2O in kaolinite during sintering cause a shift in the exothermic reaction temperature and accelerate cristobalite formation. A comparative study of mullite formation at high temperatures from kaolinite and equivalent Al_2O_3–SiO_2 mixtures in the presence of impurities, such as TiO_2, Fe_2O_3, CaO, Na_2O, and K_2O, they have shown that the kaolinite-based mixture always produces acicular mullite crystals, while the Al_2O_3–SiO_2 mixture gives rise to rectangular shapes.

Also, the presence of CaO and larger amounts of Fe2O3 added to both natural and synthetic Al_2O_3–SiO_2 mixtures causes an increase in crystal size.

Decomposed kaolinite shows poor solid-state sinter ability below ~1,200 °C. Norton (1937) added soda lime silica glass of small amounts of chemical compounds such as carbonates of alkali and alkaline-earth, iron oxide, titanium dioxides, some fluorides e.g., NaF, CaF_2, AlF_3, MgF_2, LiF approx. composition

(70 % SiO_2, 15 % CaO, and 15 % Na_2O) to clay and heat treated. He observed mullite formation at comparatively low temperature and with considerable yield in comparison to raw clay. Similar observation is noted by Budnikov who used slag or ash with clay.

Parmalee and Rodriguez (1942) have shown that ZnO, MgO, Fe_2O_3 shows positive catalytic effect by the addition of 1 % oxides.

Budnikov and Shmukler (1946) first reported that the larger the value of (r/e = ionic radius/ionic charge), the more effective is the mineralizing action.

Palmeri (1952) showed that effects of the mineralizes decreased with addition from CaF_2 to CaO to MgO. Thus, the results of the above studies seem contradictory.

Moore and Prasad (1955) fired kaolin–alumina mixture with various oxides. Maximum mullite formation occurs with addition of 1 % or less. Similar observation was noted earlier by Parmelee and Rodriguez (1942).

According to Chaudhury (1969) minor amount of CaO promote mullitization. However, addition of higher content of it breaks down mullite.

Grofcsik and Tamas (1961) used fired kaolins ad clays in the presence of various mineralizes and found that no mineralizes would increase the mullite quantity of the fired product.

The main possible source of contradictory results is the estimation of mullite quantity. This determination has been done by the following procedures.

1. Hydrogen fluoride dissolving method.
2. Plan metering under microscope.
3. Internal standard X-ray method
4. Rietveld whole pattern analysis.

Al these may involve methodological errors in estimation process. Since mullite measured during decomposition of kaolinite in the temperature range of 1,100–1,250 °C is weakly crystalline variety, all characteristic XRD peaks are not fully developed and these are broad in nature. Hence, comparison of it with highly crystalline mullite is obviously leads to error.

Moreover, study of influence of mineralizes on alumina-silica mixture or kaolinite is of theoretical importance considering radious and charge of cation used (Avgustinik et al. (1954), Prabhakaram (1968), Mackenzie (1969)). China clay used in white ware industries are usually matured at and above 1,250 °C.

Grofcsik and Tamas (1961) measured quantity of mullite by three possible methods namely, plan metering under optical microscope, HF dissolution technique and internal standard X-ray method, They concluded that quantity of mullite formed out of kaolinite did not increase with oxide additions on kaolinite. Mineralizers influenced the change in size and specific surface of the mullite crystals. According to them, oxides result in recrystallization phenomena.

Chaudhuri (1969) found that lower concentration (1 %) of oxides such as Na_2O, K_2O, and CaO and large amount (4 %) of TiO_2 and Fe_2O_3 addition are effective in mullitization out of kaolinite. The results are based on mullite estimation by X-ray technique.

How is the so-called mineralizers act? A small quantity of it may help in nucleation of mullite in mullite(A) phase as generated during the decomposition of metakaolinite. The mullite formation process proceeds as per the Eq. 21.7 as shown above.

The higher concentration of those oxides may convert cubic spinel phase to orthorhombic mullite with leaving behind some quantity of amorphous phase.

Some of the oxides may form solid solution with mullite and helps growth if it.

21.11 Effect of Iron and Titanium Oxide on Decomposition of Kaolinite to Mullite Formation

The behavior of chemical and mineralogical impurities during the thermal transformation of kaolinite on heating has been studied for electron paramagnetic resonance resonance by Djemai et al. (2001). Iron was detected in two different states in progressive heat-treated kaolinite (Fig. 21.16).

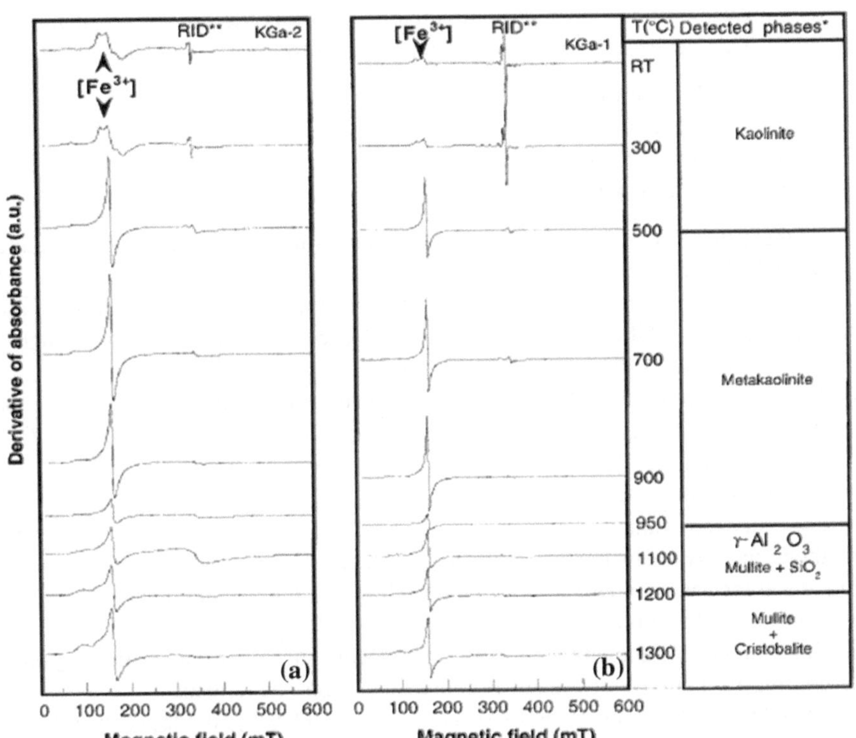

Fig. 21.16 EPR spectra of raw and heated kaolinites (after Djemai et al. 2001). Reprinted by permission of the American Ceramic Society

(1) Isolated Fe^{+3} ions at low magnetic field (B = 160 mT). This signal results from isomorphous replacement of paramagnetic Fe^{+3} for Al^{+3} ions at the octahedral Al sites of kaolinite.
(2) The resonance observed at high magnetic field = 300 mT is due to a broad resonance of superparamagetic Fe-demonstrated as RIDs.

The three steps of transformation are presented below.

(i) In K-MK transformation step (500–900 °C).

A drastic modifications in the local environment of Fe^{+3} ions occurs. The low field signal is reduced to an almost isotropic signal and high field signal due to RID is strongly reduced (Fig. 21.16). The increase in the concentration of dilute Fe^{+3} ions may be related to the oxidation of Fe^{+2} ions of starting kaolinites during collapse of kaolinite structure.

(ii) In MK – Al–Si Spinel + Mullite(A) + SiO_2(A) transformation step (950–1,100 °C).

The intensity of low field signal is strongly reduced. A broad superparamagnetic signal is observed at B = 330 mT.

The strong decrease in Fe^{+3} absorbance is related to the appearance of a strong superparamagnetic signal which is due to transformation of metakaolin to Al–Si Spinel phase, etc.

The drastic decrease of the amount of paramagnetic Fe^{+3} ions may be interpreted as the result of Fe^{+3} diffusion within the aluminosilicate framework and its subsequent exsolution outside the amorphous matrix, which leads to the formation of superparamagnetic concentrated-iron rich domains likely either as Fe^{+3} clusters within spinel phase or as Fe^{+3} oxide nano phase.

(iii) In Al–Si Spinel + Mullite(A) + SiO_2(A) to Mullite + Cristobalite Transformation step (>1,100 °C).

The low field signal again undergoes a drastic change in shape and its intensity increases. It is interpreted as a new diffusion of Fe^{+3} ions Al–Si Spinel + Mullite(A) + SiO_2(A) within the aluminosilicate framework, leading to the incorporation of iron into the mullite structure.

The resonances that are observed at high magnetic field and are due to superparamagnetic iron oxides weaken and are still exist in the mullite and cristobalite mixture.

Soro et al. (2003) studied the role of iron in the formation of mullite out of kaolinites containing minor quantity of structural iron, and kaolinite containing a larger deposit of iron prepared synthetically on basal surfaces. Calcined materials were studied for Fe environment by Mossbauer spectroscopy and for mullite stoichiometries by XRD Rietveld refinements. During nucleation process of mullite, Fe^{+3} is involved in structural reorganization and preferably occupy the octahedral site of the mullite structure by replacing Al^{+3}. This results to significant

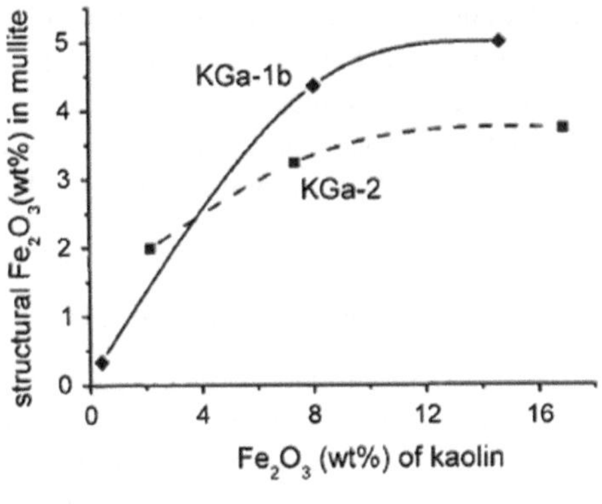

Fig. 21.17 Structural Fe quantities in mullite versus structural Fe in the initial kaolinite plus iron initially deposited (after Soro et al. 2003). Reprinted by permission of the American Ceramic Society

increase of c-axes parameter of mullite. The quantity of substitution attains a plateau at about 5 wt%.

Fe_2O_3 depending on kaolinite and its initial characteristics chosen (Fig. 21.17).

The results of Djemai et al. 2001) and Soro et al. (2001) that that added iron promotes the growth of the size of mullite conforms previous observation of Grofcsik and Tamas (1961). Moreover, iron is found to absorb during heating and formation of metakaolin. It exolves out during spinel formation stage at exothermic temperature. It further re-enters on heating to mullite lattice and forms solid solution with change of cell parameters.

The EPR and Mossbauer spectroscopic studies enlighten these two pictures. With these data in mind, one can explain the role of other oxides during mullitization process of kaolinite. It is to be predicted that may act or help in nucleation in mullite(A) generated during decomposition of metakaolinite at first exotherm.

21.12 Industrial Use of Kaolinite (See References)

(i) Utilization alkali-leached heat-treated kaolinite.
(ii) Clay-alumina mullite reaction to synthesize mullite.
(iii) Production of refractory bricks.
(iv) Production of firebricks and shapes 45 % mullite grog.
(v) Break down of mullite.
(vI) Synthesis of alkali/alkaline earth silicates.
(vII) Synthesis of composites.
(viii) White ware industries, e.g., production of earthen ware, tile, insulators, sanitary wares, etc.

21.13 Summary

Following conclusions are emerged out regarding mullitization of kaolinite which was studied by XRD and electron diffraction technique. There are two distinct steps in mullitization. Primary mullite which is formed at ~980 °C are poorly crystalline. All XRD peaks are broad and diffuse and are difficult to measure the precise value of lattice parameter. Mullite which is formed beyond second exotherm is well developed. Both cell volume and **a** parameter increases with heating schedule. There are two old views of mullite formation. According to first group, solid-state reaction of γ-Al_2O_3/SiO_2 (A) is responsible for mullitization. Topotactic transformation of Al–Si spinel accounts for secondary mullitization.

Chakraborty (1992) proposed a newer concept of mullitization. According to him, mullite formation occurs in two paths. In the first path, a polymorphic transformation of Al–Si spinel happens suddenly during the occurrence of second exotherm.

$$\underset{\substack{\text{Al-Si spinel}\\ \text{Cubic}\\ (25\text{-}30\ \text{wt\%})}}{3Al_2O_3.2SiO_2} \xrightarrow[1250\ ^\circ C]{1100\ ^\circ C} \underset{\substack{\text{Mullite}\\ \text{Orthorhombic}}}{3Al_2O_3.2SiO_2} \tag{21.7}$$

In the second path, mullite (A) which was developed during heating kaolinite at 980 °C nucleates and finally crystallizes to mullite in the lower temperature region below 1,200 °C and at temperature beyond 1,250 °C.

$$\text{Mullite(A)phase} \xrightarrow[1600\ ^\circ C]{1250\ ^\circ C-} \underset{\text{Mullite}}{3Al_2O_3.2SiO_2} \tag{21.8}$$

This is consistent with the exhibition of a predominantly large exotherm in DTA analysis (Fig. 3.1). In this process, initial formed mullite at ~1,050 °C is poorly crystallized. With the increase of temperature, broadening of the peaks decrease side by side resolution increase and strain value of mullite decreases. In the continuation of the second process, Al–Si spinel rapidly transforms at ~1,250 °C and triggers to larger quantity of mullite development.

Change in lattice constants values of mullite formed at temperature >1,200 °C versus temperature data calculated by quasar analysis agrees with that measured by Tomura et al. and others. Considering changes in width, resolution of some X-ray reflection pairs, strain values of mullite formed during heating clay, it is predicted that composition of mullite does not change on heat treatment. It is the crystallinity of mullite which increases on progressive heating of kaolinite.

I.R. spectroscopic studies shows (i) the disappearance of mostly $Al0_6$ groups in kaolinite to $Al0_4$ in metakaolinite in the first step of transformation, (ii) partial development of $Al0_6$ during 980 °C reaction in the second step of thermal reaction, and (iii) further decrease of the development of $Al0_6$ with consequent increase of $Al0_4$ during mullite formation in the third step of reaction series.

This basic information accords MAS NMR data presented by Brown et al. (1985), Meanhod et al. (1985), Sanz et al. (1988), and others.

Measured $Al0_6/Al0_4$ ratio indicates that Al sites in kaolinite heated to 1,000 °C are reorganized on further heat treatment to develop more $Al0_4$ groups connected with remaining $Al0_6$ chains which is necessary for building up of mullite structure.

Clausius-Clapeyron equation is found to be applicable to Chakraborty's 3:2 composition spinel model to 3:2 mullite transformation equation. It predicts that increase of pressure could lower down transformation temperature, which is really happen in practice.

A small amounts of chemical compounds such as carbonates of alkali and alkaline-earth, iron oxide, titanium dioxides, some fluorides have been used to study their role in influencing both first and second exothermic reactions as well as catalytic action of mullitization of kaolinite.

Various authors note that a small amount of oxides acts as mineralizer and promotes mullite formation. Higher content does not increase quantity of mullite but helps vitrification and sintering of mass and effectively increase size of mullite crystals. It is believed that some oxides helps nucleation of mullite in mullite (A) phase present in kaolinite heat treated to first exothermic temperature.

The EPR and Mossbauer spectroscopic results of Djemai et al. (2001) and Soro et al. (2001) indicated the following role of iron oxide catalyst on mullitization.

(i) Iron promotes the growth of the size of mullite conforms previous observation of Grofcsik and Tamas (1961).
(ii) Iron is found to absorb during heating and formation of metakaolin.
(iii) It exolves out during metakaolin to spinel formation stage at exothermic temperature.
(iv) It further re-enters on heating to mullite lattice and forms solid solution with change of cell parameters of mullite crystals and preferably occupy the octahedral site of the mullite structure by replacing Al + 3. This results to significant increase of c-axes parameter of mullite.

References

S.O. Agrell, J.V. Smith, Cell dimension, solid solution, polymorphism, and identification of mullite and sillimanite. J. Am. Ceram. Soc. **43**, 69–76 (1960)

A.I. Avgustinik, M.F. Nazarenko, and V.A. Sviridenko, The effect of valance and radious of the cation of mineralizers on the process of mullitizatio. Zh. Priklad. Khim. **27**, 782 (1954)

M. Bellotto, High temperature phase transformation in kaolinite: the influence of disorder and kinetics on the reaction path. Mater. Sci. Forum **166–169**, 3–20 (1994)

G.W. Brindley, J. Lemaitre, Thermal oxidation and reduction of clay minerals, in *Chemistry of Clay and Clay Minerals*, ed. by A.C.D. Newman (Longman Scientific and Technical, Essex, 1987), pp. 319–70

G.W. Brindley, M. Nakahira, The kaolinite-mullite reaction series: I, II, and III, J. Am. Ceram. Soc. **42**(7), 311–324 (1959)

I.W.M. Brown, K.J.D. MacKenzie, M.E. Bowden, R.H. Meinhold, Outstanding problems in the kaolinite–mullite reaction sequence investigated by 29Si and 27Al solid-state nuclear magnetic resonance: II, High-temperature transformations of metakaolinite. J. Am. Ceram. Soc. **68**(6), 298–301 (1985)

P.P. Budnikov, K.M. Shmukler, Effect of Mineralizers on the Process of Mullitization of Clays, Kaolins, and Synthetic Mas. Zh. Priklad. Khim. **19**, 1029 (1946)

M. Bulens, A. Leonard, B. Delmon, Spectroscopic investigations of the kaolinite–mullite reaction sequence. J. Am. Ceram. Soc. **61**(1–2), 81–84 (1978)

T.G. Carruthers, B. Scott, Reactive hot pressing on kaolinite. Trans. Brit. Ceram. Soc. **67**, 185 (1968)

T.G. Carruthers, T.A. Wheat, Hot pressing of kaolin and of mixtures of alumina and silica. Proc. Br. Ceram. Soc. **3**, 259 (1965)

W.E. Cameron, Composition and cell dimensions of mullite, ceram. Bull. **56**(11), 1003–1011 (1977)

A.K. Chakraborty, Resolution of thermal peaks of kaolinite by TMA and DTA. J. Am. Ceram. Soc **75**(7), 2013–2047 (1992)

A.K. Chakraborty, Application of TMA and DTA studies on the crystallization behavior of SiO2 in thermal transformation of kaolinite. J. Thermal Anal. **39**, 280–299 (1993)

A.K. Chakraborty, S. Das, Measurement of structural parameters of kaolinite formed in an Indian kaolinite. Clay Sci. **12**(4), 235–242 (2004)

A.K. Chakraborty and D.K. Ghosh, Interpretation on the changes of co -ordination number of Al in the thermal changes of kaolinite, Clay Science, **8**, 45–57 (1991a)

A.K. Chakraborty, D.K. Ghosh, Kaolinite–mullite reaction series. The development and significance of a binary aluminosilicate phase. J. Am. Ceram. Soc. **74**(6), 1401–1406 (1991b)

A.K. Chakraborty, D.K. Ghosh, Comment on the Interpretation of the kaolinite–mullite reaction sequence from infra-red absorption spectra. J. Am. Ceram. Soc. **61**(1–2), 90–91 (1978a)

A.K. Chakraborty and D.K. Ghosh, Re - examination of the kaolinite to mullite reaction series, J. Am. Ceram. Soc. **61**(3–4), 170–173 (1978b)

A.K. Chakraborty and D.K. Ghosh, Study of phase transformation of Al2O3 – SiO2 gel and kaolinitic clay, Trans. Ind. Ceram. Soc. **37**(5), 192–200 (1978c)

A.K. Chakraborty, New DTA on the thermal analysis of diphasic mullite gel. J. Therm. Analysis, **46**, 1413–1419 (1996a)

A.K. Chakraborty, DTA Characterisation of three types of Al2O3-SiO2 gels made from TEOS-Al(OBu)3 mixture with variation of water, Ceram. Int. **22**, 463–469 (1996b)

A.K. Chakraborty, D.K. Ghosh, P. Kundu, Comment on "Structural characterization of the spinel phase in the kaolin–mullite research series through lattice energy". J. Am. Ceram. Soc. **68**(8), C-200–C-201 (1986)

A.K. Chakraborty, S. Das, S. Gupta, Evidence for two stage mullite formation during thermal decomposition of kaolinite. Brit. Ceram. Trans. **102**(4), 33–37 (2003)

S.P. Chaudhuri, X-ray study of induced mullitization of clay. Trans. Indian Ceram. Soc **28**, 24 (1969)

J.E. Comeforo, R.B. Fischer, W.F. Bradley, Mullitization of kaolinite. J. Am. Ceram. Soc. **31**(9), 254–259 (1948)

J.J. Comer, Electron microscope studies of mullite development in fired kaolinite. J. Am. Ceram. Soc. **43**(7):378–384 (1960)

J.J. Comer, New electron-optical data on the kaolinite–mullite transformation. J. Am. Ceram. Soc. **44**(11), 561–563 (1961)

W.I. DeKeyser, Reactions at the point of contact between SiO_2 and Al_2O_3, in *Science of Ceramics*, vol. 2, ed. by G.H. Stewart (Academic Press, London, 1965), pp. 243–257

A. Djemai, G. Calas, J.P. Muller, R.A. Condrate, Role of Structural Fe(III) and iron oxide nanophases in mullite coloration. J. Am. Ceram. Soc. **84**(7), 1627–1631 (2001)

J.F. Duncan, K.J.D. Mackenzie, P.K. Foster, Kinetics and mechanism of high-temperature reactions of kaolinite minerals. J. Am. Ceram. Soc. **52**(2), 74–77 (1969)

S. Durovic, A Statistical model for the crystal structure of mullite, Kristallografiya **7**, 339 (1962)

F. Freund, Infrared spectra of kaolinite, metakaolinite, and Al-Si spinel. Ber. Deut. Keram. Ges. **44**(181), 392–397 (1967a)

F. Freund, Kaolinite-metakaolinite, a model of a solid with extremely high lattice defect concentrations. Ber. Dtsch. Keram. Ges. **44**(I), 5–13 (1967b)

J. Grofcsik, F. Tamas, *Mullite, Its Structure, Formation and Significance* (Akademiai Kiado, Publishing House of the Hungarian Academy of Sciences, Budapest, 1961), p. 109

A.F. Gualtieri, M. Bellotto, G. Artioli, J.M. Clark, Kinetic study of the kaolinite-mullite reaction sequence. II. Mullite Formation. Phys. Chem. Miner **22**, 215–222 (1995)

D.W. Hoffman, R. Roy, S. Komarneni, Diphasic xerogels, a new class of materials: phases in the system Al_2O_3- SiO_2. J. Am. Ceram. Soc **67**(7), 468–471 (1984)

S.M. Johnson, J.A. Pask, Role of impurities on formation of mullite from kaolinite and AI, O,-SiO2 mixtures. Am. Ceram. Soc. Bull. **61**, 838–842 (1982)

W.D. Kingery, *Itroduction to Ceramics*, (John Wiley & Sons, Inc., New York, 1960) p.324

A.J. Leonard, Structural analysis of the transition phases in the kaolinite–mullite thermal sequence. J. Am. Ceram. Soc. **60**(1–2), 37–43 (1977)

A.J. Leonard, M.J. Genet, J. Lemaitre, M. Bulens, B. Delmon, Reply to the Comment on "Structural analysis of the transition phases in kaolinite–mullite thermal sequence". J. Am. Ceram. Soc. **62**(9–10), 529–531 (1979)

S.T. Lundin, Studies on triaxial white ware bodies. Almqnist Wiksall, Stockholm, Sweden. **32**, 103 (1959)

K.J.D. MacKenzie, The effect of impurities on the formation of mullite from kaolinite-type minerals: I, The effect of exchangeable cations. Trans. Br. Ceram. Soc. **68**(3), 97–101 (1969)

S. Majumder, B. Mukherjee, Reply to the Comment on " Structural characterization of the spinel phase in the kaolin-mullite research series through lattice energy". J. Am. Ceram. Soc. **69**(8), C-201 (1986)

I.J. McCom, M.M. Rebbeck, M. Rachmawati ad S.M. Faeta-Boada, Development of microstructure of fired Ecuadorean clay. Brit. Ceram. Tran. **98**(50), 213–218 (1999)

T.D McGee, Contn. of fireclays at high temperatures: I, Methods of analysis; II, Mineralogical composition, III, Deformation characteristics. J. Am. Ceram. Soc. **49**(2), 83–94 (1966)

R.H. Meinhold, K.L.D. Mackenzie, I.W.M. Brown, Thermal reactions of kaolinite studied by solid state 27Al and 29Si NMR. J. Mater. Sci. Lett **4**, 163–166 (1985)

H. Moore, M.R. Prasad, The effects of various mineralizing agents in promoting recrystallization in mixtures of clay–alumina during es. J. Soc. Glass Tech. **39**, 314T (1955)

K. Murthy, F.A. Hummel, X-ray study of the solid solution of TiO_2, Fe_2O_3, and Cr_2O_3 in mullite ($3Al_2O_3.2SiO_2$). J. Am. Ceram. Soc. **43**, 267 (1960)

M. Niwa, N. Katada, Y. Murakami, Thin silica layer on alumina: evidence of the acidity in the monolayer. J. Phys. Chem. **94**(16), 6441–45 (1990)

C.L. Norton, Jr., Can.Pat.366784, 15 June 1937

K. Okada, N. Ostuka, J. Ossaka, Characterization of spinel phase formed in the kaolin–mullite thermal sequence. J. Am. Ceram. Soc. **69**(10), C-251–C-253 (1986)

K. Okada, N. Otsuka, S. Somiya, Review of mullite synthesis routes in Japan. Ceram. Bul. **70**(10), 1633–1640 (1991)

F. Onike, G.D. Martin, A.C. Dunham, Time–temperature–transformation curves for kaolinite. Mater. Sci. Forum **7**, 73–82 (1986)

V.R. Palmeri, Mullite formation by decomposition of kaolinite. J. Soc. Glass Tech **36**, 25–28N (1952)

C.W. Parmelee, A.R. Rodriguez, Catalytic mullitization of kaolinite by metallic oxide. J. Am. Ceram. Soc. **25**, I (1942)

H.J. Percival, Reply to the Comment on the " Interpretation of the kaolinite-mullite reaction sequence from infra-red absorption spectra". J. Am. Ceram. Soc. **61**(1–2), 91 (1978)

H.J. Percival, J.F. Duncan, P.K. Foster, Interpretation of the kaolinite–mullite reaction sequence from infrared absorption spectra. J. Am. Ceram. Soc. **57** (2), 57–61 (1974)

A.J. Perrotta and J. E. Young, Silica-free phases with mullite-type structure. J. Am. Ceram. Soc. **57**(9), 405–407 (1974)

P. Prabhakaram, Exchangeable cations and the high-temperature reactions of Kaolinite. Trans. Br. Ceram. Soc. **67**, 105–24 (1968)

Y. Saito, T. Takei, S. Hayashi, A. Yasumori, K. Okada, Effect of Amorphous and Crystallie SiO2 Additives on γ-Al2O3- to- α-Al2O3 Phase Transitions. J. Am. Ceram. Soc **81**(8), 2197–2200 (1998)

J. Sanz, A. Madani, J.M. Serratosa, J.S. Moya, S. Aza, ^{27}Aluminum and ^{29}Silicon-magic-angle spinning nuclear magnetic resonance study of the kaolinite–mullite transformation. J. Am. Ceram. Soc. **71**(10), C-418–C-421 (1988)

H. Schneider, T. Rymon-Lipinski, Occurrence of pseudo tetragonal mullite. J. Am. Ceram. Soc. **71**(3), C-162–C-164 (1988)

K.H. Schuller, H. Kromer, Primary mullite as a pseudomorph after kaolinite, in *Proceedings of the International Clay Conference*, Mexico City, 1975, ed. by S.W. Bailey (Applied Publishing, Wilmette, 1976), pp. 533–538

N. Soro, L. Aldon, J. Olivier-Fourcade, J.C. Jumas, J.P. Laval, P. Blanchart, 'Role of iron in mullite formation from kaolins by Mössbauer Spectroscopy and Rietveld Refinement. J. Am. Ceram. Soc. **86**(1), 129–34 (2003)

W.G. Staley, G.W. Brindley, Development of noncrystalline material in subsolidus reactions between SiO_2 and Al_2O_3. J. Am. Ceram. Soc. **52**, 616–19 (1969)

S. Tomura, Y. Shibasaki, H. Mizuta and T. Maeda, Yogo. Kyokai Shi. **98**, 917 (1986)

F. Vaughan, Energy changes when kaolin minerals are heated. Clay Miner. Bull. **2**(13), 265–274 (1955)

V. Viswabaskaran, F.D. Gnanam, M. Balasubramanian, Mullitisation behavior of South Indian Clays. Ceram. Int. **28**(5), 557–564 (2002)

K. Von Gehlen, Oriented formation of mullite from Al – Si Spinel in the transformation series Kaolinite – Mullite . Ber. Deut. Keram. Ges. **39**(6), 315–320 (1962)

F.M. Wahl, R.E. Grim, High temperature DTA and XRD studies of reactions. Twelfth national conference on clays and clay minerals, pp. 69–81 (1964)

K. Wang, M.D. Sacks, Mullite formation by endothermic reaction of α-alumina/silica microcomposite particles. J. Am. Ceram. Soc. **79**(1), 12–16 (1996)

A. Weiss, K.J. Range, J. Russow, The Al, Si-spinel phase from kaolinite (isolation, chemical analysis, orientation and reactions to its low-temperature precursors), in *Proceedings of the International Clay Conference*, Tokyo, 1969, vol. 2 (Israel Universities Press, Jerusalem, 1970), pp. 34–37

G.K. Willamson, W.H. Hall, Acta. Met. **1**, 22 (1953)

B.E. Yoldas, Thermal stabilization of an active alumina and effect of dopants on the surface area. J. Mater. Sci **11**, 465–470 (1976)

B.E. Yoldas, Effect of ultrastructure on crystallization of mullite. J. Mater. Sci **27**(24), 6667–6672 (1992)

Further Readings

H.R. Baharvandi and A.M. Hadian, Investigation on addition of kaolinite on sintering behavior and mechanical properties of B4C. J. Mater.Eng. Perform. **18**(4), 433–7 (2009)

I.M. Bakr, and S.M. Naga, Role of B2O3 in formation mullite from kaolinite and α – Al2O3 mixtures. Brit . Ceram .Trans. **101**(3), 133–136 (2002)

V. Balek and M. Murat, The Emanation thermal analysis of kaolinite clay minerals. Thermochim. Acta, 282/283, 385–97 (1996)

S.P. Banerjee and N.R. Sircar, Sintered Mullite for High Alumina Refractories. in *Proc. of the Seminar o High Alumina Refractories*, ed. by N.R. Sircar, Ind. (Refractory Makers Association, Kolkata, 1973) p. 60–64

W.E. Blodgett,High-strength alumina porcelains. Am. Ceram. Soc . Bull. **40**, 74–77 (1961)

S.P. Chaudhuri, Influence of mineralizers on the constitution of hard porcelain: I Mineralogical composition. Ceram. Bull. **53**(2), 169–71 (1974)

S.P. Chaudhuri, Influence of mineralizers on the constitution of hard porcelain: II, Microstructures. Am. Ceram. Soc. Bull. **53**(3), 251–54 (1974)

S.P. Chaudhuri, Ceramic properties of hard porcelain in relation to mineralogical composition and microstructure: VI, Thermal shock resistance and thermal expansion. Trans. Indian Ceram. Soc. **34**(1), 30–34 (1975)

S.P. Chaudhuri and P. Sarkar, Constitution of porcelain before and after heat treatment: I, mineralogical composition. J. Eur. Ceram. Soc. **15**, 1031–35 (1995)

C.Y. Chen, G.S. Lan, and W.H. Tuan, Preparation of mullite by the reaction sintering of kaolinite and alumina. J. Eur. Ceram. Soc. **20**(14–15), 2519–2525 (2000)

Y.F. Chen, M.C. Wang, and M.H. Hon, Phase transformation and growth of mullite in kaolin ceramics. J. Eur. Ceram. Soc. **24**(8), 2389–97 (2004)

F.H. Clews, H.M. Richardson, and A.T. Green, Action of alkalis on refractory materials : XVI, Cone-deformation study of Certain alkali –silica- alumina –ferric oxide mixtures. Trans. Brit. Ceram.Soc. **44**(2), 21–24 (1945)

J. Dubois, M. Murat, A. Amroune, X. Carbonneau, and R. Gardon, High temperature transformation in kaolinite: The role of the crystallinity and of the firing atmosphere. Appl. Clay Sci. **10**, 187–98 (1995)

G.M. Gad and R. Barrett, High-temperature break down of mullite and other alumino – silicates in presence of alkalis. Trans. Brit. Ceram. Soc. **49**(11), 470–91 (1950)

W.H. Hawkes, The Production of Synthetic Mullite, Trans. Brit. Ceram. Soc. **61**, 689–703 (1962)

A.W.H. Hawkes, 37–42 Symposium on High –Alumina Refractories, 37. the Production of synthetic mullite. Trans. Brit. Ceram. Soc. **61**(11), 689–703 (1962)

M. Heraiz, A. Merrouche and N. Saheb, Effect of MgO addition & sintering parameters on mullite formation through reaction sintering kaolin and alumina. Adv. Appl. Ceram. **105**(6), 285–284 (2006)

K. Kamano, Preparation of mullite ceramics from kaolin and luminous minerals (Part 1) Effects of grain size & grinding of raw materials . J. Ceram. Soc. Jap. **102**, 78–83 (1994)

S. Kawi, A High Surface – area silica – clay composite material. Mat. Lett. **38**, 351–55 (1999)

H.S. Kim and P. Nicholson, Use of mixed-rare-earth oxide in the preparation of reaction-bonded mullite at 1300°C. J. Am. Ceram. Soc. **85**(7), 1730–4 (2002)

W.E. Lee, and Y. Iqbal, Influence of mixing on mullite formation in porcelain. J. Eur. Ceram. Soc. **21**(14), 2583–86 (2001)

J. Lemaitre and B. Delmon, Effect of mineralizers on properties of kaolin bodies. Am. Ceram. Soc. Bull. 59(2), 235–38 (1980)

W. Li, K. Lu, and J. Walz, Effects of added kaolinite on sintering of freeze-cast kaolinite–silica nanocomposite I. Microstructure and phase transformation.J. Am. Ceram. Soc. **95**(3), 883–891 (2012)

W. Li, K. Lu, and J. Walz, Formation, structure and properties of freeze-cast kaolinite-silica nanocomposites. J. Am. Ceram. Soc. **94**(4), 1256–1264 (2010)

K.C. Liu , Microstructure & microanalysis of mullite processed by reaction sintered of kaolin – alumina mixture. Ceram. Today – Tomorrow's Ceram. Pt. 66A. ed. by P. Vincenzini, (Amst, Elsvr. 1991) p. 177 – 186

K.C. Liu, G.Thomas, A. Cabalero, J.S. Moya, S. de Aza, Time – temperature – transformation curves for kaolinite – α - Al2O3 . J. Am. Ceram. Soc. **77**, 1545–1552 (1994)

Y. Iqbal and W.E. Lee, Fired porcelain microstructures revisited. J. Am.Ceram. Soc. **82**(12) 3584–90 (1999)

Y. Iqbal and W.E. Lee, Micro structural evolution in triaxial porcelain. J. Am Ceram. Soc. **83**(12) 3121–27 (2000)

T. Lundin, Electron microscopy of whiteware bodies. Trans. Int. Ceram.Congr. **4**, 383–90 (1954)

S.T. Lundin, Microstructure of porcelain. Natl. Bur. Stand. (U.S.), Misc Publ. **257**, 93–106 (1964)

L. Mattyasovszky-Zsolnay, Mechanical strength of porcelain. J. Am. Ceram. Soc. **40**, 299–306 (1957)

O. Matsuda, Effect of alumina and glass on formation of needle like mullite from kaolinite. J. Ceram. Soc. Jap. **100**, 725–730 (1992)

F. Nadachowski , The Assessment of the phase composition of fireclay refractories based on ternary phase diagrams. Refractories J. **4**, 126–131 (1965)

C.L. Norton, The Influence of time on maturing temperature of white ware bodies: II. J. Am. Ceram. Soc. **14**, 192–206 (1969)

K. Okada, N. Watanabe, K.V. Jha , Y. Kameshima, A. Yasumori, and K.J.D. MacKenzie, Effect of grinding and firing conditions on $CaAl_2Si_2O_8$ phase by solid formation of kaolinite with $CaCO_3$. Appl. Clay Sci. **23**(5–6), 329–336 (2003)

A.D. Papargyris and R.D. Cooke, Structure and mechanical properties of kaolin based ceramics. Br. Ceram. Trans. **95**(3), 107–20 (1996)

D.S. Perera and G. Allott, Mullite morphology in fired kaolinite / halloysite clays. J. Mater. Sci. Lett. **4**, 1270–1272 (1985)

N.V. Pitak and R.M. Fedoruk, Formation of kaolin refractories heated to different gaseous atmospheres. Refractories, **34**, 454–457 (1993)

N.V. Pitak, Mullite based on Novosdtsa Kaolin, Commercial alumina. Inorganic Mater. **10**(4), 650–651 (1974)

N.V. Pitak, R.S. Shulyak and Z.D. Zhukov, Molten kaolin & its properties. Ogneupory, **2**, 32–39 (1977)

N.V. Pitak, Morphological character of mullite : An important factor for evaluating quality of refractories. Industrial Ceramics, **3**(8), 7–8 (1997)

Pranab Das, Role of magnesium compounds on mullitization of clay –alumina mixtures. Trans. Ind.Ceram. Soc., 39, 113–118 (1980)

P. Rado, The Strange case of hard porcelain. Trans. Br. Ceram. Soc. **71**(4), 131–39 (1971)

H. Rager, H. Schneider and H. Graetsch, Chromium incorporation in mullite. Am. Mineral. **75**, 392–397 (1990)

H.R. Rezaie, W.M. Rainforth, and W.E. Lee, Mullite evolution in ceramics derived from kaolinite, kaolinite with added α - Al2O3 and sol–gel precursors. B rit. Ceram. Trans. **96**(5), 181–87 (1997)

J.M.A. Rincon, G. Thomas and J.S. Moya, Micro structural study of sintered mullite. J. Am. Ceram. Soc. **69**(2), C-29–C-31 (1986)

A.R. Rossini, Mulitization of mixtures of kaolinite clay & aluminum hydroxides. Bol. Soc. Espan. Ceram. **9**, 579– 91 (1970)

A. Roy, Bauxite, Principle raw material for alumina & refractories. Ind. Ceram, **37**, 215–21 (1994)

C. Sane and R.L. Cook, Effect of grinding and firing treatment on the crystalline and glass content and the physical properties of whiteware bodies. J. Am. Ceram. Soc. **34**, 145 (1951)

J.E. Schroeder, Inexpensive high-strength electrical porcelain. Am.Ceram. Soc. Bull. **57**, 526 (1978)

K.H. Schuller, Reactions between mullite and glassy phase in porcelains. Trans. Br. Ceram. Soc. **63**, 102–17 (1964)

K.H. Schuller, Reactions between mullite and glassy phase in porcelains. Trans. Br. Ceram. Soc. **63**(2), 102–17 (1964)

G.P. Souza, E. Rambaldi, A. Tucci, L. Esposito, and W.E. Lee, Micro structural variations in porcelain stoneware tiles as a function of flux system. J. Am. Ceram. Soc. **87**(10), 1959–66 (2004)

E.C. Subbarao, Sintered mullite from china clay-alumina mixtures. Trans. Ind. Ceram. Soc. **37**(6), 225–235 (1978)

T.Tarvornpanich, G.P. Souza, and W.E. Lee, Micro structural evolution on firing soda–lime–silica glass fluxed white wares. J. Am. Ceram. Soc. **88**(5), 1302–8 (2005)

T. Tarvornpanich, Recycled colourless soda–lime–silica glass as an alternative flux in *white wares, Ph.D. Thesis*, (University of Sheffield, U.K., 2007)

T. Tarvornpanich, G. P. Souza, and W. E. Lee, Micro structural evolution in clay-based ceramics I: Single components and binary mixtures of clay, flux, and quartz filler. J. Am. Ceram. Soc. **91**, 2264–71 (2008)

T. Tarvornpanich, G. P. Souza, and W. E. Lee, Micro structural evolution in clay-based ceramics II: Ternary and quaternary mixtures of clay, flux, and quartz filler. J. Am. Ceram. Soc. **91**(7), 2272–80 (2008)

S. Venkataramani, Clay–Alumina Mixtures, Trans. Ind. Ceram. Soc. **39**, 113–118 (1980)

A. G. Verduch, The Formation of mullite from Serecite & its mixture with alumina & kaolin. in *Science of Ceramics*. ed. by G. H. Stewart, vol- 1, (Acad. Press for Brit. Ceram. Soc., London, 1962) p. 285

V. Viswabaskaran, F. D. Gnanam, and M. Balasubramanian, Mullite from clay-reactive alumina for insulating substrate application. Appl. Clay Sci. **25**, 29–35 (2004)

V. Viswabaskaran, F. D. Gnanama, and M. Balasubramanian, Effect of MgO, Y_2O_3 and boehmite additives on the sintering behavior of mullite formed from kaolinite-reactive alumina. J. Mater. Process. Technol. **142**(1), 275–281 (2003)

J. White, Chemistry of high alumina bauxite –based refractories. With special reference to effects of TiO2- A Review. Trans. Brit . Ceram. Soc. **81**(40), 109–111 (1982)

K.-H. Yang, J.-H. Wu, C.-S. Hsi, and H.-Y. Luw, Morphologically textured mullite in sintered tape-cast kaolin. J. Am. Ceram. Soc. **94**(3), 938–44 (2011)

Chapter 22
Reasons for First and Second Exothermic Peaks

22.1 Introduction

Metakaolinite is a poorly crystalline material. It is formed from kaolinite by absorption of heat energy and thus it is conceivable that it is metastable with high free energy content. It is obvious that this stored energy must release out during subsequent crystallization of metakaolinite. Three transformations may take place theoretically.

(i) The silica component of metakaolin may crystallize to β-cristobalite and in some instances to β-quartz. (ii) Metakaolinite may itself transform into poorly crystallized intermediate Al–Si spinel phase. (iii) It may form to mullite at any temperature.

Formation of any of the crystalline form may be the cause of 980 °C exotherm. Since the intensity and rapidity of exothermic process is certainly due to crystallization phenomenon. To reveal the real cause of it, researchers tried to analyze the products of heated clays by various physico chemical means. It is shown that the sudden evolution of energy is associated with shrinkage or increase in density, decrease in Al_2O_3 solubility etc. XRD analysis observed that in major cases, both spinel and mullite are found to coexist in fired kaolinites around 1,000 °C. In certain cases quartz or cristobalite are noted. Question likely arises: which phase is responsible for the first exothermic peak? Is it quartz or cristobalite and/or is it spinel or mullite or both? Thus, there exist different views on the cause of 980 °C exotherm.

22.2 Earlier Thoughts

22.2.1 γ-Al_2O_3 Formation May be a Cause of 1st Exotherm

Based on diffraction study, earlier researchers believed γ-Al_2O_3 formation is the cause of 980 °C exotherm. Among them Insley and Ewell (1935) first noted that

A. K. Chakraborty, *Phase Transformation of Kaolinite Clay*,
DOI: 10.1007/978-81-322-1154-9_22, © Springer India 2014

950 °C exotherm coincides with the formation of crystalline γ-Al_2O_3. But Colegrave and Rigby pointed out that hydrous aluminum oxide begin to produce γ-Al_2O_3 at temperature as low as 565 °C and DTA curve shows no sudden exothermic reaction at al. The retardation of γ-Al_2O_3 formation in case of kaolinite was explained by Insley and Ewell. They believed that SiO_2 (A) hinders γ-Al_2O_3 formation in case of kaolinite to same extent. Meanwhile temperature rises and when it occurs, it does so rapidly and in consequence the heat of crystallization is liberated with suddenness to produce a peak.

Newer group of researchers namely Percival et al. (1974) by I.R. study; Leonard et al. (1977) by RED study, Lemaitre et al. (1975) by density measurement, Brown et al. (1985) by MAS NMR studies. Sonuparlak et al. (1987) by HREM studies supported the old γ-Al_2O_3 hypothesis to be responsible for 980 °C exotherm. The so called γ-Al_2O_3 phase has been designated as Al–Si Spinel, the composition of it is shown to be composition analogous to the composition of 3:2 mullite (Chap. 20).

The possibility of γ-Al_2O_3 phase formation as a cause of 980 °C has been indicated theoretically by Mazumdar and Mukherjee (1983) through lattice energy calculation. Table 12.3 shows that $\Delta H_{1250\ K}$ for Eq. 12.4 is (+) ve when $\Delta H_{298\ K}$ for.γ-Al_2O_3 was taken from Avgustinik's data and it is found to be (−) ve by using Kroger's or Shieltz and Soliman's data. The numerical values, however vary to a great extent in all the three cases of above mentioned Eqs. 12.2, 12.4 and 12.7. The different $\Delta H_{1250\ K}$ values obtained for Eq. 12.7 is not exactly tally with the value (−54, 39 kJ/mole) given by Mazumdar and Mukherjee for their Eq. 5. In fact, the measured heat of reaction values obtained by different workers vary to a great extent from −5 to −15.5 kcal/mole as reviewed by Blair and Chakladar (1972). The acceptable value is due to Nicholson and Fulrath (1970) which is 40 kJ/mol. On comparison of this measured value with the calculated values for the Eqs. 12.4, 12.7 and 12.17, it is very difficult to ascertain which reaction would actually occur at 980 °C. Therefore, any forecast regarding formation of γ-Al_2O_3 at 980 °C exothermic peak would be erroneous. Further, $\Delta H_{1250\ K}$ values for Eq. 12.7 and 12.17 do not seem to differ much when a particular data was chosen as the basis of silica crystallization at 980 °C.

22.2.2 Mullite Formation May be a Cause of 1st Exotherm

With the view towards establishing the significance of various observed thermal effects, Bradley and Grim (1951) stated that sudden sharp exothermal effects may be due to rapid evolution of a new hard, dense phase of large articulated units from the parent structures without catastrophic rearrangement within the units. Where as soft irregularly constituted phases certainly cannot evolve detectable energy in slow development and probably do not in moderately fast precipitation. Subsequently, Roy et al. (1955) believed that formation of γ-Al_2O_3 is not the cause of this peak. Since, samples show γ-Al_2O_3 formation as low temperature as at 650 °C for halloysite. Even then this preheated halloysite exhibited exotherm. Even on heating several samples of kaolinite and halloysite at 800 °C which contained γ-Al_2O_3 but they noted exotherm unaffected. As such they excluded γ-Al_2O_3 formation to the intensity of exotherm. They thought and argued that metakaolinite is a metastable high free energy phase and always tends to revert to the stable assemblage with evolution of energy as heat. The first of these steps accompanied by any appreciable evolution of heat is the one that gives rise to 950 °C exotherm. It is most unlikely that such a large and pronounced heat effect would be caused by the formation of a metastable phase γ-Al_2O_3 at a temperature which is nearly the limit of its metastable existence under atmospheric conditions. Therefore, exothermic peak probably result of final stable assemblage, mullite and cristobalite. Mullite is reported in some cases just above the exothermic peak and in many cases 100 °C or less above this peak, it is reasonable to expect that the exothermic peak is caused by the crystallization of some mullite. The mullite formed around exothermic peak temperature seldom yields an X-ray diffraction pattern owing to its small particle. The crystals become large enough (by heating for longer periods or at higher temperature) to give powder X-ray pattern. Comeforo et al. (1948) opined that mullite is a comparatively hard, dense phase and more stable than the non crystalline 'metakaolin' and its crystallization into mullite nuclei of articulated units derived by collapse of a skeletal anhydride would account for both the observed heat effect and the observed bulk shrinkage of kaolinite bodies fired to approximately 1,000 °C. They further believed γ-Al_2O_3 is not a factor in the mechanism. In earlier days it is impossible to ascertain the structural configuration of metakaolinite due to lack of sufficient diffraction data. They suggested the persistence of a "uniquely crystalline anhydrous phase related structurally to kaolinite". They further suggested that in case of metakaolinite, dehydroxylation leads to co-ordinate strings of alumina octahedral as occurs in case of pyrophillite. Since, distribution of octahedral Al ions in kaolinite is like that of pyrophyllite.

John (1953) assumed that transition from the higher energy state of metakaolinite octahedral chain (face shared) to lower energy state of mullite chain (edge shared) would result a sudden release of thermal energy and a more stable structure. In case of well order kaolinite proper Al ion distribution occurs, the mullite crystallization is observed. They interpreted that exothermal effect as the result of energy involved primarily in the rearrangement of oxygen network to a network approaching that of mullite. They pointed that the DTA is most sensitive in revealing the first change which takes place, which represent the rearrangement of packing of the oxygen network. Subsequent reorganization of cat ions follows thereafter and is detected by the development of appropriate diffraction characteristic say mullite. The intensity in DTA is more directly correlative with the mullitization process and the Al_2O_3 appears as an incidental phase with regard to the observed thermal effects. However, Slaughtor and Keller (1959) argued that the formation of extremely small mullite nuclei (not detectable by X-ray) should preclude a very large heat of formation for metakaolin to mullite formation. However, mullite was detected in almost all clays fired in DTA furnace to the completion of exothermic reaction and thus they concluded that at least part of the

980 °C exothermic reaction of kaolinite must be attributed to mullite formation. They preheated fire clay of Missouri which contents 90 % kaolinite to two different temperatures. X-ray analysis showed abundant quantity of spinel type phase at 950 °C and the absence of any spinel at 900 °C/24 h fired samples. Subsequent DTA analysis of these samples preheated at two temperatures followed by X-ray analysis of DTA run residues revealed different observation. Sample preheated at 900 °C developed a strong 980 °C exothermic reaction with formation of spinel and mullite phases. However, samples preheated to 950 °C developed no exothermic reaction at 980 °C and neither more spinel phase nor any detectable mullite. With these observation they concluded formation of spinel phase was responsible for the reaction than the reaction should have been noted regardless of previous formation of spinel phase.

22.2.3 *β-Quartz Formation May be a Cause of 1st Exotherm*

β-quartz hypothesis as a cause of 980 °C peak comes up later on. The possibility of formation of *β*-quartz in thermal evolution of kaolinite was first proposed by Soliman (1961). According to Schieltz and Soliman (1966) the energy that produces the peak is enthalpy. Based on accurate thermodynamic data such as heat of formation, specific heat, entropy of formation etc. of kaolinite and its high temperature products they calculated the enthalpy change and change in free energy of six possible transformation of metakaolinite at 980 °C (1250 °K). The following equation

$$Al_2O_3 \cdot 2SiO_2 \rightarrow 1/3\,\underset{\text{Crystalline}}{(3Al_2O_3 \cdot 2SiO_2)} + 4/3\,\underset{\text{Cristobalite}}{SiO_2} \tag{12.8}$$

shows maximum $\Delta H_{1250\text{ K}}$ and minimum $-\Delta G$ values and thus it may be the most stable reaction thermodynamically. Thereafter, they calculated energy of crystallization of the products of 980 °C transformation of kaolinite. Calculated values of energy of crystallization of γ-Al_2O_3, silica and mullite are −36,513, −7,189 and −336,180 cal/mol respectively. Comparing the above values, they showed that energy of crystallization of mullite is about nine times that of γ-Al_2O_3. Accordingly, they suggested that the 980 °C exothermic peak is neither due to crystallization of γ-Al_2O_3 and SiO_2, nor to mullite, γ-Al_2O_3 and SiO_2 but most probably due to crystallization of mullite and SiO_2. The percentage contribution mullite and silica to the exothermic energy was also calculated by them are 92 and 8 % respectively.

Based on the calculations on exothermic energies of Schieltz and Soliman (1966), Nicholson and Fulrath (1970) attempted to prove the fist exothermic peak is due to *β*-quartz crystallization. Their argument was based on the following observation.

(i) They cited the works of Roy et al. (1955) and Comer (1961) that both spinel and mullite forms at lower temperature and thus they emphatically stated that there was no reason why the spinel phase which appeared continuously over a long range of temperature after Brindley and Nakahira (1959) would give rise to the sudden sharp exothermic reaction observed at 980 °C. The only phase remaining was the SiO_2(A). This SiO_2 had been assumed by them to be continuously discarded during spinel and mullite crystallization and it was possible that the accumulation of silica at 980 °C favored a sudden crystallization.

(ii) They measured the exothermic heat released at 980 °C of various kaolinites namely API 4, API 9, API 17 and DRG by differential thermal calorimeter and found to be −9 kcal/mol compared to heat of solution measurement by calorimetric technique which was approximately 16–25 cal/gm. To compare with; they also calculated theoretically the enthalpy of crystallization of 1 mol of amorphous SiO_2 to β-quartz at 1,000 °C as follows.

$$\Delta H_{298\,\mathrm{K}} \text{ for } \beta\text{-quartz} = -209{,}900 \text{ cal/mol. } C_P = 14.41 + 1.94 \times 10^{-3}\mathrm{T}.$$

$$\Delta H_{298\,\mathrm{K}} \text{ for } \mathrm{SiO_2(glass)} = -202{,}000 \text{ cal/mol, } C_p = 13.38 + 3.68 \times 10^{-3}\mathrm{T}$$

$$-\,3.45 \times 10^{-5}\mathrm{T}^{-2}.\ \Delta H_{1250\,\mathrm{K}} = \Delta H_{298\,\mathrm{K}} + \mathrm{INT}_{298 \text{ to } 1250}\,\Delta C_p \cdot dT$$

$$_{1250\,\mathrm{K}}\ (\mathrm{SiO_2 \text{ glass}} \rightarrow \beta\text{-quartz}) = 9.1\mathrm{K \ cal/mol}.$$

The calculated value agrees closely with that observed in the DTC measurement. Thus, they tentatively suggested that metakaolinite decomposed according to the following equation.

$$Al_2O_3\ 2SiO_2 \rightarrow Al_2O_3\ SiO_2 + SiO_2(A). \tag{22.1}$$

At 980 °C, SiO_2(A) so librated would be converted to one mole of quartz and in consequence −9 kcal/mol of exothermic heat would be released.

(iii) They were of the opinion that after the dehydroxylation of kaolinite to metakaolinite was complete, SiO_2 would be gradually discarded from metakaolinite structure with the rise of temperature up to the starting point of the first exothermic peak. Based on this assumption they treated 950 °C heated Zettlitz kaolinite with 10 % NaOH for 7 days as to remove the released portion of SiO_2. On DTA studies, the exothermic peak was found to be present in case of untreated sample, while the same was absent in case of NaOH treated sample. With the above observation they came to the conclusion that the exothermic peak was eliminated due to removal of accumulated SiO_2(A) by NaOH.

Blair and Chakladar (1972) and latter Chakladar (1976) put forwarded the view that spinel formation is not the cause of 980 °C exotherm. They applied Clausius Clapeyron equation on the basis of following assumptions.

(i) The silica (A) liberated during the transformation of metakaolin would remain in the spinel lattice i.e., spinel phase is supersaturated with SiO_2 while the transformation at the exothermic peak temperature would be as follows.

A. $$\text{Metakaolin} \rightarrow \text{Spinel with Supersaturated Silica} \tag{22.2}$$

B. $$\begin{array}{l}\text{Spinel with Supersaturated}\\ \text{Silica} \rightarrow \text{Spinel} + SiO_2(A)\end{array}$$

C. $$SiO_2(A) \rightarrow \beta\text{-quartz} \tag{22.4}$$

Equation (22.2) corresponds to one component ($c = 1$) and two phase ($p = 2$) which satisfied the requirement for application of Clausius Claypeyron equation. The integrated form of the equation is given below.

$$\ln \frac{T}{T_o} = \frac{V_o}{\Delta H_o}(P - P_o) \tag{22.5}$$

Measured value of exothermic energy in case of metakaolin to spinel transformation is −5 to −15.5 kcal/mol i.e., ΔH is negative. Change in molar volume (ΔV) for metakaolin to spinel transformation was calculated using data of Brindley and Nakahira (1959), Freund (1967) and Range et al. (1970) and reported by them to be −79.13 cm^3/g atom. As both ΔV and ΔH are negative, the application of an external pressure as per Eq. 22.5 should increase the transformation temperature of metakaolinite (T should be greater than T_0) which is exactly opposite to that experimentally observed. Thus, Chakladar (1976) believed that ΔV is definitely negative, the enthalpy of metakaolin to spinel transformation should be positive. Secondly, there might be a possibility of favored SiO_2(A) to β-quartz transformation. Since it has (−ve) ΔH and (−ve) ΔV for this transformation. With these arguments they emphasized β-quartz crystallization hypothesis for 980 °C exotherm.

22.3 Factors Affecting 980 °C Exotherm

Some of the factors namely heat treatment schedule, external use of some oxides, associated impurity oxides present in clays and application of external pressure affect both the occurrence and intensity of the 980 °C exotherm.

22.3.1 Non Equilibrium (or Dynamic) Heating

It is observed in Table 22.1 that no crystalline phases are formed on heating Bhandak kaolinite when it is heated below its exothermic peak temperature by dynamic heating schedule. But it forms both spinel and mullite after the exotherm. Similar observation was noted earlier by Insley and Ewell (1935) using Zettlitz kaolinite and Glass (1954).

All of them noted the formation of spinel phase just after the occurrence of exothermic peak. In addition to spinel and mullite crystallization, Chakraborty (1976) observed that liberation of free SiO_2(A) does not take place from Bhandak kaolinite heated below 980 °C during dynamic heating, but occur only at the exotherm (Table 22.2).

Therefore, in non equilibrium soaking at different temperature below 980 °C, neither spinel and mullite phase nor any amount of SiO_2 is liberated out of the metakaolinite structure. But the formation of spinel and mullite phases and exolution of silica takes place at exothem itself. Thus, there exist an interrelationship between three such occurrences i.e. silica liberation, spinel formation and incipient crystallization of mullite out of decomposition of metakaolinite.

Table 22.1 Crystalline phases formed from Bhandak kaolinite at various temperature during equilibrium and non-equilibrium soaking by heating at @ of 10 °C/min (after Chakraborty 1978)

Non equilibrium soaking			Equilibrium soaking at 900 °C	
Temperature °C	600–900	1000	12 h	24 h
Phases identified by XRD	No crystalline phases are formed	Al–Si spinel + Mullite phase	Al–Si spinel + Mullite phase	Al–Si spinel + Mullite phase

Table 22.2 Percentage of free silica estimated from bhandak kaolinite heated to different temperatures for different periods of soaking (after Chakraborty 1978)

Heat treatment temperature (°C)	Time of soaking (min)	Amount of free silica (A) estimated (%)
900	180	2.00–3.00
720	11.00	
1440	32.00	
980	0	33.50
35	34.90	
120	32.07	
600	37.98	

22.3.2 Static Heating

The situation is entirely different from the heating effect of kaolinite noted in case of dynamic heating. Figure 3.9 shows intensity of exotherm decreases with pre-treatment which confirms Insley's findings. Both spinel and mullite are noted much below 980 °C e.g. they are just discernable after 8 h of heating Bhandak kaolinite at 900 °C. Marked improvement in theirs intensities are observed at 12–24 h soaking respectively. Amount of free SiO_2 liberated gradually with soaking period (Table 22.2). It, therefore, shows that below exothermic peak temperature, spinel and mullite crystallization, liberation of SiO_2(A) occur by continued heat treatment which is usually formed at exothermic peak temperature on dynamic heating with no equilibrium soaking time. This observation is in close agreement with that of Insley and Ewell (1935), Roy et al. (1955) and Comer (1961). Insley and Ewell obtained spinel phase from Zettlitz kaolin, heated at 880 °C for 74 h soaking and by 6 h soaking at a bit higher temperature 910 °C. Glass (1954) noted the difference in phase sequences during firing under non equilibrium conditions at a constant heating rate versus equilibrium heating. For example, poorly crystallized kaolinite shows only γ-Al_2O_3 at 1,100 °C when air quenched. Where as a complete mullite with minor cristobalite development occur when sample soaked 4 h at 1,100 °C i.e. prolonged firing lowers the temperature at which phases form decreases the total temperature interval in which a complete firing sequence occurs. Roy et al. noted the appearance of spinel phase from halloysite at as low as 650 °C and from kaolinite at 850 °C respectively at static long time heating of 12–200 h. Comer (1961) noted the formation of mullite at 850 °C from a poorly crystallized kaolinite similarly at high soaking time e.g. 20 h. On comparing the results of equilibrium heated products with those of non equilibrium heat treated materials it is found that the formation temperature of spinel, mullite and SiO_2(A) are different. Accordingly, interpretations in these two cases will also be varied. It is not however, worthwhile to correlate the nature and amount of phases formed, the temperature of solid state reaction by reaction by equilibrium vs. non equilibrium soaking transformation sequence. It is found to be gradual for static heating and proceed continuously over a long range of temperature, but intermittent and sharp one as it occurs only at one temperature when heating rate follows dynamic. Therefore, the quotations on the work of Roy et al., Comer, and Brindley and Nakahira etc. are inapplicable in explaining DTA exotherm. In order to explain the cause behind it, one should have to analyze the heated products critically before and just after the peak temperature of kaolinite heated dynamically in the same rate as is used in DTA without allowing any soaking time at the respective temperature i.e. long static heat treatment should be avoided. As shown above by static heating at lower temperature than 980 °C, both spinel and mullite will form which is scheduled to form at 980 °C. Similarly long heat treatment at and above 980 °C, liberated SiO_2 will crystallize to cristobalite. Likely the interpretation that cristobalite is responsible for exotherm as conjectured by Mackenzie (1971) will be also untenable.

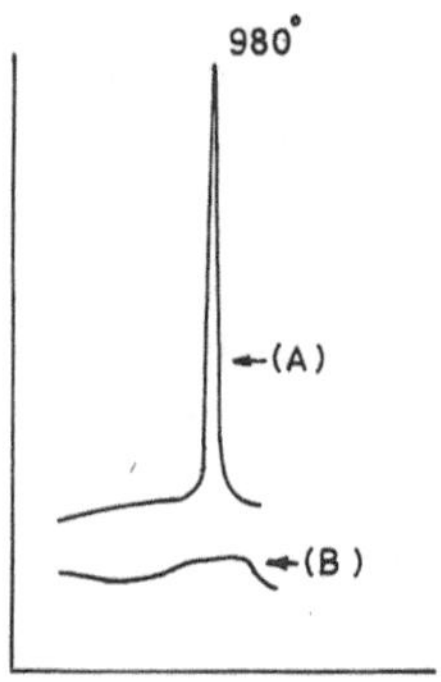

Fig. 22.1 Portions of DTA curves for bhandak kaolinite, (*A*) heated to 850 °C/2 h, (*B*) subjected to the same heat treatment and then extracted with 10 % NaOH for 1 week (after Chakraborty and Ghosh 1977). Reprinted by permission of the American Ceramic Society

Furthermore Chakraborty and Ghosh (1977, 1978) repeated the DTA study of alkali treated metakaolinite, the result confirms the earlier observation of Nicholson and Fulrath (Fig. 22.1).

But their assumption on the removal of accumulated SiO_2 from 850 °C heated kaolinite is not found to be true. Instead of loss in weight due to dissolution of associated SiO_2(A) from metakaolinite as per their assumption, an increase in weight approx. 25.8 % is observed experimentally. This result suggests that metakaolinite transforms into new materials by reacting with alkali although no X-ray pattern is noted. However, the DTA analyzed sample showed the presence of lines comparable to zeolitic phase instead spinel and mullite phases as are normally noted (Table 17.1). Therefore, NaOH treatment of 850 °C heated Bhandak kaolinite converted the metakaolinite into a zeolitic phase and the exothermic peak at 980 °C is eliminated as a result of the formation of this phase and not the removal of any SiO_2(A) by NaOH. Therefore, the contention that β–quartz is responsible for 980 °C exotherm is largely eliminated and moreover cyrstallization of SiO_2 is not supported by X-ray data.

22.3.3 R.H.P.

Chaklader and Mckenzie (1964, 1966) first studied the effect of pressure on the Hedvall effect. Later on Carruthers and Wheat (1965) reported that after pressing kaolinite at >5 tonf/in^2 and heated at temperatures at 750 °C, phases were produced which, at atmosphere pressure are not normally formed below 1000 °C. It was also shown that once these phases are formed, the material no longer shows the exhibition of strong exothermic reaction at ~980 °C. Blair and Chaklader (1971, 1972) indicated that there was an enhanced compaction during each stage of phase transformation in the consecutive reaction series from kaolinite. They showed three DTA patterns (Fig. 22.2) and observed that (i) the height of the exothermic peak for raw kaolin noticed at 1,020 °C was more or less same for heated kaolin at 800 °C; (ii) exothermic peak for RHP specimen was smaller for preheated kaolin and or raw kaolin.

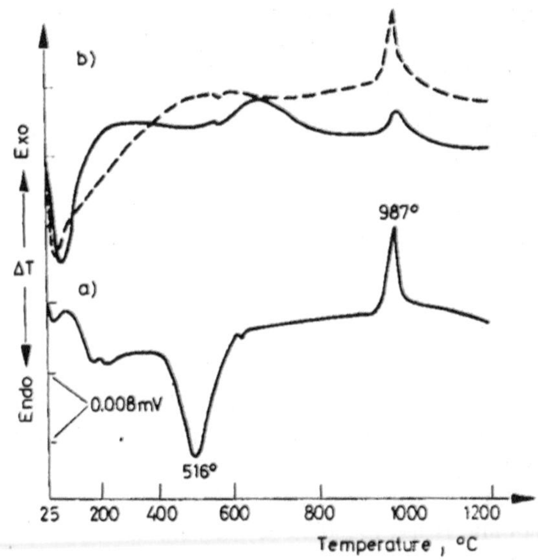

Fig. 22.2 DTA plots in air of Georgia kaolin. **a** Raw clay; **b** clay RHP to 800° (solid); **c** clay fired to 800° (broken) (After Blair et al. 1972) with kind permission from Maney Publishing, UK

X-ray diffraction pattern of RHP kaolin showed a weak pattern resembling spinel phase. This suggested that hot pressing induced the formation of spinel phase at lower temperature than found on normal firing.

Thus, the enhanced compaction (transformation plasticity) associated directly with the metakaolinite to spinel transformation precedes the exothermic peak in DTA. They also observed that the densification was encountered as low as 900 °C although DTA traces did not show any reaction until a temperature of 980 °C was reached at which temperature the transformation plasticity was almost over. Correlating compaction data with DTA plots, they proposed that transformation of metakaolinite to spinel occurs before the reaction temperatures usually encountered in DTA. Contrary, it is suggested that the reaction which is occurring at the 1st exotherm at normal atmospheric pressure might occur at lower temperature by the hot pressing method.

Carruthers and Wheat (1965) reported that after pressing kaolinite at >5 tonf/in^2 and heated at temperatures at 750 °C, phases were produced which, at atmosphere pressure, are not normally formed below 1,000 °C. Thereafter, Wheat (1967) had shown that once these phases are formed, the material no longer shows the exhibition of strong exothermic reaction at ~980 °C.

Chaklader (1976) indicated that there is enhanced compaction during each stage of phase transformation in the consecutive reaction series from kaolinite. The enhanced compaction (transformation plasticity) associated directly with the metakaolinite to spinel transformation precedes the exothermic peak in DTA. Correlating compaction data with DTA plots, he proposed that transformation of metakaolinite to spinel occurs before the reaction temperatures usually encountered in DTA. Therefore, it is concluded that the reaction which is occurring at the 1st exotherm may occur at lower temperature by the hot pressing method. Obviously, the conjecture that crystallization of SiO_2(A) to cristobalite as the reason the 980 °C exotherm is untenable.

22.4 Criteria for Exhibiting Exotherm

Before pointing out the real cause of 980 °C exotherm it is necessary to focus first the guideline i.e. the minimum necessary and sufficient conditions regarding the appearance/exhibition of a DTA peak in general. The following are to be considered for giving reasonable explanation of exothermic peaks occurs at 980, 1,200 °C and last exotherm at ~1,300 °C respectively.

(i) As the exothermic reaction is due to crystallization reaction, then which crystal phase is forming at the exotherm is to be ascertained first.
(ii) The quantitative yield of crystallization reaction must be sufficiently high such that a sufficient or minimum quantity of liberated heat energy would generate differential emf in a thermo-couple circuit of a most sensitive DTA apparatus to be detectable.
(iii) Whatever be the nature of crystallization reaction, it will lead to evolution of energy. But to produce an exotherm during DTA the rate of crystallization reaction must be sufficiently rapid such that crystallization would take place in the same heating schedule as usually followed in DTA analysis (e.g. rate of heating, 2, 5, 10, 15 °C etc. per minute with non equilibrium packing).

Therefore, the present aim is to sought out the phase which forms just at the exothermic peak during heating kaolinite at the same rate or condition as used in DTA analysis. According to the above views, the amount of phase form at just exothermic peak temperature of Bhandak kaolinite followed by the analogous heat treatment schedule was carried out. Figure 20.1 shows that quite an appreciable amount of spinel forms suddenly as indicated by its step rise in its intensity at 980 °C during non equilibrium heating at the rate of 10 °C/min. This result verifies the earlier observation of Insley and Ewell and Glass that rapid formation of spinel coincides with the T_m of the exotherm. In this contest, the combine X-ray diffraction and differential thermal analysis experiment carried out by Tszuki et al. (1969) is of much convincing. In this experiment one of the junctions of the differential thermocouple was inserted into the sample holder and was mounted properly on the goniometry of X-ray diffractometer. The goniometer was scanned oscillatory in the range of diffraction angle 2θ, 65.5–66.7° in which 440 reflection of spinel is included. This spinel which was conceived earlier as γ-Al_2O_3 is now designated as Al–Si spinel as per the discussion presented in Chap. 20.

(i) Various workers observed to form mullite with spinel at and around exothermic peak temperature. Glass (1954) was of the opinion that the relative amount of the formation of two phases depend upon the origin of the clay mineral and its associated impurity content. Some authors e.g. Keler and Leonov etc. noted more spinel and little mullite where as Mikheev and Stulov observed little spinel but more mullite.

(ii) Semi quantitative yield of spinel as noted by, Okada et al. (Fig. 14.2) noted comparatively a significant amount of spinel than mullite. Tsuzuki et al. showed that in the high temperature runs, the rate of crystallization of spinel is the largest at the beginning and then it shows down with time. At the lower temperature runs, crystallization begins after an induction period and then the rate becomes large following a sigmoid shape of the curve.

In comparison to spinel formation, mullite formation as shown in case of Bhandak kaolinite at 1,000 °C is insignificant (Fig. 20.1). Latter develops only around 1,200–1,250 °C with decomposition of former phase. Therefore, the mullite formation may be the reason for exotherm at 1,250 °C and less likely for 980 °C exotherm. As per theoretical calculation of crystallization energy of γ-Al_2O_3 and mullite phase by Schieltz and Soliman (1966) it is noted that the value for mullite is nine times as that of γ-Al_2O_3. Although the amount of mullite phase forms in little amount, but its high crystallization energy (as conceived by Roy et al. 1955; Comefore et al. 1948; John 1953) will contribute some enthalpy during exotherm. Further, spinel forms in quite appreciable amount and with a great speed (as shown) so it's crystallization enthalpy to 980 °C exotherm contributes a major role. Therefore, the formation of both spinel (major) and mullite (minor) constitute essentially the crystalline part of the exothermic reaction. It was shown that both spinel and mullite are polymorphic phases having same chemical composition, differ only in crystallographic form. Probably, there is no wrong in accounting mullite with spinel as a contributor of exothermic heat energy at 980 °C. Accordingly, Eq. 20.2 is designated as most probable 980 °C exothermic reaction by Chakraborty and Ghosh (1978).

$$\begin{aligned} \text{Metakaolinite} \xrightarrow[\text{Exotherm}]{980^\circ} & \text{(i) Al-Si spinel phase, 20-30 \%} \\ & \text{(ii) Mullite phase, 5 \%} \\ & \text{(iii) Mullite(A), 30-40\%} \\ & \text{(iv) } SiO_2 \text{ rich(A) phase, 36 \%} \end{aligned} \tag{20.2}$$

Figure 22.3 shows that metakaolinite to spinel/mullite transformation is a very sharp reaction as it occurs within a very short range of temperature interval i.e. within the beginning and completion of exothermic peak. Table 22.2 noted that the amount of free SiO_2(A) as liberated at 980 °C remains more or less constant even a long soaking is allowed at the exothermic peak temperature. This shows that reaction is complete with the completion of the exothermic peak and no further transformation is produced with increase of soaking at that temperature.

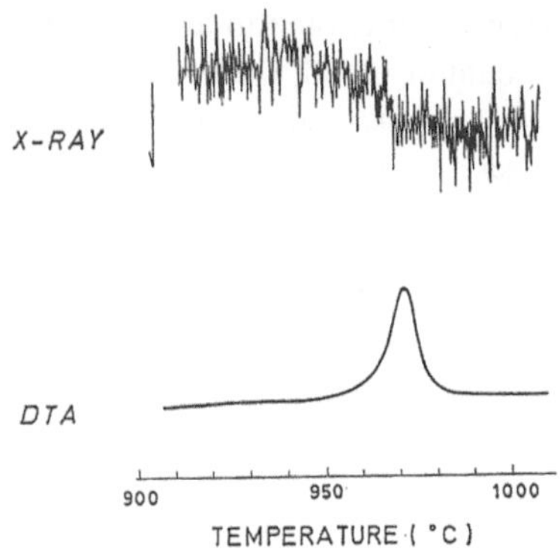

Fig. 22.3 Indicates that the increase in spinel content is coincident with the exothermic peak on the differential thermal curve (After Tszuki et al. 1969). Reproduced by permission of The Clay Science Society of Japan

According to Lee et al. (2001) the main origins of the first exothermic reaction were proposed to be the extraction of silica (A) from metakaolinite and the gradual nucleation of the mullite phase. At ~1,020 °C i.e., after the completion of the 1st exotherm three information are revealed.

(1) Centre of the amorphous broad back ground shifts to a lower angle as per XRD pattern which indicates segregation of SiO_2(A).
(2) Electron diffraction intensity of spots of spinel phase increased considerably in comparison to the faint spots of the same of kaolinite sample heated to 940 °C.
(3) Spinel phase is started to form before exothermic reaction, It is however evident from XRD patterns that both traces of weakly crystalline mullite and major amount of spinel phase are existed concurrently. Therefore, it is concluded that crystallization of mullite and spinel phase are responsible for the occurrence of exotherm which is unlike to what Lee et al. (2001) stated.

22.5 Correlation Between Phases Forming with Exotherm

Insley and Ewell (1935) showed the dependence of the size of the exothermic effect upon the preliminary heat treatment of Zettlitz kaolin. It was shown apparently that a decrease in the size of the peak height with increase in the temperature of the preliminary heat treatment. The amount of acid soluble alumina in heat treated sample, including both samples heated at constant temperatures and sample air-quenched upon reacting a given temperature shows a relation to the heat treatment which parallels that of the intensity of the exothermic effect and the appearance of spinel phase. A similar observation was noted by Chakraborty and Ghosh (1978). For example, geometry of the exothermic peak decreases with increase of soaking time during heating Bhandak kaolinite at 900 °C. At 24 h soaked samples, the peak vanishes completely (Fig. 3.9). Where as the converse is true for the formation of spinel, mullite and exolution of free SiO_2(A) with increase of preheating time (Tables 22.1, 22.2). Therefore, the gradual formation of spinel, mullite and liberation of SiO_2(A) with consequent diminution of intensity of exotherm are correlated. When the preheat treatment time is zero, the amount of spinel, mullite formation is likely to be zero and liberation of SiO_2(A) is also to be negligible then the intensity of 980 °C exotherm is maximum and vice versa.

22.6 Application of Clausius–Clapyron Equation on 1st Exotherm

Still there are some questions remain to be answered. 1. Whether Clausius Clapyron equation is applicable in the decomposition process of metakaolinite The Eq. 20.2 as shown above corresponds to component (C) = 5 instead of 1 which is essential for the applicability of the above thermodynamic equation. Secondly, the reaction is not probably reversible.

2. Why 1st exothermic peak temperature increases first and then decreases again with the increase in application of pressure as noted by Stone ad Rowland (Fig. 3.10), Carruthers and Scott in their (Fig. 16.1)?

3. Why the spinel formation is favored by application of pressure as noted by Weiss et al. (1970) and Chaklader (1976)

22.7 Endo-Exo Region of 980 °C Reaction of Metakaolinite

During decomposition of metakaolinite, a sharp differential contraction peak accompanied by an asymmetry at the beginning of the peak has been noted in the DTMA curve. The temperature at which this contraction peak occurs corresponds with the 980 °C exothermic peak in DTA. The asymmetric position indicates that contraction versus slow decomposition of metakaolinite begins much before the 980 °C crystallization reaction. The mechanism can be understood from the information gathered from a more sensitive TMA run (Fig. 3.4). The appearance of first small differential contraction peak is quite analogous to the endothermic dip (Fig. 3.1) occasionally found just before 980 °C during DTA runs of some kaolinites. This dip was explained earlier to be caused by the removal of the last traces of –OH groups present in the metakaolinite structure as had been observed in the TGA experiment and IR studies by various researchers.

It is assumed that in the final dehydroxylation region, metakaolinite sheets come closer to each other and contract with slow removal of part of the –OH groups and fully dehydrate during the endothermic dip. The sheet structure then collapses followed by the liberation of 35–37 wt% SiO_2(A) forming a highly porous skeleton of the anhydrous aluminosilicate(A) unstable phase. It crystallizes suddenly during the ongoing densification process to cubic mullite along with a small amount of orthorhombic mullite, leaving behind 30–40 wt% mullite (A) phase as a residue. Cubic mullite is the densest phase in the kaolinite transformation series and as a consequence of its sudden formation, the second differential contraction peak and the concurrent peak in DTA, which occur at the 980° exotherm, is very sharp. Thus, two phenomena occur sequentially. First is the final dehydroxylation of metakaolinite followed by removal of the last traces of –OH and expulsion of SiO_2 with the liberation of an anhydrous aluminosilicate (A) unstable phase. The second is the densification of the later mass and crystallization of a major amount of cubic mullite and a minor amount of its orthorhombic form. The small exothermic peak which occurs just before the large 980 °C DTA peak needs an explanation. The intensity of this small exothermic peak it more pronounced in Bhandak kaolinite than in Rajmahal kaolinite. Occasionally, this peak had been noticed in other kaolinites. XRD results showed that at 980 °C Bhandak kaolinite developed more orthorhombic mullite than Rajmahal kaolinite. Therefore, it can be concluded that the origin of this small exothermic peak occurring prior to the large 980 °C exotherm (Fig. 3.1) is related to partial and preferential crystallization of small amount of orthorhombic mullite from the decomposed metakaolinite phase. Shrikrishna et al. (1990) noted the existence of some single crystals of mullite in addition to spinel phase in their alkali leached sample heat treated previously.

22.8 Second Exothermic Peak

Mullite is a quite hard, dense and stable crystalline phase in the binary Al_2O_3–SiO_2 system. Most likely its formation out of the transformation of Al–Si spinel is highly expected to exhibit an exotherm in DTA trace. Now question is: Where? Usually the velocity of mullite formation from static heat treated Bhandak kaolinite is slow around 1,000 °C but it increases very suddenly at 1,150 °C (Fig. 14.5). After that it still increases slowly up to 1,400 °C and attains almost constant. Various researchers claimed the rapid rise of mullite formation is the cause of 2nd exotherm at about 1,250 °C in DTA. The following observations noted by various investigators show how and approximate temperature interval at which mullite forms during heating kaolinite as in same rate used in DTA. Instead of reaching a conclusion from static heat treated samples,

Glass (1954) choose to heat various kaolinites dynamically at the rate of 10 °C/ min and showed that the major formation of mullite take places (by X-ray) at about 1,200–1,250 °C where the temperature of 2nd exothermic peak generally lie.

In an extension to this explanation, continuous high temperature X-ray diffraction and DTA studies on well crystallized, poorly crystallized and Halloysite by Wahl and Grim (1964) could be cited here. They noted that initial appearance of mullite was found on the diffractogram at 1,180 °C which coincides with the beginning of a 2nd exotherm at approximately 1,200 °C in DTA curve (Figs. 3.7 and 5.6).

Similarly, Bellotto et al. (1994) also noted rapid mullitization at ~1,200 °C in their dynamic heated sample (Fig. 5.4).

Moya et al. (1985) confirmed the exhibition of 2nd exotherm after removal of free SiO_2(A) by alkali.

The semi quantitative amount of orthorhombic mullite developed in the dynamic heating schedule of Rajmahal kaolinite was estimated by measuring the heights of the 0.537 nm peaks from XRD recordings is as shown in Fig. 22.4. It shows that in between 1,230 and 1,270 °C, the intensity of peak related to mullite suddenly increases at the cost of disappearance of Al–Si Spinel phase. This result also supports 2nd exotherm is the cause of mullitization. A new concept of mullitization is put forward below. The formation of orthorhombic mullite obtained from static heat treatment of Bhandak kaolinite is shown in (Fig. 14.5). These curves show two distinct temperature regions of mullite formation, which verifies an earlier observation.

In the present study, the DTA curve in temperature range 1,100–1,400 °C exhibits two exotherms. These results indicate that those two peaks are related to the two paths of mullite formation from the two precursor phases e.g., cubic mullite and residual mullite (A) phase according to the following evidence.

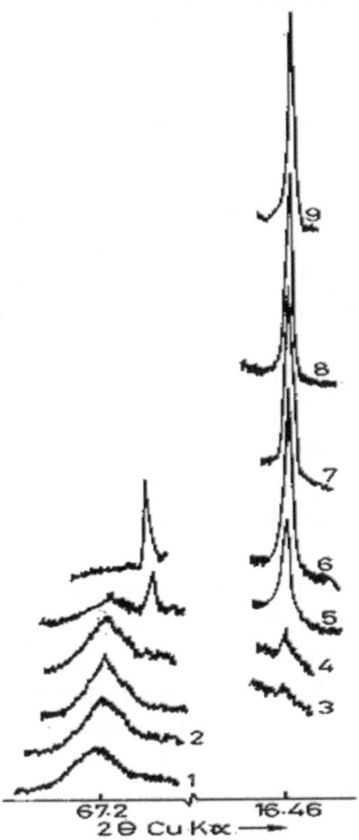

Fig. 22.4 Portions of XRD recordings of DTA—analyzed Rajmohol kaolinite sample collected at the following successive increases of temperature: **1** 980°, **2** 1050°, **3** 1100°, **4** 1230°, **5** 1270°, **6** 1305°, **7** 1345°, **8** 1392°, and **9** 1,420 °C (After Chakraborty 1992). Reprinted by permission of the American Ceramic Society

Transformation of Cubic Mullite

When cubic mullite (density of 3.5 g/cm^3) transforms to orthorhombic mullite (density of 3.17 g/cm^3) its density will decrease and theoretically an expansion will take place. In practice, the contraction, curve of English china clay (Fig. 3.4) shows how the TMA profile changes its slope or brakes at three points (marked A, B and C) which corresponds with the 1,250 °C differential expansion peak in DTMA. The breaks on the smooth contraction profile between 1,100 and 1,400 °C when compared to the differential expansion peak in DTMA suggest that mullitization, as evidence in the first cyrstallization region shown in Fig. 3.6 occurs with expansion. Therefore, exothermic peak maximum at about 1,250 °C in the DTA and the same peak at the same temperature in DTMA can be explained by the polymorphic transformation of cubic to orthorhombic form of mullite. Regarding the nature of this transformation, it is observed that cubic mullite transforms over a range temperature, e.g., ~1,270 °C, in contrast to its sharp formation at 980 °C exothermic peak in DTA, a broad exotherm is observed in the vicinity of 1,250 °C in DTA. The same phenomenon occurs in the differential expansion peak in DTMA (Chakraborty 1993).

Crystallization of Mullite(A)Phase

Besides the mullitization process described above, a further mullite formation takes place on dynamic heating of Rajmahal kaolinite beyond 1,300 °C (Fig. 22.5).

For example, at ~1,400 °C the crystallization of mullite takes place and then stops at a point which distinctly separates two crystallization regions. This suggests that nucleation of mullite has taken place earlier in the aluminosilicate phase formed from 40 wt% of the collapsing of metakaolinite. It was shown by Freund (1960) that this collapsed mass was full of lattice defects and pores. As a consequence, the mass subsequently contracts as the temperature increases. Partial contraction occurs around the 980 °C endothermic-exothermic region and the rest of the large contraction is completed between 1,100 and 1,400 °C where a big and

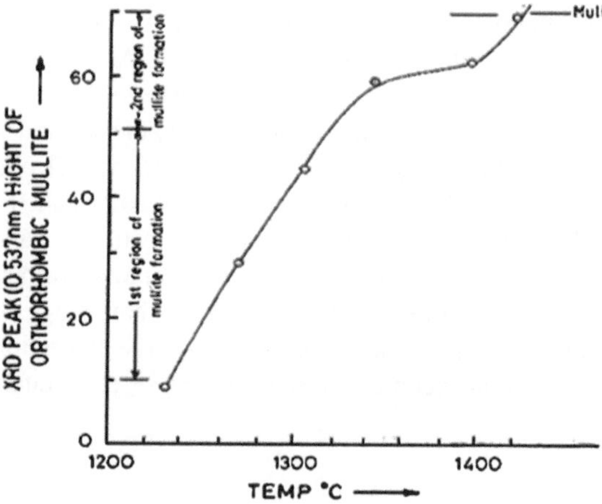

Fig. 22.5 Formation of orthorhombic mullite versus temperature of heating of rajmohol kaolinite (Chakraborty 2003)

broad differential contraction peak occurs in the DTMA. The inherent alkali, alkaline earth oxides. TiO_2 and Fe_2O_3 occurring as ~3.5 wt% in the Rajmahal kaolinite play a significant role in the sintering of the mass as a whole and also in the nucleation and growth of mullite as a mineralizer or catalytic agent. As shown, mullite formation occurs over a wide range of temperatures (from about 1,100–1,400 °C) during the dynamic heating schedule. This mode of crystallization corresponds exactly with the broad exothermic peak in the same temperature range. This indicates that mullitization takes place early in mullite(A) phase and proceeds continuously showing maximum crystallization in 1,300–1,400 °C range where the second exotherm (1,330 °C) also lies. Meanwhile, at ~1,250 °C the associated cubic mullite transforms and adds to the growth of orthorhombic mullite. Thus, during the heating of kaolinite, two mullitization process occur simultaneously, which confirms earlier work.

22.9 Summary

Metakaolinite is a metastable phase with high free energy content. It must exhibit few exotherms during subsequent crystallizations.

Theoretical consideration point of view, the possible reasons for occurrence of 1st exotherm are first of al discussed considering crystallization either of the following phases e.g., mullitization, Al–Si spinel formation and finally β-quartz formation.

The first exothermic peak is found to be very sensitive and is affected by heating schedule, preheat treatment condition and by pressing during DTA run. The criteria for exhibition of an exotherm is discussed.

The crystalline phase which forms just at the exothermic peak during heating kaolinite at the same rate or condition as used in DTA analysis is sorted out. Results of Insley and Ewell and Glass that rapid formation of spinel coincides with the T_m of the exotherm is convincing. The combine X-ray diffraction and differential thermal analysis carried out by Tszuki et al. (1969) is most worthy. The increase in intensity of XRD peak of Al–Si Spinel phase coincides with the exhibition of exotherm. Various workers observed to form weakly crystalline mullite is also crystallize along with spinel at and around exothermic peak temperature. Relatively, the yield of Al– Si Spinel phase formation is significantly larger than mullite as noted both by Okada et al. (Fig. 14.2), and Tsuzuki et al.

Chakraborty and Ghosh (1991) also estimated an appreciable amount of Al–Si spinel phase and traces of mullite are found to form suddenly at 980 °C during non equilibrium heating @ 10 °C/min in a short range of temperature. Obviously, the decomposition of metakaolinite vis-a-vis exhibition of 1st exothermic reaction (Eq. 20.2) has been established which verifies much earlier findings.

$$\text{Metakaolinite} \xrightarrow[\text{Exotherm}]{980^\circ} \begin{array}{l}\text{(i) Al-Si spinel phase, 20-30 \%}\\ \text{(ii) Mullite phase, 5 \%}\\ \text{(iii) Mullite(A), 30-40\%}\\ \text{(iv) SiO}_2\text{ rich(A) phase, 36 \%}\end{array} \quad (20.2)$$

Finally, the occurrence of the 1st exotherm is correlated with nature of preheat treatment or soaking time, crystallization behavior of Al–Si spinel, mullite and exolution of siliceous phases. The Clausis Clapeyron equation seems inapplicable to this exothermic reaction (20.2)

The reason of the formation of endothermic dip due to final dehydroxylation of kaolinite is substantiated by corroborating DTA and DTMA curves. Endothermic dip is analogous to exhibition of first differential contraction peak in DTMA. In comparison to slow crystallization of mullite during the occurrence of 980 °C exotherm, predominant mullitization occurs in the broad temperature range 1,100–1,400 °C. In this range mullite formation occurs in two stages.

In the first stage, Al–Si rapidly polymorphic ally transforms at ~1,250 °C as per Eq. (21.7) and accounts for 2nd exotherm. The 2nd exothermic peak maximum occurs at ~1,250 °C in the DTA is analogous to the same at the same temperature in DTMA. A gradual mullitization occurs by nucleation/crystallization process out of mullite (A) phase in the entire temperature range as stated above as per Eq. (21.8). This accounts for a broad exotherm over the same temperature range in DTA.

Thus, two mullitization processes exhibit two exotherms in DTA. The energy of crystallization of silica was calculated by Schieltz and Soliman (1966) to be −7,189 cal/mol. Based on this data, several arguments are put forward by Nichoson and Fulrath (1970), Blair and Chakladar (1972) and later Chaklader (1976) to explain β-quartz formation as cause of 1st exotherm. By standardizing an alkali leaching for estimation of free silica(A) content of kaolinite heat treated to 980 °C and by repeating the DTA study of alkali treated metakaolinite for 7 days Chakraborty and Ghosh (1977, 1978) discard β-quartz hypothesis.

References

M. Bellotto, High temperature phase transformation in kaolinite: the influence of disorder and kinetics on the reaction path. Mater. Sci. Forum 166–169, 3–20 (1994)

G.R. Blair, A.C.D. Chaklader, Firing versus reactive hot-pressing. J. Therm. Anal. **4**, 311–322 (1971)

G.R. Blair, A.C.D. Chaklader, Kaolinite-mullite series: firing versus reactive hot-pressing. J. Therm Anal **4**, 311–322 (1972)

W.F. Bradley, R.E. Grim, High-temperature thermal effects of clay and related materials. Am. Mineral. **36**(3/4) 182–201 (1951)

G.W. Brindley, M. Nakahira, The kaolinite-mullite reaction series: I, II, and III,. J. Am. Ceram. Soc. **42**(7), 311–324 (1959)

I.W.M. Brown, K.J.D. MacKenzie, M.E. Bowden, R.H. Meinhold, Outstanding problems in the kaolinite–mullite reaction sequence investigated by 29Si and 27Al solid-state nuclear magnetic resonance: II, high- temperature transformations of metakaolinite. J. Am. Ceram. Soc. **68**(6), 298–301 (1985)

T.G. Carruthers, T.A. Wheat, in *Hot Pressing of Kaolin and of Mixtures of Alumina and Silica.* Proceedings Britain Ceramic Society, vol. 3, p. 259 (1965)

A.C.D. Chaklader, L.G. McKenzie, Am.Ceram.Soc.Bull. **43**, 892 (1964)

A.C.D.Chaklader, L.G. McKenzie, J. Am. Ceram. Soc. **49**, (1966)

A.C.D. Chaklader, Flow properties of clay minerals during phase transformation. Cent. Glass. Ceram. Res. Bull. **23**(1), 5-14 (1976)

A.K. Chakraborty, D.K. Ghosh, Re-examination of the kaolinite to mullite reaction series. J. Am. Ceram. Soc. **61**(3–4), 170–173 (1978)

A.K. Chakraborty, Resolution of thermal peaks of kaolinite by TMA and DTA. J. Am. Ceram. Soc. **75**(7), 2013–2047 (1992)

A.K. Chakraborty, Application of TMA and DTA studies on the crystallization behavior of SiO2 in thermal transformation of kaolinite. J. Therm. Anal. **39**, 280–299 (1993)

A.K. Chakraborty, New data on thermal effects of kaolinite in the high temperature region. J. Therm. Anal. **71**, 799–808 (2003)

A.K. Chakraborty, D.K. Ghosh, Re-examination of the decomposition of kaolinite. J. Am. Ceram. Soc. **60**(3–4), 165–166 (1977)

A.K. Chakraborty, D.K. Ghosh, Comment on the Interpretation of the Kaolinite-Mullite Reaction Sequence from Infra-Red Absorption Spectra. J. Am. Ceram. Soc. **61**(1–2), 90–91 (1978)

A.K. Chakraborty, D.K. Ghosh, Re-examination of the kaolinite to mullite reaction series. J. Am. Ceram. Soc. **61**(3–4), 170–173 (1978)

A.K. Chakraborty, D.K. Ghosh, Kaolinite-mullite reaction series. The development and significance of a binary aluminosilicate phase. J. Am. Ceram. Soc. **74**(6), 1401–1406 (1991)

A.K. Chakraborty, D.K. Ghosh, On the origin of the exothermic peak in the thermogram of a kaolinitic clay : spinel versus Ô quartz crystallization, Cent. Glass & Ceram. Res. Instt. Bull. **23**(1), 38–40 (1976)

J.E. Comeforo, R.B. Fischer, W.F. Bradley, Mullitization of kaolinite. J. Am. Ceram. Soc. **31**(9), 254–259 (1948)

J.J. Comer, New electron-optical data on the kaolinite–mullite transformation. J. Am. Ceram. Soc. **44**(11), 561–563 (1961)

F. Freund, Die Deutungder Exothermen Reaktio Des Kaolinite Als Reaktio Des Aktiven Zustandes. Ber. Deut. Keram. Ges. **37**, 209–218 (1960)

F. Freund, Kaolinite-Metakaolinite, a model of a solid with extremely high lattice defect concentrations. Ber. Dtsch. Keram. Ges. **44**(I) 5–13 (1967)

H.D. Glass, High-temperature phases from kaolinite and halloysite. Am. Mineral. **39**, 193–207 (1954)

I.H. Insley, R.H. Ewell, Thermal behavior of the kaolin minerals. J. Res. Natl. Bur. Stand. **14**[S], 615–627 (1935)

W.D. Johns, High-temperature phase changes in kaolinites. Mineral. Mag. **30**(222) 186–198 (1953)

S. Lee, Y.J. Kim, H.J. Lee, H-S. Moon, Electron-beam-induced phase transformations from metakaolinite to mullite investigated by EF-TEM and HRTEM. J. Am. Ceram. Soc. **84**(9) 2096–2098 (2001)

J. Lemaitre, M. Bullens, B. Delmon, in *Influence of Mineralizers on the 950°C Exothermic Reaction of Metakaolinite*, ed by S. W. Bailey. Proceedings of the International Clay Conference, Mexico City, Mexico, July 1975 (Applied Publishing Ltd., Wilmette, 1975), pp. 539–544

A.J. Leonard, Structural analysis of the transition phases in the kaolinite–mullite thermal sequence, J. Am. Ceram. Soc. **60**(1–2), 37–43 (1977)

K.J.D. MacKenzie, Comment on differential thermal calorimetric determination of the thermodynamic properties of kaolinites. J. Am. Ceram. Soc. **54**(3), 174 (1971)

S. Majumder, B. Mukherjee, J. Am. Ceram. Soc. **69**(8), C-201 (1986)

J.S. Moya, C.J. Serna, J.E. Iglesias, On the formation of mullite from kandaties. J. Mater. Sci. 20, 32–36 (1985)

P.S. Nicholson, R.M. Fulrath, Differential thermal calorimetric determination of the thermodynamic properties of kaolinite. J. Am. Ceram. Soc. **53**(5), 237–240 (1970)

H.J. Percival, J.F. Duncan, P.K. Foster, Interpretation of the kaolinite–mullite reaction sequence from infrared absorption spectra. J. Am. Ceram. Soc. **57**(2), 57–61 (1974)

R. Roy, D.M. Roy, E.E. Francis, New data on thermal decomposition of kaolinite and halloysite. J. Am. Ceram. Soc. **38**(6), 198–205 (1955)

K. Range, J.Russow, G.Oehinger ad A.Weiss, Ber. Dt. Keram. Ges. **47**, 545 (1970)

N. C. Schieltz, M. R. Soleman, in *Thermodynamics of various high temperature transformation of kaolinite*, ed. by E. Ingerson. Proceedings of 13th National Conference on Clays, Madison, Wisconsin, (Pergamon press, Monograph, 1966) No. 25, p. 419–425

M. R. Soliman, Thermodynamics of various high temperature reactions of kaoinite , In *Clays and Clay Mineras*, Pro.12th Natl. Conf.19, Editor –in- Chief, Earl Ingerson, (Pergamon Press, Oxford ,N.Y., 1961) p. 247

B. Sonuparlak, M. Sarikaya, I.A. Aksay, Spinel phase formation during the 980°C exothermic reaction in the kaolinite-to-mullite reaction series, J. Am. Ceram. Soc. **70**(11), 837–842 (1987)

K. Srikrishna, G. Thomas, R. Martinez, M.P. Corral, S. Aza, J.S. Moya, Kaolinite–mullite reaction series: a TEM study. J. Mater. Sci. **25**, 607–612 (1990)

Y. Tsuzuki, K. Nagasawa, A transitional stage to 980°C exotherm of kaolin minerals. Clay Sci. **3**(5), 87–102 (1969)

F.M. Wahl, R.E. Grim. High temperature DTA and XRD studies of reactions. Twelfth National Conference on Clays and Clay Minerals, pp. 69–81 (1964)

K.J. Weiss, Range, J. Russow, in *The Al,Si-Spinel Phase from Kaoinite (Isolation, Chemical Analysis, Orientation and Reactions to Its Low-Temperature Precursors)*. Proceedings of the International Clay Conference, vol. 2. Tokyo, 1969, (Israel Universities Press 1970) p. 34–37

T. A. Wheat, Ph.D. Thesis, (Uiversity of Leeds, U.K., 1967)

Chapter 23
Cristobalite Phase

23.1 Introduction

23.1.1 Role of SiO_2 Component of Kaolinite During Heating

At the first stage of transformation of kaolinite to metakaolinite at ~600 °C, XRD shows a broad diffuse asymmetrical reflection at 0.43 nm. It is attributed to noncrystalline nature of metakaolinite instead of free SiO_2(A).

At the second stage, exothermic peak at 980 °C, the X-ray intensity of 0.43 nm band reduces and shifts Lee et al. (1999). It is assumed that this band is related to liberation of SiO_2(A). In this paper, the crystallization behavior of this released silica on further heat treatment of kaolinite has been discussed. From theoretical calculations, Schieltz and Soleman (1966) pointed out that the reaction.

$$\mathrm{Al}_2O_3.2SiO_2 \rightarrow 1/3\,(3Al_2O_3\,2SiO_2) + 4/3\,\mathrm{SiO_2(crist)} \qquad (12.8)$$

showed a maximum decrease of free energy. They calculated the energies of crystallization of mullite, γ-Al_2O_3 and SiO_2 are −336.180, −36.513, and −7.199 kcal mol^{-1} respectively. The energy contribution of SiO_2 to the 980 °C exotherm as in the above equation was −9.6 kcal mol^{-1} in agreement with the value measured experimentally with a differential calorimeter. However, neither β–quartz nor cristobalite has ever been reported in the XRD patterns of kaolinite heated at 980 °C other than by them. This might be due to the presence of quartz as impurity in the original kaolinite. The apparent elimination of this exotherm on alkali treatment of metakaolinite with a view to extracting the free silica prior to DTA supported the β–quartz hypothesis. However, the actual reason behind the removal of the 980 °C peak was the formation of a zeolitic phase (Table 17.1) by the direct reaction between metakaolinite and NaOH solution, in contrast with the observation of Flank (1979). Furthermore, silica could not be extracted by NaOH due to the nonavailability of free SiO_2 in the metakaolinite structure. The contentions of Roy et al. (1955) and Blair and Chaklader (1972) that neither spinel nor the spinal and mullite phases together were responsible for the 980 °C exotherm

A. K. Chakraborty, *Phase Transformation of Kaolinite Clay*,
DOI: 10.1007/978-81-322-1154-9_23, © Springer India 2014

because of their earlier development proved untenable. The phase transformations occurring either on equilibrium heating or under reactive heating conditions are entirely different from those taking place during nonequilibrium heating as in case of DTA. Accordingly, the concept of a Al–Si spinel with mullite-like composition gained ground and it was found to be a feasible 980 °C phase in the K–M reaction series. Semi-quantitative amounts of this spinel phase together with other phases that developed at ~980 °C were estimated by Okada et al. (1986), Tszuki et al. (1969) and Chakraborty and Ghosh (1991). The formation of these phases from the sharp decomposition of metakaolinite is associated with an enthalpy change and those should not be ignored, as was done by Blair and Chaklader (1972). In contrast, application of the *Clausius–Clapeyron* equation to the transformation of metakaolinite was found to be untenable as discussed. On thermodynamic and experimental reasoning, the views of all earlier authors were re-examined so far as silica crystallization is concerned as the source of the 980 °C exotherm in DTA. Ultimately, it was concluded that, as the formation of ~20 wt% Al–Si spinel and ~5 wt% mullite, together with ~40 wt% mullite (A) phase and silica (A) phase ~3 5–37 wt% SiO_2 occurred very suddenly at 980 °C, the most probable cause of the 980 °C exotherm would be the formation of the latter phases as per the Eq. 20.2.

At the third stage, Al–Si spinel and mullite (A) phase transformed to orthorhombic mullite with exotherms at about 1,250 °C and at ~1,330 °C in DTA analysis. The reduced 0.43 nm band decreases further with increase in intensity of mullite (Fig. 5.1). Chakraborty (1992) assumed that mullite (A) phase crystallizes to mullite instead of γ-Al_2O_3/SiO_2(A) solid state reaction as proposed earlier. K–M reaction series thus constitutes three stages of transformation. Now question is, what happens to silica phase exolved out of decomposition of metakaolinite at ~980 °C. Obviously, it is expected that it would crystallize during heat treatment.

23.2 Earlier Views of Last Exotherm

XRD study by several researchers noted ß-cristobalite development at ~1,200 °C and marked quantity of it crystallizes in the temperature range 1,300–1,400 °C. The majority of earlier workers e.g., (i) Glass (1954), Gernad-Hirne and Menaret (1956); (ii) Johnson et al. (1982) attribute the presence of third exothermic peak as the crystallization of cristobalite from a large amount of siliceous phase (A) accumulated from the breakdown of metakaolinite structure. (iii) The major formation of cristobalite was found to take place at about third exothermic peak temperature, when the kaolinite heated continuously (Fig. 3.5). Chakraborty (1993) showed that a contraction and or a differential contraction peak (Fig. 3.6) coincide with exotherm in the temperature range 1,410 °C. This rapid contraction is theoretically expected for crystallization of siliceous phase (A) as per the equation below.

23.3 Cristobalite Development in Kaolinite

Chakraborty (1993) measured the amount of β-cristobalite formed in Rajmohol kaolinite heated to $\sim$1,400 °C. By quantitative XRD technique this value is found to be 20–2 wt%. This result supports the concept of silica crystallization as the cause of last exotherm in the DTA trace of kaolinitic clay. Other corroborative evidences are outlined below.

$$\begin{array}{ccc} \mathrm{SiO_2\ rich} & \xrightarrow[\text{Contraction}]{1420^\circ\mathrm{C}} & \mathrm{SiO_2(\alpha - Cristobalite)} \\ \mathrm{phase(A)} & & \mathrm{(Density\ -\ 2.20\,m/cc)} \\ \mathrm{Density\ -} & & \\ \mathrm{2.22\,gm/cc} & & \\ & & \downarrow^{\text{Cooling}} \\ & & \mathrm{SiO_2(\beta - Cristobalite)} \\ & & \mathrm{(Density\ -\ 2.38\ -\ 2.34\,gm/cc)} \end{array} \tag{23.1}$$

(i) Effect of alkali treatment on DTM.
When the SiO_2 rich phase liberated during the 980 °C reaction of metakaolinite had been removed by alkali leaching, the DTMA run of the leached sample showed an increment in the area (shaded portion of the differential expansion peak at $\sim$1,250 °C(B) in comparison to the area for the unleached sample (A) (Fig. 18.7). Second, the large differential contraction peak (e′) decreases on leaching. These two observations confirm the effectiveness of silica removal by NaOH leaching. Third, it was also shown that due to silica removal, differential contraction peak at 1,430 °C is completely eliminated and the XRD recording of the DTMA run sample shows only mullite and the absence of β-cristobalite. Finally, he showed that DTA of leached residue does not show exhibition of the 1,420 °C exotherm. Moya et al. also confirmed this view.

(ii) Mode of cristobalite formation.
Chakraborty (1993) showed that the XRD analysis of dynamic heat-treated Rajmahal Kaolinite, the growth patterns of β-cristobalite occur as a faint 0.404 nm Bragg diffraction peak at $\sim$1,300 °C. The intensity of this peak increases slightly at $\sim$1,320 °C, but at $\sim$1,420 °C a marked increases in its intensity is noted. This observation indicates that cristobalite crystallizes slowly before the occurrence of the differential contraction peak at 1,420 °C. It may crystallize partially during the DTMA run, in the portion marked (g) (Fig. 3.6) and at or around the peak temperature, i.e., at 1,420 °C, a considerable amount of cristobalite formation takes place. This observation is corroborated by recording the cooling curves during TMA and DTMA runs of Rajmahal kaolinite (Fig. 23.1).

Fig. 23.1 Cooling curves (*A1*) TMA and (*A*) DTMA of Rajmohol kaolinite run up to 1,420 °C in the region of $\alpha \leftrightarrow \beta$ cristobalite inversion compared with the same *B* and *B1* of the same kaolinite run up to 1,350 °C. Sample size 15 mm, sensitivity 100 ± μm, cooling rate: normal cooling in ordinary atmosphere, instrument: shimadzu thermo mechanical analyzer system (after Chakraborty 1993)

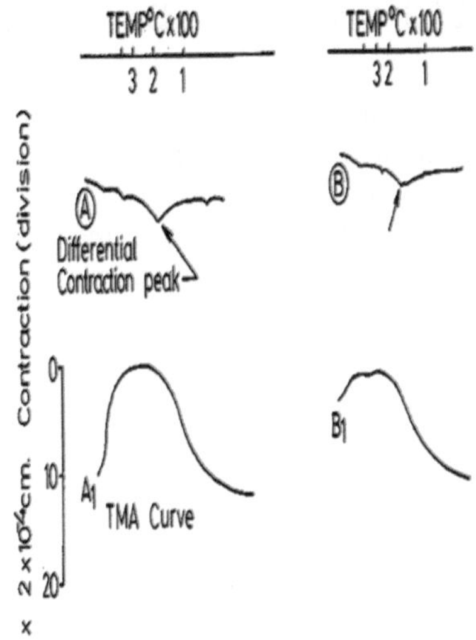

In the cooling cycle, kaolinite run previously up to 1,420 °C shows a contraction in the TMA trace (A1) and correspondingly a differential contraction peak in the DTMA record (A).

This phenomenon is due to the reversible transformation of α-cristobalite formed earlier during the heating cycle of kaolinite run to its β-modification at ~240 °C during the normal cooling run, as shown in Eq. (23.1). This inversion is also recorded in kaolinite heated to 1,350 °C. i.e., a temperature less than the peak temperature of cristobalite formation, usually occurs at 1,420 °C. However, in this run (B.B1) the deflection of the peak (B) and the amount of contraction are less than in the previous case. The amount of contraction and the geometry of the DTMA peak are related to the quantity of cristobalite formed during the heating cycle in the kaolinite sample. This contraction behavior indicates that a fraction of the cristobalite is formed prior to the differential contraction peak at ~1,420 °C in DTMA. This is also evident from the semi-quantitative XRD estimations of cristobalite out of kaolinite heated to 1,350–1,420 °C (Fig. 23.2).

At ~1,420 °C, the heat liberated due to the transformation of residual SiO_2 rich (A) phase (Eq. 23.1) to a major amount of β-cristobalite is enough for the exhibition of a marked contraction in the TMA plot and a corresponding exothermic peak in the DTA record.

The formation of cristobalite before the commencement of this exotherm was also noted by Wahl and Grim (1964) and Johnson et al. (1982) in their poorly crystallized kaolinite and impurity-mixed kaolinite, respectively. Therefore, the earlier explanation put forward by Chakraborty and Ghosh (1977) that SiO_2(A) liberated at 980 °C did not crystallize at the temperature and that the 980 °C exothermic peak was not due to the crystallization of either β-quartz or β-cristobalite holds good. At the same time, the corroborative evidence presented above definitely suggests that the transformation of ~37 wt% siliceous(A) phase to ~20

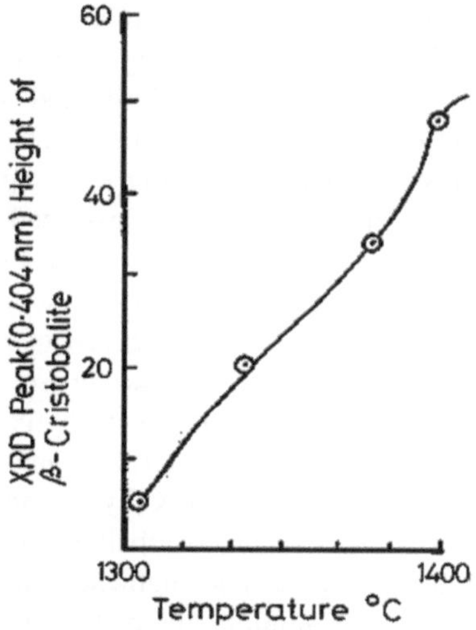

Fig. 23.2 XRD intensity (peak height) of β-cristobalite versus temperature of heating of Rajmohol kaolinnite under dynamic heating conditions (rate of heating 20 °C/min) with no soaking time (after Chakraborty 1993)

wt% β-cristobalite is responsible for the 1,420 °C exotherm in the DTA trace and the contraction peak in the DTMA record. Cristobalite formation according to Bellotto (1994) possibly appears to be nucleation controlled.

23.4 Changes in Lattice Parameter of Cristobalite and its Explanation

The relative changes in lattice strain, crystal size, and changes in lattice parameter of cristobalite formed during heating kaolinite at elevated temperatures have been determined by Quasar Analysis using the same software program. The results are shown by Chakraborty (2003a, b, c, d) in Table 23.1.

Present result shows that L.C. values of cristobalite of the order of 0.502 nm at ~1,350 °C and is comparable with data published by Tomura et al. (1986). However, lattice constant values seem higher than those noted by Verduch (1958), Yamada and Kimura (1962) while preparing cristobalite from silica gel.

Lattice constant values of cristobalite obtained by Chakraborty (0.502 nm at ~1,350 °C) is found to be much higher than the same synthesized from other sources. Verduch heat-treated silicic acid for his kinetic studies and showed that

Table 23.1 Changes in size, strain, and lattice constant of cristobalite formed on heating english kaolinite

Heat treatment Temperature (°C)	Size (μm)	Strain (rms)	Lattice constant of cristobalite (nm)		
			a	b	c
1,250	73	0.4497	0.49820	0.49820	0.73540
1,350	434	0.3208	0.50168	0.50168	0.70910

cristobalite attained maximum perfection during heating silicic acid at ~1,425 °C and its interplaner spacing attained to 0.4042 nm.

According to Chakraborty and Ghosh (1978), the silica (A) which is generated out of the evolution of the metakaolinite structure contains 5–6 wt% Al_2O_3. It is presumed that this alumina should play a definite role in crystallization behavior of silica. It may further be assumed that Al atom will remain in the cristobalite structure in the tetrahedral site by replacing some Si atom. As a result, both crystallization phenomenon and lattice constants value will vary and possibly increase since the atomic size of Al is greater than that of Yamauchi and Kato (1943) claimed that occurrence of foreign cations in the cristobalite lattice may interrupt the regular layer rhythms resulting to imperfection. Tamela (1949) reviewed the earlier literature and showed the effect of heating a series of silica alumina gels to 1,160 °C by examining on X-ray analysis. In presence of alumina, a weak pattern of cristobalite is obtained. With 2 wt% alumina, pattern of cristobalite is greatly strengthened and full cristobalite pattern is visible. With more increase of alumina, the pattern of cristobalite is lost together. It may be believed that structure of cristobalite is much close to silica gel. Again cristobalite has the same structure as potassium aluminate without potassium ion in the center of the ring of aluminum tetrahedral. In co precipitated alumina silica hydrogel interfolding of silicon and aluminum ions occurs. This is occurred by the condensation of the hydroxyls existing at the interface between silica particles. The tetrahedral structure of silica being extremely stable will influence the coordination of the oxygen around the aluminum ion. The aluminate structure in its turn affects the orientation of the silica tetrahedral adjoining the aluminate structure. At the inter particle interface it is expected that well-defined cristobalite rings exist. This cristobalite nuclei act as seeds for the crystal growth observed when hydrogel of low alumina content is heated to 1,150 °C analogous to synthetic silicate gel containing low percentage of alumina. The amorphous silica which is exolved out of the decomposition of kaolinite is also containing some alumina inherently as noted during alkali extraction process as noted by Chakraborty and Ghosh (1978) in Fig. 17.1b, c.

Therefore, the reason of increase of **a** and **c** value may be due to crystallization of silica incorporated with alumina to form cristobalite on heating process.

It is explained that due to inheritance of some amount of alumina in amorphous silica, the lattice parameter of cristobalite may increase.

23.5 Summary

The possibility of crystallization of silica is not the source 980 °C exotherm, is discussed in detail from the thermodynamic and experimental point of views. Several researchers namely Glass (1954), Gernad-Hirne and Menaret (1956), Wahl and Grim (1964), Johnson et al. (1982) noted ß-cristobalite development at

>1,200 °C by XRD study and conjectured the occurrence of last exotherm is due to formation of cristobalite.

The concept of silica crystallization is supported by the following results.

(i) Chakraborty (1993) measured the amount of β-cristobalite formed in Rajmohol kaolinite heated to ~1,400 °C by quantitative XRD technique this value is found to be 20–21 wt%.
(ii) The exhibition of a contractile and a differential contraction peak in DTMA coincides with exothermic peak at 1,410 °C in DTA.
(iii) The conformation of crystallization of SiO_2(A) to β-cristobalite is shown by its mode of its formation before and after the 1,400 °C exotherm by TMA and DTMA analysis of kaolinite. The exhibition of $\alpha \leftrightarrow \beta$ cristobalite inversion at ~240 °C confirms crystallization of free silica vis–a–vis exhibition of fourth exotherm of DTA analysis of kaolinite.
(iv) Complete elimination of exhibition of the 1,420 °C exotherm, and contraction peak in DTMA vis–a–vis effectiveness silica extraction process supports the above view.
(v) Semi-quantitative XRD estimations of cristobalite of dynamic heat treated samples showed that maximum amount of cristobalite formation occur at the temperature of exotherm. However, it starts to form slowly from ~1,300 °C. Accordingly, TMA curve shows a continued contraction with rise of heating and finally a sharp contraction at 1,400 °C.

L.C. values of cristobalite of the order of 0.502 nm measured by Tomura et al. (1986) and Chakraborty (2003a, b, c, d) are found to be much higher (0.4042) when it is synthesized from silica gel (Verduch 1958). The reason of increase of **a** and **c** value may be due to crystallization of silica (A) incorporated with alumina as noted during alkali extraction process.

References

M. Bellotto, High temperature phase transformation in kaolinite: the influence of disorder and kinetics on the reaction path. Mater. Sci. Forum 166–169, 3–20 (1994)

G.R. Blair, A.C.D. Chaklader, Kaolinite-mullite series: Firing vs. reactive hot-pressing. J. Therm. Anal. **4**, 311–322 (1972)

A.K. Chakraborty, D. K. Ghosh, Re-examination of the decomposition of kaolinite. J. Am. Ceram. Soc. **60**(3–4), 165–166 (1977)

A.K. Chakraborty, D. K. Ghosh, Re-examination of the kaolinite to mullite reaction series. J. Am. Ceram. Soc. **61** 3–4, 170–173 (1978)

A.K. Chakraborty, D.K. Ghosh, Kaolinite-mullite reaction series. The development and significance of a binary aluminosilicate phase. J. Am. Ceram. Soc. **74**(6), 1401–1406 (1991)

A.K. Chakraborty, Resolution of thermal peaks of kaolinite by TMA and DTA. J. Am. Ceram. Soc.**75**(7), 2013–2047 (1992)

A.K. Chakraborty, Application of TMA and DTA studies on the crystallization behavior of SiO_2 in thermal transformation of kaolinite. J. Therm. Anal. **39**, 280.299 (1993)

A.K. Chakraborty, Cristobalite formation of an Indian kaolinite investigated by rietveld technique. Presented at the Seminar XTRI, held at DMRL, Hyderabad, India (2003a)

A.K. Chakraborty, DTA study of preheated kaolinite in the mullite formation region. Thermochima Acta. **398**(1–2), 203–209 (2003b)

A.K. Chakraborty, Effect of packing on the nature and intensities of high temperature exotherms of kaolinite. Clay Sci. **12**(2), (2003c)

A.K. Chakraborty, New data on thermal effects of kaolinite in the high temperature region. J.Therm. Anal. **71,** 799–808 (2003d)

W.H. Flank, Behavior of kaolinite pellets at elevated temperature. Clays Clay Minerals **20**(1), 1–18 (1979)

M.J.Gerad-Hirne, M.J.Meneret, Les reactions thermiques a haute temperature des kaolis at argiles kaoliniti ques. Bull. Soc. Franc.Ceram.No.30. **25** (1956)

H.D. Glass, High-temperature phases from kaolinite and halloysite. Am. Mineral. **39**, 193–207 (1954)

S.M. Johnson, J.A. Pask, J.S. Moya, Influence of impurities on high-temperature reactions of kaolinite. J.Am.Ceram.Soc. **65**(1), 31–35 (1982)

S. Lee, Y.J. Kim, H-S. Moon, Phase transformation sequence from kaolinite to mullite investigated by an energy-filtering transmission electron microscope. J. Am. Ceram. Soc. **82**(10), 2841–2848 (1999)

K. Okada, N. Ostuka, J. Ossaka, Characterization of spinel phase formed in the kaolin-mullite thermal sequence. J. Am. Ceram. Soc. **69**(10), C-251–C-253 (1986)

R. Roy, D.M. Roy, E.E. Francis, New data on thermal decomposition of kaolinite and halloysite. J. Am. Ceram. Soc. **38**(6), 198–205 (1955)

N.C. Schieltz, M.R. Soleman, Thermodynamics of various high temperature transformation of kaolinite, in *Proceedings of the 13th National Conference on Clays*, ed. by E. Ingerson (Pergamon Press, Madison, 1966), pp. 419–425. Monograph, No. 25

S.Tomura, Y.Shibasaki, H.Mizuta, T.Maeda, Yogo.Kyokai Shi. 98,917 (1986)

Y. Tsuzuki, K. Nagasawa, A transitional stage to 980°C exotherm of kaolin minerals. Clay Sci. **3**(5), 87–102 (1969)

A.g. Verduch, Kinetics of cristobaite formation from silicic acid. J.Am.Ceram.Soc. **41**(11), 427–432 (1958)

F.M. Wahl, R.E. Grim, High temperature DTA and XRD studies of reactions in *Twelfth National Conference on Clays and Clay Minerals*. pp. 69–81(1964)

T. Yamauchi, S. Kato, Thermal analysis of raw clays. J. Jpn. Ceram. Assn. **50**, 303 (1943)

H. Yamada, S. Kimura, Studies on coprecipitates of alumina and silica gels and their transformation at higher temperature. J.Ceram.Assn. Japan, **70**(3) 65–71 (1962)

A. Yamuna, S. Devanarayana, M. Lalithambika, Phase-pure mullite from kaolinite. J. Am. Ceram. Soc., **85**(6) 1409 (2002)

Chapter 24
Topotaxy IN K-MK-AL/SI Spinel-Mullite Reaction Series

Grim (1951) first indicated that high temperature transformation of a material is some cases proceeds in an orderly manner. The structure of the new phase inherits directly from the parent material. It is assumed that oxygen network of the parent material may slightly be modified during heating. Only the cations diffuse and rearrange to develop a new crystal lattice.

Bradley and Grim (1951) showed that pyrophyllite crystals retain a crystal form and yield a characteristic X-ray diffraction diagram after its dehydration. The aluminum layer for anhydrous pyrophyllite consists essentially at a serial of zig zag chain octahedral extending parallel to the a-axis or to the approximate equivalent (110) and $(1\bar{1}0)$. Octahedral chains of similar nature running parallel to the c-axis are a prominent feature of the mullite crystal structure. At the early state of heating pyrophyllite, mullite developed is more asymmetric. It is suggested that long strings of alumina, co-ordination octahedral are carried over directly from the one structure to the other.

Lotgering (1959) was first suggested the term "Topotaxy" to define all solid state transformation in which each single crystal of starting material will change into a single crystal of the product. Thus, topotaxy refers to the dimensional and structural correspondences in three axial directions of the different phases involved in three parts of transformations, e.g., a geometrical part, a textural part, and a chemical part.

24.1 Geometrical Part

It directs the way in which one crystal lattice can be transformed into another without disturbing the over all three-dimensional arrangement very much. Interrelationship of the thermal decomposition products of kaolinite was first postulated by Brindley and Nakahira (1959a, b, c) using single crystal X-ray diffraction technique and thereafter by Comer (1961), McConnel and Fleet (1970) by electron diffraction technique.

A. K. Chakraborty, *Phase Transformation of Kaolinite Clay*,
DOI: 10.1007/978-81-322-1154-9_24, © Springer India 2014

In these studies, the same kaolinite crystal is heated in stages to consecutive higher temperature and sequential phase development and its decomposition to subsequent phase/s are noted by X-ray/electron microscopy. The preferred crystallographic orientation of the newly developed spinel phase with respect to parent kaolinite is first of all noted and then atomic co-ordinates related between them are established. Fourth, what should be the probable mode of cationic and anionic migration or rearrangement of atomic species so that one can realize the cause of change of atomic co-ordinates between them.

Kaolinite on heating at 400 °C dehydroxylated to form metakaolinite and thus lost its three-dimensional regularity. They showed diffraction maxima of metakaolin to be those of only hko of the parent kaolinite, i.e., two-dimensional regularity was maintained in metakaolinite. On heating at 950 °C, they further showed that the product of heating at a temperature of 950 °C was a well-oriented spinel phases which had (111) perpendicular (001) of the parent kaolinite and (110) parallel to the original b-axis of the kaolinite. On heating at temperatures above 1,000 °C mullite appeared as a final crystalline phase. Brindley and Nakahira did not observe any preferred orientation.

Udagawa et al. (1969) depicted the continuance of oxygen packing during phase evolution of kaolinite on heating. Johns (1965) reviewed the topotactic crystallization of high temperature phases derived out of kaolinite. He summarized the earlier results of the layered sequence of atoms perpendicular to the (001) of kaolinite or metakaolinite and parallel (111) direction in the spinel phase. They showed that metakaolinite can condense by the expulsion of 5.1/3 oxygen's and 2.2/3 Si per 48 oxygen, producing a layered sequence, which with lateral migration of the remaining Si and Al produces the spinel phase.

John (1965) summarizes the topotactic transformation of kaolinite in preferred orientation. Figure 24.1 shows that at ~500 °C kaolinite experiences dehydroxylation and forms queasy crystalline metakaolinite phase whose **a** and **b** axial dimensions and orientation are essentially the same as kaolinite, but with a collapse in the **c** direction. At ~925 °C, metakaolinite layers condense to form a Al–Si defect spinel phase which also exhibits preferred orientation with its (111) direction perpendicular to (001) of kaolinite and with its (110) directions parallel to **b** of kaolinite. Above 1,050 °C the spinel phase transforms to mullite. This phase likewise is preferentially oriented with **c** axis parallel to the (110) directions of the spinel phase. He further tries to relate three sets of mullite crystals form with their c-axis aligned parallel to the (110) spinel directions. In the structure of sillimanite Fig. 24.1 the chains of Al–O octahedral and Si.Al–O tetrahedral extend parallel to the c-axis. Similarly, articulated chains run in the (110) directions in spinel. He assumed that the topotactic development of mullite from spinel involves incorporation of similar linearly disposed structural elements.

In addition to provide orientation between kaolinite to metakaolinite and to spinel phase, Comer (1961) also put forward further evidence of strong preferred orientation between spinel to mullite from single crystal electron diffraction study. He showed from Fig. 6.2 that both Al–Si spinel and mullite phase co-exist. Since the strong-oriented (002) reflections of mullite phase is found just inside the broad

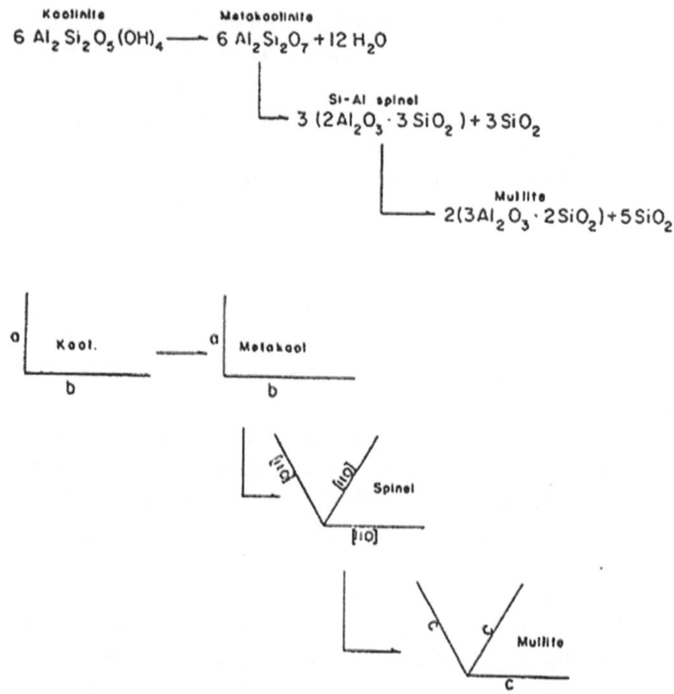

Fig. 24.1 Crystallization sequence from kaolinite showing orientation relation of phases formed (After John, 1965). "Used with permission. *ACerS Bulletin*, The American Ceramic Society"

weather (440) reflections of Al–Si spinel phase. Thus, Comer extended the axial relationship among the entire phases.

Von Gehlon (1970) measured the grain orientation of mullite by means of an X-ray Geiger counter pole figure goniometer in fired test pieces of kaolinite. It has been found that (111) plane of Al–Si spinel is in best analogy with (310) of mullite.

(111) Al–Si spinel phase ——— > (310) mullite phase

The oxygen arrangement in this (310) plane of mullite is very similar to that in the (111) plans of Al–Si spinel phase. This oriented transformation preserves/conserves essential parts of the primary crystal lattice.

Using Weissenberg and precession photographs of single crystals, Weiss et al. (1969) showed that this orientation is only one out of three, each of which occurs in six equivalent forms, differing by a rotation of 60° about the normal to the layers of the precursing kaolinite. Experimental statistics (26 crystals, precession photographs) indicate that the orientation of Brindley and Nakahira has only 45 % probability. Naturally the orientation of the mullite phase, which is formed at the next stage of the thermal treatment, is much more complicated than assumed hitherto.

Latter on by SAD study on heated kaolinites Tsuzuki et al. (1969) and McConnell et al. (1970) noted strong orientation effect between parent kaolinite and spinel phase.

Srikrishna et al. (1990) observed to grow mullite crystallites perpendicular to the plane of metakaolinite/kaolinite sheets.

Onike et al. (1986) noted that an electron diffraction pattern obtained from kaolinite sample soaked for 5 h at 965 °C showed large lens-shaped spots, which were indexed as the (220), (440), and (222) reflections of spinel phase. On orientation of the sample, they showed that the (110) reflection of spinel in electron diffraction patterns is parallel to the b-axis of the original kaolinite. Thus, preferred orientation between kaolin and spinel is conformed. On further heating of kaolinite sample soaked for 48 h at 965 °C, the lath-shaped mullite crystals shows a preferred orientation at 120° to each other.

McConville et al. (1998a, b) fired dense extruded kaolinite at 1,000 °C and showed (440) ring diffraction pattern. They concluded that texture development depends on the shaping technique used. Textured mullite morphology is obtained out of calcined kaolin powder compact obtained by extrusion and by die-pressing techniques.

Lee et al. (1999) found the orientation between spinel and mullite.

Chen et al. (2000) made powder kaolin compact by die-pressing technique. In this technique, kaolin flakes noted to lie down on the plane which is perpendicular to the die-pressing direction. On calcination the long axes of the mullite needles grow and tend to be perpendicular to the die-pressing direction. Mullite needles show preferred orientation, it is related to the preferred orientation in kaolin powder compact (Fig. 24.2).

24.2 Textual Part

During the electron microscopic study, Comefero et al. (1948) showed the followings.

(i) The retention of hexagonal kaolinite outlines at elevated temperatures indicating the continuance of structural order in anhydride of kaolinite.
(ii) Development of mullite needle in an oriented manner.

They noted that the growth of mullite has progressed sufficiently so that mullite needless is recognizable in association with undisturbed kaolinite hexagonal grain relicts. In some, such needless even appear to be growing out of a particular relict. In these cases, the elongation of the needle is parallel with an edge of the relict hexagon. In more incompletely mullitized instances, e.g., after firing at 1,300 °C, groups of mullite needless in associated growth simulate hexagonal aggregations. With these two observations they concluded that a directional dependence likely persist through the over all transition from original kaolinite grain to the mullite aggregates.

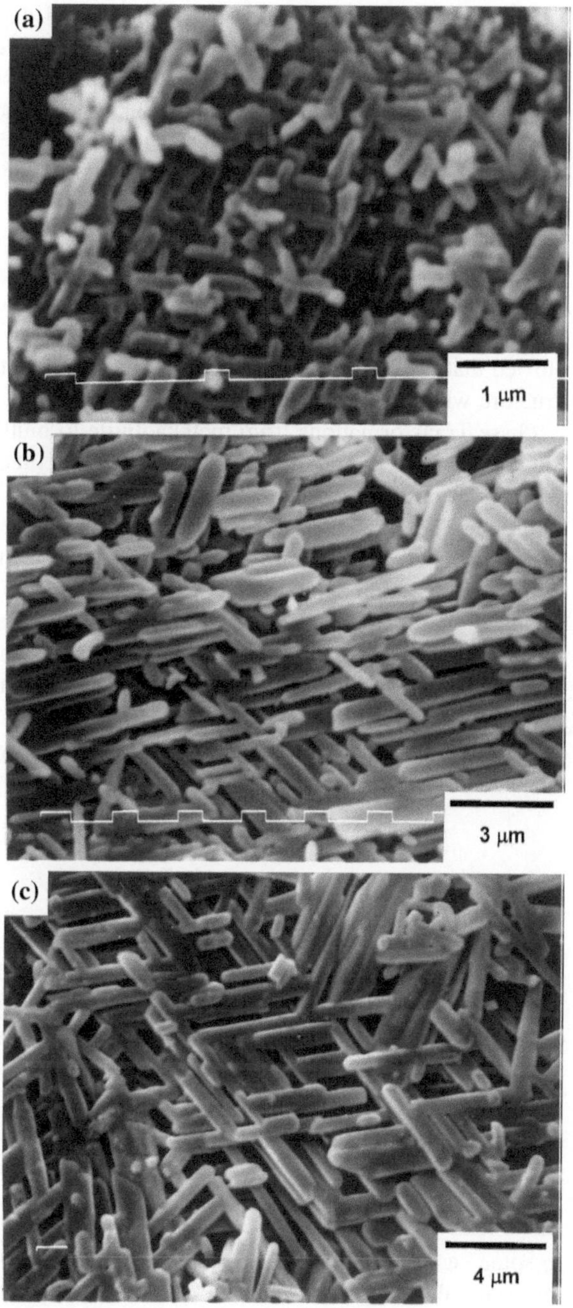

Fig. 24.2 The morphology of mullite grains on the plane perpendicular to the die-pressing direction in the specimens sintered at **a** 1,400 °C, **b** 1,500 °C, and **c** 1,600 °C for 1 h. The glassy phase was removed by acid etching (After Chen, Lan ad Tuan, 2000). Used with permission from Ceram. International

Scheuller and Comer (1975) by SEM showed that original sheet structure of kaolinite is preserved during firing after noting the electron micrograph of kaolin FP80 containing well-developed and idiomorphic crystals. Some of the aggregates with booklet textures with parallel orientation of the platelets. These stacks of crystals are retained when clay is heated to 1,200 °C or even 1,300 °C when mullitization occurred. They concluded that primary mullite as a pseudo morph after kaolinite.

By electron microscopic studies of carbon replica of heated kaolinites, Comer (1960) suggested that the growth of mullite occurs along ordered alumina octahedral chains. According to Comeforo et al. (1948) these chains are the octahedral alumina layer present in original kaolinite. Brindley and Nakahira (1959a, b, c) showed that octahedral chains also prevailed in the intermediate Al–Si spinel type structure which thereafter proceeded to mullite on heating.

Phase Transformations from metakaolinite to mullite investigated by EF-TEM and HRTEM was carried out by Lee et al. (2001). Electron beam heating gives rise to a similar effect as furnace heating on phase transformation from metakaolinite to mullite. They predicted that beam heating gave rise to production of structurally more disordered silica(A). Both diffraction and EDS studies qualitatively indicates the formation of spinel phase and evolution of silica. It is difficult to determine the composition of spinel phase as it sits on unconverted metakaolinite and lots of free silica using a probe of 4 nm diameter. According to them there is a possibility that the spinel-type phase has lower symmetry than the suggested cubic symmetry as suggested by Brindley and Nakahira (1959a, b, c). HRTEM images of mullite crystals demonstrate their random orientations with respect to the parent metakaolinite.

24.3 Chemical Part

Chakraborty and Ghosh (1978a, b, c) first established the persistence of Si–O–Al bonds in the entire reaction series of K–Mk–Al–Si spinel–Mullite by their alkali extraction study. This view was confirmed by Moya et al.(1985), Rincon et al. (1986), Okada et al. (1986), and finally by Srikrishna et al. (1990).

24.4 Summary

In some cases of solid state transformation, sometime single crystal of a starting material will change into a single crystal of the product. Following interrelationship of the thermal decomposition products of kaolinite was observed. First postulation is that of Brindley and Nakahira (1959a, b, c) using single crystal

X-ray diffraction technique and which is supported by McConnel and Fleet (1970) by electron diffraction technique by Onike et al. (1986).

(i) Kaolinite–Metakaolinite–Al–Si Spinel.

It was apparent that transformation of kaolinite to metakaolinite to cubic spinel phase proceeds in an orderly manner. Structural studies indicate that metakaolinite maintaining two-dimensional regularity. It shows that the cubic axis (111) is perpendicular to (001) of kaolinite and cubic face diagonal (110) is parallel to b of kaolinite. However, the experimental statistics indicate that the probability of the above orientation is only 45 % (Weiss et al. 1970).

Second interrelationship was put forward by Comer (1961).

(ii) Al–Si Spinel–Mullite

Reflection (002) of mullite is found parallel to (440) of spinel phase. The topotactic transformation of kaolinite in preferred orientation was reviewed the literatures were summarized by John (1965).

SEM study indicates primary mullite as a pseudo morph after kaolinite (Scheuller and Kromer, 1975).

TEM study shows the development of mullite in an oriented manner. The growth of mullite occurs along ordered alumina octahedral chains (Comer, 1960). These chains are the octahedral alumina layer present in original kaolinite Comeforo et al. (1948). The octahedral chains also prevailed in the intermediate Al–Si spinel type structure which thereafter proceeded to mullite on heating (Brindley and Nakahira, 1959a, b, c).

Mullite needles show preferred orientation, and it is related to the preferred orientation in kaolin powder compact made by extruding or die-pressing technique (McConville et al. 1998a, b) and (Chen et al. 2000). However, in HRTEM study mullite crystals demonstrate their random orientations with respect to the parent metakaolinite (Lee et al. 2001).

Finally, alkali extraction study shows the persistence of Si–O–Al linkages among the phases in the entire reaction series of Kaolinite–Metakaolinite–Al–Si Spinel–Mullite, and strengthens the interrelation aspect.

References

W.F. Bradley, R.E. Grim, High-temperature thermal effects of clay and related materials. Am. Miner. **36**(3/4), 182–201 (1951)

G.W. Brindley, M. Nakahira, The kaolinite–mullite reaction series: I, a survey of outstanding problems. J. Am. Ceram. Soc. **42**(7), 311–314 (1959a)

G.W. Brindley, M. Nakahira, The kaolinite–mullite reaction series: II, metakaolin. J. Am. Ceram. Soc. **42**(7), 314–318 (1959b)

G.W. Brindley, M. Nakahira, The kaolinite–mullite reaction series: III, the high-temperature phases. J. Am. Ceram. Soc. **42**(7), 319–324 (1959c)

A.K. Chakraborty, D.K. Ghosh, Study of phase transformation of Al_2O_3–SiO_2 gel and kaolinitic clay. Trans. Ind. Ceram. Soc. **37**(5), 192–200 (1978a)

A.K. Chakraborty, D.K.Ghosh, Comment on the interpretation of the kaolinite–mullite reaction sequence from infra–red absorption spectra. J. Am. Ceram. Soc. **61**(1–2), 90–91 (1978b) (Reply by H.J. Percival. ibid., 61 (1–2) 91 (1978))

A.K. Chakraborty, D.K. Ghosh, Re-examination of the kaolinite to mullite reaction series. J. Am. Ceram. Soc. **61**(3–4), 170–173 (1978c)

C.Y. Chen, G.S. Lan, W.H. Tuan, Micro structural evolution of mullite during the sintering of kaolinite powder compacts. Ceram. Int. **26**(7) 715–720 (2000)

J.J. Comer, Electron microscope studies of mullite development in fired kaolinite. J. Am. Ceram. Soc. **43**(7), 378–384 (1960)

J.J. Comer, New Electron-optical data on the kaolinite–mullite transformation. J. Am. Ceram. Soc. **44**(11), 561–563 (1961)

J.E. Comeforo, R.B. Fischer, W.F. Brandley, Mullitization of kaolinite. J. Am. Ceram. Soc. **31**(9), 254–259 (1948)

D.R. Dasgupta, Topotactic transformations. Ind. J. Earth Sci. **1**(1), 60–72 (1974)

R.E. Grim, *Clay Mineralogy*. (McGraw-Hill Book Co., New York, 1953)

W.D. Johns, High-temperature phase changes in kaolinites. Mineral. Mag. **30**(222), 186–198 (1953)

S. Lee, Y.J. Kim, H-S. Moon, Phase transformation sequence from kaolinite to mullite investigated by an energy-filtering transmission electron microscope. J. Am. Ceram. Soc. **82**(10), 2841–2848 (1999)

S. Lee, Y.J. Kim, H.J. Lee, H-S. Moon, Electron-beam-induced phase transformations from metakaolinite to mullite investigated by EF-TEM and HRTEM. J. Am. Ceram. Soc. **84**(9), 2096–2098 (2001)

F.K. Lotgering, J. Inorg. Nucl. Chem. **9**, 113 (1959) (in J.D. Bernal; Topotaxi. Schw. Archiv. Jharb., 26, 69)

J.D.C. McConnell, S.G. Fleet, Electron optical study of the thermal decomposition of kaolinite. Clay Miner. **8**, 279–290 (1970)

C.J. McConville, W.E. Lee, J.H. Sharp, Micro structural evolution in fired kaolinite. Br. Ceram. Trans. **97**(4), 162–168 (1998a)

C. McConville, W.E. Lee, J.H. Sharp, Comparison of micro structural evolution in kaolinite powders and dense clay bodies. Br. Ceram. Proc. **58**, 75–92 (1998b)

J.S. Moya, C.J. Serna, J.E. Igesias, On the formation of mullite from kandites. J. Mater. Sci. **20**, 32–36 (1985)

K. Okada, N. Ostuka, J. Ossaka, Characterization of spinel phase formed in the kaolin–mullite thermal sequence. J. Am. Ceram. Soc. **69**(10), C-251–C-253 (1986)

F. Onike, G.D. Martin, A.C. Dunham, Time–temperature–transformation curves for kaolinite. Mater. Sci. Forum **7**, 73–82 (1986)

J.M.A. Rincon, G. Thomas, J.S. Moya, Micro structural study of sintered mullet. J. Am. Ceram. Soc. **69**(2), C-29-C-31 (1986)

K.H. Schuller and H. Kromer, Primary mullite as a pseudomorph after kaolinite. in *Proceedings of the International Clay Conference* (Mexico City, 1975). ed. by S. W. Bailey, (Applied Publishing, Wilmette, IL, 1976) p. 533–538

K. Srikrishna, G. Thomas, R. Martinez, M.P. Corral, S. Aza, J.S. Moya, Kaolinite–mullite reaction series: A TEM study. J. Mater. Sci. **25**, 607–612 (1990)

Y. Tsuzuki, K. Nagasawa, A transitional stage to 980°C exotherm of kaolin minerals. Clay Sci. **3**(5), 87–102 (1969) (Kaolin Minerals)

S. Udagawa, T. Nakada, M. Nakahira, Molecular structure of allophane as revealed by its thermal transformationin, in *Proceedings of the International Clay Conference*, ed. by Editor-in-chief Lisa Heller bisa Heller, vol. 1 (Israil University Press, Gerusalem, 1969), p. 151

K. Von Gehlen, Oriented Formation of Mullite from Al–Si spinel in the transformation series Kaolinite–Mullite. Ber. Deut. Keram. Ges. **39**(6), 315–320 (1962)

A. Weiss, K.J. Range, J. Russow, *The Al,Si-Spinel Phase from Kaoinite (Isolation, Chemical Analysis, Orientation and Reactions to Its Low-Temperature Precursors)*, In Proceedings of the International Clay Conference, Tokyo, 1969, vol. 2 (Israel Universities Press, 1970), pp. 34–37

A. Weiss, K.J. Range, J. Russow, The Al,Si-Spinel phase from kaoinite (Isolation, chemical analysis, orientation and reactions to its low-temperature precursors), In *Proceedings of the International Clay Conference*, Tokyo, 1969, vol. 2 (Israel Universities Press, 1970), pp. 34–37

Chapter 25
Dehydroxylation Mechanism

25.1 Introduction

There are two types of mechanism for dehydroxylation of kaolinite. In homogeneous dehydroxylation, the elements of water are lost more or less uniformly from all over the crystal. In inhomogeneous dehydroxylation, no oxygen is lost from those parts of the crystal where oriented conversion to a crystalline product occurs; the oxygen for the expected water comes directly from other regions of the crystal which are wholly or partly converted into pores, while various migration of cations occurs (Taylor 1962). In this mechanism, there are three main constitute of processes: (a) cation migration; (b) formation of water molecules in the donor regions and their subsequent expulsion; and (c) repacking of the oxygen framework within the acceptor regions.

With aluminum minerals, all the above three processes do not occur simultaneously. For example, cation migrations and water loss occur easily, but repacking of oxygen framework occurs with difficulty. As a result, metastable intermediate phases are formed over a wide temperature range. Bohemite, whose structure is based on a framework of oxygen is approximately in cubic close packing. At 500 °C it is dehydroxylated; the product is γ-Al_2O_3. A metastable intermediate phase in which the oxygen is still in cubic close packing. Between 500 and 1,000 °C, further movements of cations occur and probably also minor changes in oxygen packing. Above 1,000 °C, there is a major changes in the hexagonal close packing, and corundum (α-Al_2O_3.) is formed as a stable phase. Analogous dehydroxylation mechanism may occur in case of kaolinite since metastable intermediate phase, e.g., metakaolinite is formed and persisted over a temperature range of several hundred degrees.

According to De Keyser (1963e), the metastable metakalinite phase is stabilized by OH groups. Similar way of stabilization was noticed with addition of small quantity of Na_2O. Dehydration/dehydroxylation of kaolinite consists of four stages: (i) Removal of physically adsorbed water, (ii) principal dehydration, (iii) progressive dehydration of partial OH groups over a long range of temperature, and (iv) sudden dehydroxylation of residuary water. He noted an endothermic peak at ~950 °C in thermogravimetric differential analysis just before the occurrence of crystallization exotherm (Fig. 25.1). There is a direct relationship

A. K. Chakraborty, *Phase Transformation of Kaolinite Clay*,
DOI: 10.1007/978-81-322-1154-9_25, © Springer India 2014

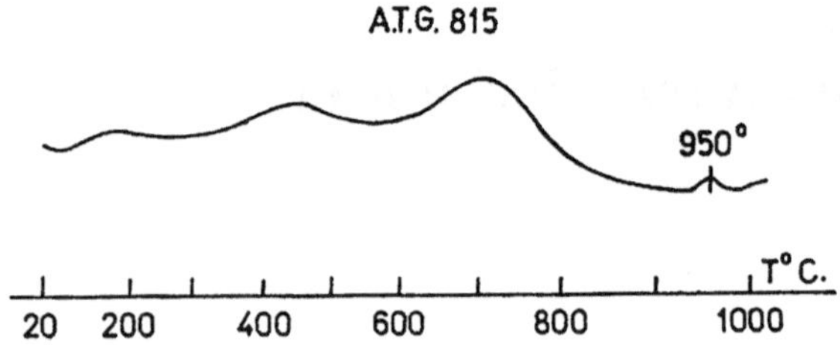

Fig. 25.1 Differential thermogravimetric analysis of Zettlitz kaolin after heating 650 °C during 14 h (after DeKeyser 1963d). Reproduced with kind permission of The Clay Minerals Society, publisher of Clays and Clay Minerals

between exhibition of this endothermic phenomenon and disappearance of the 3620 and 3420 cm^{-1} bands immediately after exothermal reaction (Fig. 7.2). This is most important clue point in dehydroxylation mechanism cum phase transformation process of kaolinite.

Pampuch (1958) laid emphasis on the role of mobile protons in dehydroxylation of kaolinite. Brindley and Nakahira (1959b, c) interpreted a homogeneous dehydroxylation based on single crystal study. Later on Taylor (1962) suggested an inhomogeneous mechanism of full dehydroxylation of kaolinite. To understand the possible mechanism of dehydroxylation, it is essential now to look into some important questions:

(i) The nature of hydroxyl groups linked to octahedral aluminum.
(ii) The products of dehydroxylation.
(iii) To locate where actually dehydroxylation taking place.
(iv) Whether it occurs sharply or slowly?
(v) To note the magnitude of water loss.

According to Pampuch (1958), the mobility of protons is most important which influence dehydroxylation reaction. There are two layers of –OH group in kaolinite. First, the top layer, contains only hydroxyl (4OH/molecule). Second, the layer (inside) contains only oxygen and hydroxyl (2O and 2OH/molecule). “Outer” hydroxyl is situated at 0.189 nm apart from the aluminum and its fundamental stretching mode gives rise to an absorption band at 3692 and 3648 cm^{-1} in I.R. spectrum. “Inner” hydroxyls are situated at 0.195 nm apart from the aluminum and give rise to an absorption band at 3620 cm^{-1}. Question is which of two hydroxyl groups would take part in the dehydroxylation first?

The ideal Al–O distance is 0.191 nm, the shortened distance between the “Outer” hydroxyls and aluminum must be ascribed to a strong polarization of the O–H bond and an increasing negative charge in the direction of Al. Due to this, strength of O–H bonds is weakened and under the influence of thermal vibrations ¾ of the protons associated with the “Outer” hydroxyl are more apt to separate from the oxygens than the remaining ¼ proton present in “Inner” hydroxyls. Pampuch shows that preferentially the intensity of 3692 cm^{-1} band associated with “Outer” O–H bonds decreased in contrast to 3620 cm^{-1} band persisted in sample heated to temperatures above 450–550 °C (Fig. 19.3). Owing to the persistence of ¼ of the OH groups to higher temperatures, Pampuch thought that only ¾ of the protons of kaolinite may take part in counter diffusion during main dehydroxylation process above 450 °C. This deficiency is very acute in kaolinite, since a counter diffusion of Al^{+3} demands more protons flowing in the opposite direction to maintain electro neutrality. The lack of mobile protons hinders, therefore, the formation of stable dehydroxylation products in kaolinite Fig. 25.2. Owing to the limited number of mobile protons, the actually formed metastable phases are also only poorly crystalline.

Pampuch (1958) describes the dehydroxylation of kaolinite as inhomogeneous one and it consists of two steps. In the first step, proton migrates from the acceptor regions to donor regions. For the maintenance of electrical neutrality, counter diffusion of Al^{+3} from donor to acceptor regions takes place as a result of formation of water along with the liberation of SiO_2(A).

After first stage of dehydroxylation when ¾ of protons take part, dehydroxylation temperature increases from 450–550 to 750 °C. In the second step, residual ¼ protons take part in dehydroxylation and form Al–Si spinel with no further liberation of SiO_2.

To prove importance of protons in dehydroxylation of kaolinite, he carried out decomposition of kaolinite in the excess of protons and showed dehydroxylated product at earlier temperature (Fig. 25.3).

He introduced additional small amount of mobile protons (of the order of 0.5 %) by exchanging the protons for potassium. H-kaolinite thus obtained showed an earlier formation of Al–Si spinel (X-ray) as demonstrated by the lack of a slower dehydroxylation between 600 and 700° which is observed in natural kaolinite. This is associated with the formation of metastable semihydrophase (metakaolinite).

The catalytic effect of additional amounts of proton on dehydroxylation of kaolinite was further substantiated by hydrothermal experiment. In this case, the onset of dehydroxylation is shifted to lower temperature, and dehydroxylation products are crystalline. This effect is ascribed to the more availability of sufficient amounts of mobile protons in contrast to usual finding that increase of water pressure results to an increase of the temperature of dehydroxylation.

Acceptor region — Donor region

$2Al_4[Si_4O_{10}](OH)_8$ — $2Al_4[Si_4O_{10}](OH)_8$

-------- $12H^+$ ----------> 450 – 550°C

<------- $4Al^{3+}$ ----------- (83.3% dehydration)

$2Al_6[Si_4O_{10}]O_6(OH)_2$ + $4SiO_2$ + $10H_2O$

$Al_6[Si_4O_{10}]O_6(OH)_2$ — $Al_6[Si_4O_{10}]O_6(OH)_2$ 750°C (100% dehydroxylation)

--------- $2H^+$ --->

<-------- 2/3 Al^{3+} ----

$9/_{16}.Si_{7.1/3}\,Al\,11.2/_3\,[\,]_5\,O32$ — $1/_2Si_8Al_{10.2/3}[\ ]_{5.1/3}O_{32}$ + H_2O

Fig. 25.2 Mechanism of thermal decomposition of kaolinite (Pampuch 1958)

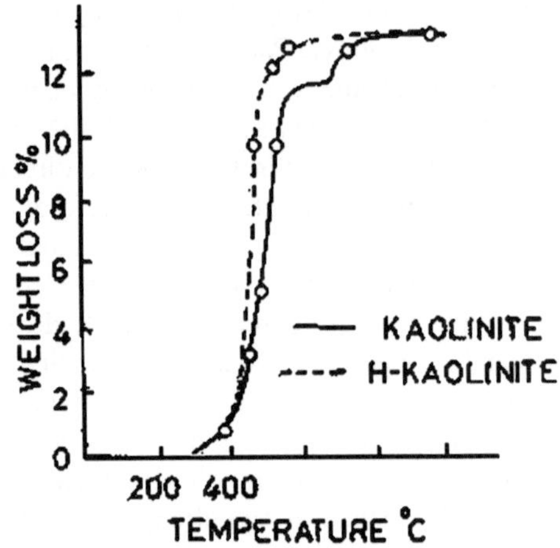

Fig. 25.3 Weight losses during the dehydroxylation of natural kaolinnite and H-kaolinite (After Pampuch 1958)

Brindley and Nakahira (1959) proposed homogeneous dehydroxylation mechanism for kaolinite to metakaolinite transformation as follows: (Fig. 25.4).

Taylor (1959b, c) has proposed inhomogeneous mechanism for metakaolinite formation. The essential feature is that the crystalline order in metakaolin resides chiefly in the oxygen packing. No oxygen is lost from the acceptor region and the type of oxygen packing in these regions tends to remain the same as in the original kaolinite. Additional cations are incorporated into the structure to make up for the protons that have been lost. All the cations, both Si^{+4} and Al^{+3} could occupy tetrahedral sites, but these are distributed in a largely random manner among the available tetrahedral interstices in the oxygen framework. *Taylor* assumed that both Al^{+3} and Si^{+4} ions migrate from donor to acceptor regions, so that the donor regions are completely destroyed. The process was represented as follows:

---- 6O		----6O		-----6O	
				3Si	
--- 6O+6OH		---- 2O		---- 6O	
4Al	500°C	4Al	925°C	4Al	1075°C
--- 6O+2H	-----→	---- 6O	------→	---- 6O	------→ $(Al_9Si)_6O_{9.6\text{-}10.0}$
4Si		4Si		3Si	
					Mullite like phase
----- 6O		---- 6O		--- 6O	+ SiO_2 (cristobalite)
				4Al	
---- 6O +6OH		----- 2O		--- 6O	
Kaolinite		Metakaolinte + H_2O		Al – Si spinel + SiO_2	↓
					$Al_{4.8}\,SiO_{1.2}\;O_{9.6}$ Mullite + SiO_2 (crist)

Fig. 25.4 Thermal transformation of kaolinite as suggested by Brindley and Nakahira (Referred by Taylor 1962). Reproduced with kind permission of The Clay Minerals Society, publisher of Clays and Clay Minerals

Accepter regions :

$$7Al_2Si_2O_9H_4 \dashrightarrow 7Al\,2.^4/_7\,Si\,2.^4/_7O_9\,H_4$$

$$4Al^{+3} + 4Si^{+4} \uparrow\downarrow 28H^+$$

Donor regions : $$2Al_2Si_2O_9H_4 \dashrightarrow 18H_2O$$

Taylor (1962) assumed metakaolin so formed fully dehydrated into an anhydrous phase. Sufficient cation migrations occur with sufficient determination into Si- and Al-rich regions at the endothermic reaction at ~600 °C with loss of water at the donor regions. Sufficient energy for cation migration, probably occur when metakaolin forms, but not for any sufficient change in oxygen packing. At the exothermic reaction (950 °C), sufficient oxygen packing changes to give Al–Si spinel. But the change is limited in that the structure is still based on nearly close packed layers. Above 1,000 °C more profound changes in oxygen packing can occur and mullite is formed.

25.2 Newer Explanation on Dehydroxylation Mechanism

Based on the results of DeKeyser (1963e) as shown above, *Chakraborty* suggested a different picture of dehydroxylation of kaolinite as shown below. In fact, some of the conceptions made by Pampuch (1958) and Taylor (1962) do not concede with experimental results as given below:

(i) Silica (A) is not liberated during dehydroxylation reaction as happens usually at ~500–600 °C as shown earlier (Fig. 17.1a). Structure of intermediate metakaolin phase is stabilized by the presence of some quantity of water molecule.
(ii) When structural coherency and orientation relationship are maintained from parent kaolinite to metakaolinite it is questioned, how the assumption of *Pampuch* that Al atoms take part in the diffusion process during dehydroxylation of kaolinite? Taylor (1962) conceived the migration of both Al and Si ions during dehydroxylation process at ~500 °C from donor to the acceptor region, then the original sequence like –O–Si–O–Al–O–Si would be changed instead of remaining intact which is most essential requirement for the development of oriented metakaolinite and or Al -Si spinel phase. Experimentally, it is noted that at ~500–600 °C when metakaolinite is formed, it is not fully dehydrated as conceived by these authors.

Further, *he* differentiated the anhydro metakaolinite phase into Si- and Al-rich regions between ~500 and 600 °C prior to spinel formation. If it really happens then a portion of silica would dissolved out preferentially from Si-rich portion of metakaoinite. This is not occurring in practice in NaOH extraction process of silica out of heated kaolinites. Silica (A) liberates only once at 980 °C exothermic reaction process by structural decomposition of metakaolinite by expulsion of last portion of OH groups (Fig. 17.1b, c). This is the second most important clue point in phase transformation process of kaolinite.

Dehydroxylation at Endo – Exo Assembly Region

Accepter region	Donor region
$[(Si_4O_{10})Al_4O_{3.5}OH]$	$[(Si_4O_{10})Al_4O_{3.5}OH]$
Metakaolin (Hydrous)	Metakaolin (Hydrous)
(Containing ~1.3 % H20)	(Containing ~1.3 % H20)

i) 1st step , Proton migration begins with (at endothermic dip just before 980°C)

----------->   H2O

<------------

ii) 2nd step , Al^{+3}, Cat ion diffusion starts in opposite direction.

iii) 3rd step , Si^{+4} starts diffusion with

------------------> -

counter diffusion of Al^{+3} in reverse direction .
Redistribution of cat ions ensues.

<--------------------

iv) 4th step, phase separation occurs

--------------- 6O
4Al
------------- 6O

Al /Si (Mixed cat ions)
----------- 6O
4Al
---------- 6O
Al/Si (Mixed cat ions}
----------- 6O

Aluminosilicate(A) phase (Anhydrous & Unstable)
It immediately starts repacking of oxygen . AlO_4 changes to AlO_6.
(Composition = 3:2 mullite)

Silica(A) phase (Present in domains and develops phase phase boundary)

↓

Crystallization at 1st exotherm.

i) Al – Si spinel (Observed as cube/dots in matrix of silica)
ii) Mullite (trace, weakly crystalline)
iii) Mullite (A) (Residual phase)

Based on dehydroxylation of kaolinite, alkali leaching study regarding the detection of extent of silica liberation different temperature of its heating are concerned, Chakraborty conceived two types of dehydroxylation mechanism which may occur consecutively in kaolinite.

(i) Up to ~87 % dehydroxylation, the mechanism may be homogeneous. It is assumed that each particle of kaolinite has reacted to the extent of ~87 % with formation of a partially dehydrated phase during the formation of first endotherm at ~500–600 °C.
(ii) The mechanism of rest ~1.3 % dehydroxylation may be inhomogeneous. Final dehydroxylation is started at ~950 °C (endothermic dip) where actually proton migration starts at the first step from acceptor to the donor region and Al ion may migrate in the opposite direction for electro neutrality point of view.

In the second step a large scale diffusion of Si ion and counter diffusion of Al ion may start resulting to generation of two separate phase boundary regions. The anhydrous aluminosilicate (A) phase is unstable. It immediately starts repacking of oxygen. Some of AlO_4 groups changes to AlO_6. Its composition becomes equivalent to 3:2 mullite. Phase separation becomes complete. The donor region contains silica (A) phase and the other phase starts crystallization into stable cubic Al–Si spinel phase.

The above assumption is based on TMA curve of a kaolinite clay. It shows two major contractions at the endo–exo region (Figs. 3.1, 3.4 and 3.7). First one (B′) may be due to collapsing of the metakaolinite layers followed by final dehydroxylation into silica rich and alumina rich phases simultaneously. The second (C′), the more steep contraction would be due to consecutive crystallization of major quantity of Al–Si spinel and minor amount of weakly crystalline mullite phase out of alumina rich phase. At the end, silica rich phase and residual mullite(A)phase are remaining as major quantity of amorphous phases and for that both X-ray and electron microscopy studies show dark and diffuse background.

The essential feature of the transformation process is that SiO_2 liberation is taking place after final stage of dehydroxylation. Secondly, Al–Si spinel phase formation is inevitable. Question is: why not mullite crystallizes to a large extent here and instead of Al–Si spinel and why there is abnormality?

Because $(Si_2O_5)^{-2}$ rigid chain which was present in metakolinite phase is most probably a controlling factor for Al–Si spinel phase nucleation. During formation of spinel of mullite like composition, large concentration of available Si would surely tend to fill up 4.9 sites and makes 3.1 sites for Al out of 8 tetrahedral sites of cubic spinel structure. It is expected that Si–O–Si bonds existing in metakaolinite structure should continue up to spinel structure. On the other hand, Si–O–Al bondings are prevailing in mullite structure. Thus, it is necessary to heat further to convert Al–Si spinel phase to mullite where more of Al substitutes Si and as a result $Al0_6$–$Al0_4$ would decrease. Therefore, the existence of Si–O–Si linkages and/or presence of silica sheet hinders direct formation to mullite out of kaolinite. According to thermodynamic calculation of free energy and lattice energy mulllite should crystallize out. However, crystallization takes place in aluminous region and a little quantity of weakly crystalline mullite forms as noticed by a small exotherm (Fig. 3.1) and a large quantity of Al–Si spinel following a sharp exotherm subsequently. The siliceous regions remain intact. This indicates the hindrance of crystallization of siliceous phase to cristobalite in presence of alumina in the silica rich phase.

It is postulated by Chakraborty and Ghosh (1991) that as bonding was originally present between cations like Si to Al via oxygen as in kaolinite, then on heating such compound, high melting constituents, e.g., Al_2O_3 would not crystallize out. On the other hand, it will transform into other binary compounds, e.g., intermediary silicon bearing spinel phase and to mullite finally.

25.3 Summary

Dehydroxylation processes of kaolinite have been argued among various authors. Mechanisms of these processes may be partly or fully homogeneous or inhomogeneous.

Homogeneous dehydroxylation mechanism for kaolinite to metakaolinite transformation was proposed by Brindley and Nakahira (1959b, c). However, the structure of it is fully dehydrated.

Contrary, Pampuch (1958) describes the dehydroxylation of kaolinite as inhomogeneous one and it consists of two steps. About ¾ of protons take part in the first stage of dehydroxylation at the temperature range from 450–550 to 750 °C. In the second step, residual ¼ of protons take part in dehydroxylation and form Al–Si spinel with no further liberation of SiO_2 which seems unlikely considering estimation of silica by Chakraborty and Ghosh (1978).

Inhomogeneous mechanism for metakaolinite formation was proposed by Taylor (1962) based on the assumption that both Al^{+3} and Si^{+4} ions migrate from donor to acceptor region, the donor region is lost and the structure is fully dehydrated.

A new dehydroxylation mechanism is put forward below by Chakraborty:

There are three steps of dehydroxylation processes. The "outer hydroxyl" layers amounts to ¾ part of the protons first takes part in the dehydroxylation process during the occurrence of endotherm in the temperature range 500–650 °C. This accounts to ~87 % dehydroxylation. The mechanism may be homogeneous. Since no cation migration leading to formation of silica takes place and structural order is still continued in this dehydration stage.

The rest ~13 % dehydroxylation occurs from remaining "inner hydroxyl" layer amounts to ¼ part of the protons. Nature of dehydroxylation consists of two remaining steps:

A gradual dehydration starts from the endothermic temperature (~650 °C) to the beginning of endothermic dip (~950 °C) as per observation of TGA trace.

Last part of dehydroxylation occurs sharply during endothermic dip noted just before first exotherm.

At the endo–exo region, dehydroxylaton mechanism of the last part of removal of OH groups may be inhomogeneous where cation migration preferably Al^{+3} initiates to move to the donor region in exchange of movement of 3 protons (H^{+}) in the opposite direction to maintain electro neutrality. Just after dehydroxylation, a profound change in oxygen packing may occur. It is further assumed that even after completion of final stage of dehydroxylation, diffusion of two cations continued in the changed oxygen packing structure. It is expected that diffusion of Si^{+4} and counter diffusion of Al^{+3} may lead to generate two separate anhydrous regions namely siliceous region at the acceptor side and aluminous region at the donor side after the final stage of dehydroxylation of kaolinite.

It is probable that some sorts of crystallization of a phase/phases may become imminent at this stage in the two separate regions those would correspond to lowest lattice energy values. Obviously, mullite is the favored one out of alumina rich side and crystallization of quartz or cristobalite in the other, i.e., silica rich regions. In practice, a trace of weakly crystalline mullite is noted, none of silica bearing phases is crystallized. On the contrary, a considerable quantity of Al–Si spinel phase is generated. These results indicate that likely the acceptor part becomes more silica rich phase instead of free silica as assumed by various researchers which contain some amount of alumina. This alumina hinders crystallization of silica at ~1,000 °C. The alumina rich phase contains silica and the global composition may correspond to composition nearly equal to composition of 3:2 mullite. Perhaps, it may contain some Si–O–Si bonds as it is developing out of metakaolinite containing $(Si_2O_5)^{-2}$ rigid hexagonal chain structure. This favors crystallization of Al–Si spinel instead of mullite. Since mullite contains Si–O–Al bondings in its structure. Thus, following sequential events are taking place during the final stage of dehydroxylation process.

At the final stage, dehydroxylation occurs by movement of proton and counter movement of aluminum. DTA shows endotherm marked as D in DTA trace (Fig. 1.3). So mechanism is inhomogeneous. Collapse of sheet structure occurs. Oxygen packing changes. The matrix becomes fully disordered. Phase separation begins to form two separate regions, one siliceous(as shown in leaching study) and other is aluminous region. Thus, dehydroxylation at endo–exo assembly region is explained.

References

G.W. Brindley, M. Nakahira, The kaolinite–mullite reaction series: II, metakaolin. J. Am.Ceram.Soc. 42(7), 314–318 (1959b)

G.W. Brindley, M. Nakahira, The kaolinite–mullite reaction series: III, the high-temperature-phases. J. Am. Ceram. Soc. 42(7), 319–324 (1959c)

A.K. Chakraborty, D.K. Ghosh, Re-examination of the Kaolinite to Mullite Reaction Series. J. Am. Ceram. Soc. **61**(3–4), 170–173 (1978)

A.K. Chakraborty, D.K. Ghosh, Kaolinite–Mullite Reaction Series. The Development and Significance of a Binary Aluminosilicate Phase. J. Am. Ceram. Soc. **74**(6), 1401–1406 (1991)

W.L. De Keyser, R. Wollast and L. De Laet, Contribution to the study of OH groups in kaolinminerals, Intl. Clay Conf., Pergamon Press. 75–86 (1963d)

W.L. De Keyser, Note Concerning the Exotherm Reaction of Kaolinite and Formation of Spinel Phase Preceding that of Mullite. International Clay Conference, (Pergamon Press 1963e), pp. 91–96

R. Pampuch, Mechanism of topotaxial thermal decomposition reactions of layer lattice silicates and hydroxides. Proceedings of the 9th Conference Silicate India, Budapest, 1958, pp. 144–151

H.F.W. Taylor, "Homogeneous and Inhomogeneous Mechanisms in the Dehydroxylation of Minerals", Clay Miner. Bull. 528045-55 (1962)

Chapter 26
Final Conclusion on the Thermal Effects of Kaolinite

Before setting out to a conclusion of the critical analysis of structural transformation of kaolinite as described in Chaps. 19–25, it is important to recapitulate the past and present scenario on thermal decomposition of it. Kaolin at high temperature finally transforms to yield stable mullite phase and β-cristobalite as per the prediction of equilibrium phase diagram of the SiO_2-Al_2O_3 system. Many investigations explained the fundamentals of the breakdown of and the subsequent recrystallization of it to mullite. Although intermediate steps of reaction have been extensively studied the interpretations are full of controversies. It is reviewed that even with the application of many recent experimental techniques to resolve one of the most interesting crystallo–chemical problems of K-M reaction series, a few areas are yet to be solved or there is absence of general agreement on various issues.

1. Earlier problems

(i) Metakaolin, its nature and structural characteristics. It is amorphous to X-ray and electron diffraction. Standard methods are inapplicable to analysis of its structure. It is forced to rely upon indirect evidences for its elucidation.
(ii) Crystal structure and chemistry of cubic spinel phase. There is a great disagreement over the nature of it and the temperature at which transformation takes place.
(iii) The radial Patterson Function is suited for studying the structural aspects of weakly crystalline transitional spinel phase. Secondly, systematic determination of relative amount of AlO_4 group throughout the thermal evolution of kaolinite by XRF will provide a complementary check of RED result. Unfortunately, the coefficients of 980°C exothermic reaction products of kaolinite are unavailable.
(iv) Composition of primary mullite and whether the ratio of Al and Si in mullite changes with increasing temperature is a question. Determination of lattice parameter of primary mullite may be erroneous. Formation of mullite of tetragonal variety (t-mullite) at $\sim$1000°–1050°C requires complementary proof.

A. K. Chakraborty, *Phase Transformation of Kaolinite Clay*,
DOI: 10.1007/978-81-322-1154-9_26, © Springer India 2014

(v) Reason of 980°C exothermic event. Identification of reaction products that results in exothermic reaction is an essential step. Is it due to spinel phase? Whether a fraction of mullite phase also contributes to this exotherm. Or is nucleation of mullite solely responsible for the exotherm? Interpretation of it is still a matter of speculation. It is also believed that γ-Al_2O_3 is not a factor in the mechanism. It is even conjectured that evolution of energy with exhibition of characteristic exotherm, with concurrent and marked decrease in solubility, are indicative of stable crystallization phenomenon of mullite. Exothermic event is also demonstrated as of two-fold origin, i.e., both nucleation of mullite and formation of γ-Al_2O_3 is the reason behind exotherm. Moreover, there is absence of accurate and reproducible value of enthalpy data of 980°C exotherm.

(vi) Endothermic dip (D) preceding first exotherm plays an important role in the transitional stage to exothermic reaction. This is generally ignored. General DTA and TGA curves do not provide convincing information in this matter.

(vii) The second and third exothermic peak temperatures are sensitive to the presence of impurities available in the origin of kaolinite. The fluxing effect of interlayer cations, e.g., potassium, sodium present in mica and illite inhibit the crystallization of cristobaite. Thus, naturally occurring kaolinites contain chemical and mineralogical impurity oxides which influence the thermal behavior of clays and likely change the DTA traces. Due to this reason, a large variation of results are seen in the literature.

(viii) Characteristics of natural kaolinites and their differences in their structural order play a role in their transformation behavior. Accordingly, variations are noted in the occurrence of phases in stability range of temperature.

(ix) Based on topotatic relationship between kaolinite and its transformed phases, IR, and DTA curves, two mechanisms of dehydroxylation (either homogeneous or inhomogeneous) are suggested. There are no evidences in either of the mechanisms.

These problems hinder to arrive at a definite general conclusion in interpretations of K-M reaction processes. It is hoped that the pertinent problems in this reaction series are chronologically presented in the individual chapter of the book.

2. Predictions by scientific communities

There are a lot of predictions in the K-M literature. These are not always a matter of speculation. Some of the findings are experimentally based. Yet, one theory does not explain other established results of the methods applied in the characterization process.

(i) It may be simple γ-Al_2O_3 by XRD but pioneering single crystal study predicts Al-Si spinel.

(ii) It is also predicted that the composition of spinel may be analogous to composition of 3:2 orthorhombic mullite (o-mullite) and it is designated as cubic mullite (c-mullite), i.e., it contains as high as 28 wt % SiO_2.

(iii) Contrarily, spinel phase was found to contain SiO_2 to the extent of 8–11 wt % only by analytical TEM-EDS analysis and HRTEM study respectively. Therefore, the question is how much silicon is present in the spinel phase?
(iv) In most of the repetitive alkali leaching techniques, the end point of SiO_2(A) – alkali reaction is not maintained properly, which may lead to erroneous prediction in determination of silica content in spinel phase by energy dispersive X- ray spectroscopy.
(v) MAS NMR was applied to characterize the disordered solids like metakaolinite and spinel phase to study the local atomic environment of Al and Si and it is predicted that 980°C exotherm is not associated with formation of a particular phase.
(vi) To explain the reasons behind the cause for exothermic reaction in DTA, which is purely a dynamic heating process, some researchers analyzed heat treated clay samples of different thermal history which misleads the actual reason behind exotherm.
(vii) Mullitization process may occur through a degree of segregation of silica and alumina by way of decomposition of metakaolinite and then a more extensive diffusional process of Al and Si elements. Author theory predicts simple structural transformation of orderly mixed spinel phase. Thus, contrast between segregation-recombination path versus. continuous structural transitional path is still to be a subject of study.

3. Difficulties encountered by distinguished authors

The main reasons and difficulties in characterization of phases are the following.

(i) Important and landmark studies in phase transformation claim that powder X-ray pattern of metakaolin is amorphous and first crystalline product of it is also a poorly crystalline γ-Al_2O_3 followed by mullite.
(ii) Nature of spinel phase, whether it is Al-Si spinel or Al spinel (γ- Al_2O_3), is rather difficult to determine, because it is poorly crystalline, small crystal size and showing 2-3 number XRD peaks, are very broad and diffuse, and lastly even on heating for as long as 28 hr at ~950°C doesn't develop into a well crystallized state. Now the reader will realize the tough problem of its identification. Thus, the grain size of spinel phase in the nanometer range makes the structural analysis most challenging. Not only this, the problem is most critical considering its composition of Al- Si spinel, i.e., the content of Si in spinel phase.

Other fundamental difficult problems are the following.

(i) Crystallographic relationship between kaolinite to metakaoliite to spinel phase is evaluated . But there exist uncertainties on the clear axial relationship between spinel to mullite, i.e., between kaolinite to mullite as a whole.

(ii) IR studies suggest three stages of dehydroxylation of kaolinite. The large principal and small final dehydroxylation peaks are relatively sharp. Those OH groups which stabilize the metakaolin structure release gradually over a range of temperature between ~700°C to temperature just below first exotherm. This result indicates the possibility of varying hydroxyl groups in structural units of metakaolin.

(iii) During phase separation process at final stage of dehydroxylation of metakaolinite, a large quantity of siliceous phase SiO_2(A) to the extent of ~35–37 wt % is liberated. This SiO_2(A) part poses a great thrust to characterization of Al-Si spinel phase either by XRD or in TEM.

(iv) Besides liberation of SiO_2(A), a large quantity of mullite in amorphous state also remains as a coexisting phase in 980°C decomposition product of kaolinite. Now the reader must realize a very difficult task of identification of spinel phase whether it is γ-Al_2O_3 or Si bearing spinel due to its low content (~20–30 wt %), low crystallinity, and it is admixed with an average 70–80 wt % amorphous phase mixtures. Due to these reasons, composition analysis of spinel phase even by EF-TEM study is found inconclusive.

(v) Interpretations on the kinetics and mechanism of phase transformation of kaolinite studied either by isothermal or by non-isothermal process are varying. Thermal transformation of kaolinite is governed by both thermodynamic and kinetic laws. However, most of the studies in this process are done from a kinetic point of view. It is necessary to present thermodynamic and kinetic contributions separately.

(vi) The growth of mulite needles sometimes occurs in preferred orientation to hexagonal orientation to kaolinite flakes and supports the idea of epitaxial type of transformation. But it is not always noted in electron microscopic studies.

4. Investigation methods chosen by various researchers

Use of thermal analysis method for phase transformation of kaolinite has shown much and widespread interest among ceramists and mineralogists. Numeros efforts by them on the investigation on the heat effects of clays of various sources are obviously desirable. A short summary of the various investigation approaches are shown in Chap. 2 vide its Table 2.1. Detail investigations out of a large number of researchers over a period of more than 100 years are serially mentioned. Experimentally noted figures of some researchers are presented after obtaining their kind permission for reproduction along with the inclusion of author's own newer data of several studies. These are chronologically and elaborately presented in Chaps. 3–18. Experimental results of various researchers and scientific personnel are comparable and imagined to different kinds of flowers blooming in their individual investigative fields of study. The present author just tried to pluck various types of colorful flowers from their clay –mullite research area and thereafter sorted out these flowers in different gradations considering their shape, size, and colors for future use. This way of presentation of the earlier literature along with the incorporation of new conceptions of present author are done in chapter wise mode with a view to develop lucid pictures of K-M reaction processes which are likely analogous to conglomeration of flowers kept in different vases such that these most adorable and would be acceptable globally by students, professors, researchers, and technocrats.

5. Visualization and envisagement

Based on the pertinent observations and experimental results of eminent scientists, the present author elaborately addresses the various issues e.g., (i) on the various earlier problems, (ii) difficulties encountered by distinguished authors, and (iii) predictions by scientific communities on the decomposition of kaolinite as stated above. These critical reviews are presented in Chaps. 19–25 with proper cross references from previous chapters. This service is considered as an equivalent to formulation of flower bouquet or fabrication of flowers garland, etc., with some insertion of his own conceptions and visualization regarding the problems and mechanisms of K-M reaction series out of sorted flowers that are kept in vases. It is largely expected that these sorts of critical exercises as submitted here may evoke a melody in the mind of a reader when they attentively go through this book. The present author with his bowed head dedicates these newly made garland and two bouquets to the feet of his Revered Sadhan-Siddha Sadguru Swami Nigamananda Pramhansa and ends the book.

6. Conclusion drawn by the present author

Reaction sequences of decomposition of kaolinite are chronologically enumerated and is shown in Fig. 26.1.

Finally, thermal events of kaolinite are explained and summarized below.

1. At 100–150 °C. Physically adsorbed moisture is removed.
2. At the 1st endothermic region at ~500–650 °C.

First step is kaolinite to metakaolinite transformation with loss of crystalline structure. Loss of structural OH groups amounting to ~87 % and forms metakaolinite with residual OH groups ~13 %.

3. At the endo–exo peak region at the temperature 950–1,000 °C.

Second step is the rapid decomposition of metakaolinite at ~1,000 °C Loss of last traces of OH groups at the endothermic dip followed by phase separation into siliceous and aluminous regions. A part of the aluminous region crystallizes to weakly crystalline primary mullite (minor) with exhibition of a small exotherm just before 980° and other part crystallizes to Al–Si spinel phase (major) at the first exothermic peak. The significant amount of residual, two separate, amorphous phases e.g., siliceous region (SiO_2(A)) and residual aluminous region (Mullite(A)) resulting to depict globally a very weak X-ray pattern and a band.

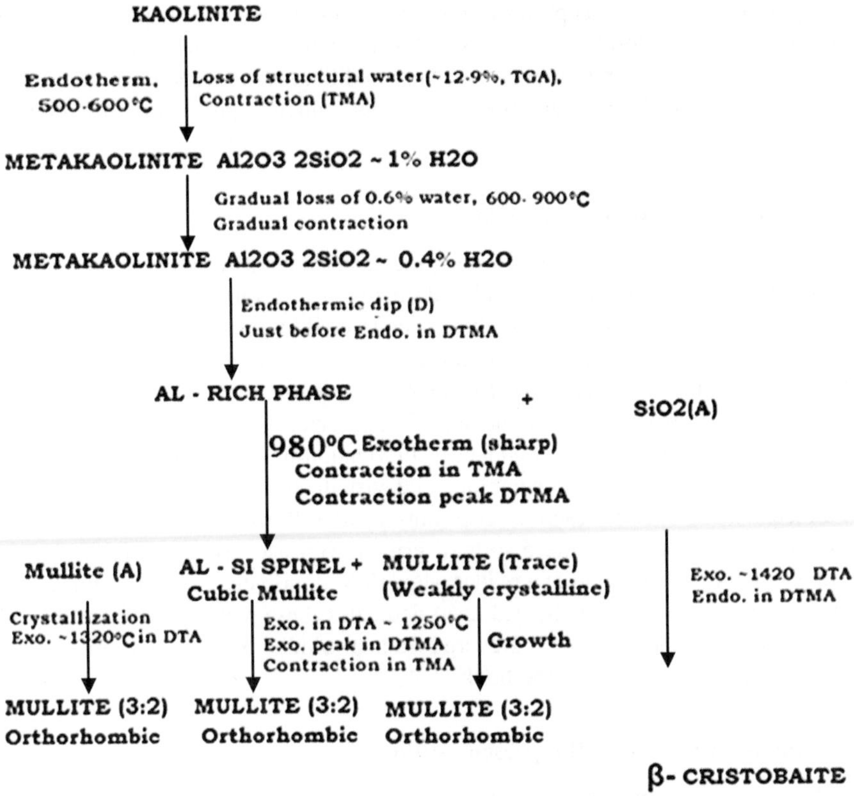

Fig. 26.1 Reaction Sequences of Decomposition of Kaolinite (After Chakraborty)

4. Two exothermic regions of mullite formation in the temperature range 1,100–1,400 °C.

Third step is the mullitization out of Al–Si spinel and amorphous aluminosilicate phases (mullite (A)) over a long range of temperature from $\sim$1,050 °C to $\sim$1,350 °C. During the entire region, two growth steps of mullite formation occur (one starts from 1,050° to $\sim$1,150 °C and other sharply at $\sim$1,250 °C with continuous increase up to 1,350 °C). A broad exotherm is due mullitization out of residual aluminous phase (Mullite(A)) by nucleation–crystallization mechanism. A small exotherm is always noticed by various researchers may be due to phase transformation of Al–Si spinel phase to 3:2 mullite. Both sharpness and resolution of mullite peaks increase with continued heating up to 1,550 °C.

5. Final exothermic region of cristobalite formation at 1,400–1,430 °C.

Concurrent to mullite formation, the remaining silica-rich alumina phase crystallizes at above 1,200 °C to β-cristobalite. The lattice parameter study shows a axis data of cristobalite is much more than that of the value noted in cristobalite samples obtained from pure silica gel this data indicates the presence of Al^{+3} in cristobalite structure. Thus, by lattice parameter data one can predict the formation of silica-rich aluminous phase during the decomposition of metakaolinite. Crystallization of cristobalite is very sharp at $\sim$1,400 °C and cause a small exotherm in DTA.

Appendix

Table A.1 Chemical analysis of kaolinites of different origins

Clay	SiO_2	Al_2O_3	TiO_2	Fe_2O_3	CaO	MgO	K_2O	Na_2O	SO_3	S	L.O.I.
Zettlitz	46.90	37.40	0.18	0.65	0.29	0.27	0.84	0.44	0.03	–	12.95
Georgia KGa-1 (ordered)	44.20	39.70	1.39	0.08	ND	0.03	0.002	0.05	–	–	13.78
Georgia KGa-2 (disordered)	43.90	38.50	2.08	0.098	ND	0.03	0.65	–	–	–	13.77
Rajmohol	43.62	40.69	1.06	0.06	0.08	Trace	0.06	0.08			12.80
Bhandak	43.41	40.74	1.30	0.49	0.22	Trace	0.02	0.07			13.33
Kolloid (Karlovy vary)	98.9% clay mineral				0.87	0.24	0.02	0.15			
English China	44.90	40.84	–	0.90	0.07	0.49	–	–	–	–	13.50
Halloysite (Mexico)	44.01	38.89	0.20	0.15	0.42	0.20	0.27	0.07	–	–	15.58
Dickite	40.80	35.70	0.02	0.04	3.96	Trace	0.02	0.02	–	–	17.68
Fireclay (Scottish)	48.74	30.93	1.18	3.50	0.48	0.37	0.96	0.20	–	–	13.78

A. K. Chakraborty, *Phase Transformation of Kaolinite Clay*,
DOI: 10.1007/978-81-322-1154-9, © Springer India 2014

Appendix

Author Index

A. K. Chakraborty, *Phase Transformation of Kaolinite Clay*,
DOI: 10.1007/978-81-322-1154-9, © Springer India 2014

Subject Index

A. K. Chakraborty, *Phase Transformation of Kaolinite Clay*,
DOI: 10.1007/978-81-322-1154-9, © Springer India 2014